Calculus

Exercises

DMK | Deutschschweizerische Mathematikkommission des VSMP
(Verein Schweizerischer Mathematik- und Physiklehrkräfte)

Calculus

Exercises

Editor: DMK Deutschschweizerische Mathematikkommission des VSMP
(German-Swiss Mathematics Commission of the Swiss Society of Mathematics
and Physics Teachers), dmk.vsmp.ch

Authors: Baoswan Dzung Wong, Marco Schmid, Regula Sourlier, Hansjürg Stocker, Reto Weibel
Translation: Marc Schmidlin
Overall management: Barbara Fankhauser

1st edition 2023
ISBN 978-3-280-04226-7 print
ISBN 978-3-280-09270-5 e-book

Typesetting and graphics: Marco Schmid, Baar
Cover illustration: Atelier Tschachtli, Bern

Orell Füssli Verlag
© 2023 Orell Füssli AG, Zürich
All rights reserved

This publication is catalogued in the German National Library (Deutsche Nationalbibliothek);
detailed bibliographic information can be accessed on the internet: www.dnb.de.

Includes e-book
Use the access code provided on the inside of the book cover.

 Extra material is available for download
at www.hep-verlag.ch/calculus.

Other works in the series:
Analysis – Aufgaben, 978-3-280-04200-7
Analysis – Ausführliche Lösungen, 978-3-280-04201-4

Orell Füssli Verlag Lernmedien
info@hep-verlag.ch
www.hep-verlag.ch/ofv

Remarks Concerning the Book's Use

This book of exercises is the English translation of the German book «Analysis», which was developed for the instruction of calculus in the basic mathematics course and covers the subject matter in calculus treated in the four-year Swiss upper secondary school as required by the «Kanon Mathematik» (math.ch/kanon). Both versions can be used as a continuation of the book «Algebra 9/10».

- Each chapter, apart from chapter X, starts with introductory questions or exercises. The exercises following the blocks of theory span from questions of comprehension to diverse applications. Apart from chapter X at the end of the book, each chapter ends with the subchapters «Miscellaneous Exercises» and «Review Exercises».

- The blocks of theory in the green boxes are concise and cannot and should not replace the development of the mathematical connections during lessons. Therefore, detailed derivations have been omitted.

- The book contains a quite extensive and thematically diverse collection of exercises. It is therefore unavoidable that a suitable subset of exercises tailored to each class will have to be selected.

- The subchapters «Further Topics» address topics and questions that exceed the basic mathematics course and can be adressed selectively with interested pupils as well as in focus and elective mathematics courses. For these subchapters there are neither «Miscellaneous Exercises» nor «Review Exercises».

- Chapter X «Functions» at the end of the book reviews, and extends on, the specific topics from «Algebra 9/10» that are addressed in the preceding chapters. The complete review of this chapter should not in general be necessary.

- Apart from proofs and justifications, as well as certain figures, the answers and numerical results of all the exercises are compiled at the end of the book. Detailed solutions are available in a separate volume (only in German).

Special Label

An exercise sheet is available for this exercise. The exercise sheets can be downloaded from hep-verlag.ch/calculus.

Feedback on the book is welcome, no matter if it concerns errors, additions, praise or criticism. Messages should be sent to info@hep-verlag.ch using «Calculus» as the subject line.

April 2023   The Authors

Remarks Concerning the Translation

The translation aims to be a one-to-one translation of the German text and is typeset not only to correspond exercise by exercise but also page by page, so that it can be used in parallel with the German version. In addition, the translation also retains the mathematical notation used in the German text, as this is the notation commonly used in mathematics in German-speaking Swiss schools. Specifically, this concerns the decimal numerals (e.g. 12'345.67 instead of 12,345.67) and the notation used for long and polynomial division as well as single-letter variable names, which derive from the first letter of the German terminology (e.g. costs are $K \mathrel{\widehat{=}}$ «Kosten»).

April 2023   The Translator

Contents

Acknowledgement

The following colleagues have supported the compilation of the book by testing individual chapters in their lessons and giving us feedback and indicating improvements:

Lucius Hartmann (KZO Wetzikon), Regula Hoefer (Gymnasium Kirchenfeld, Bern), Nora Mylonas (Alte Kanti Aarau), Andrea Peter (Kantonsschule Sursee), Rolf Peterhans (Kantonsschule Zug), Patrizia Porcaro (Gymnasium am Münsterplatz, Basel), Aline Steiner (Gymnasium Bäumlihof, Basel), Angela Vivot (Kantonsschule Kollegium Schwyz), Salome Vogelsang (Kantonsschule Frauenfeld), Josef Züger (Bündner Kantonsschule, Chur)

We would also like to thank Hansruedi Künsch (ETH Zürich), who reviewed the book as the representative of the university and gave us feedback as well as thank our colleagues Markus Egli (Kantonsschule Uetikon am See) and Richard Schicker (Kantonsschule Zug), who also acted as expert editors and gave us helpful suggestions.

Baoswan Dzung Wong, Marco Schmid, Regula Sourlier-Künzle, Hansjürg Stocker, Reto Weibel as well as Barbara Fankhauser (project leader)

Editor's Preface

The bilingual German/English Matura has become established in many Swiss upper secondary schools (Gymnasien) and prepares pupils for their university studies. In most schools offering a bilingual Matura, the subject of mathematics is taught immersively in English. For this reason, the DMK (German-Swiss Mathematics Commission) considers it very important to have teaching materials in English available, consistent with Swiss upper secondary school curricula and the «Kanon Mathematik» (math.ch/kanon). We are proud to provide mathematics teachers with a collection of exercises which corresponds precisely to the DMKs teaching aid «Analysis». This new English-language calculus book covers all of the calculus material relevant for the Matura and also includes an introductory approach to differential equations.

We have decided not to translate the book of solutions. Since the English translation of the exercise book retains the single letter variable names of the German original, the German solutions can be used directly.

We would like to thank the authors, namely Baoswan Dzung Wong, Regula Sourlier-Künzle, Marco Schmid, Hansjürg Stocker and Reto Weibel. Marco Schmid was also responsible for the typesetting, for which he deserves an extra Thank You. We also would like to thank the translator Marc Schmidlin for his excellent work. Furthermore, we would also like to thank Baoswan Dzung Wong and Robert Geretschläger who double-checked the translation from a linguistic point of view. A special Thank You goes out to the project manager, Barbara Fankhauser, for all her dedication and coordination work and to Fabio Mussi from the learning media team at Orell Füssli Verlag for their support.

Finally, we would like to thank Martin Haefner and the Kurt and Silvia Huser-Oesch Foundation for their generous financial support, without which it would not have been possible to realise this translation.

For the DMK
Andrea Peter, President

1 Sequences and Series

1. *How do we know how it continues?* What might be the next two terms for each of the given number sequences below?

a) 1, 10, 100, 1000, 10'000, ...   b) 5, 7, 9, 11, 13, 15, 17, ...   c) 0, 3, 8, 15, 24, 35, ...

d) 3, 1, 4, 1, 5, 9, 2, 6, 5, ...   e) 1, 0, 2, 0, 0, 3, 0, 0, 0, 4, 0, ...   f) 1, 1, 2, 3, 5, 8, 13, ...

g) $\frac{1}{2}, \frac{1}{6}, \frac{1}{12}, \frac{1}{20}, \frac{1}{30}, \ldots$ h) 10, 16, 17, 32, 33, 35, ...   i) 1, 2, 4, 7, 11, 16, ...

j) 1, 2, 4, 8, ...   k) 1, 5, 12, 22, 35, 51, ...   l) 31, 28, 31, 30, 31, ...

m) 1, 5, 0, 4, 1, 7, ...   n) 0, 0, 0, 6, 24, ...   o) 2, 3, 5, 7, 11, 13, 17, ...

p) 2, 1, 4, 3, 6, 5, 8, ...   q) 2, 7, 15, 26, ...   r) 1, 2, 5, 10, ...

s) 4, 4, 4, 4, 4, 5, 6, 4, 4, ...   t) 0, 0, 4, 1, 4, 4, 4, 6, 6, 7, 7, ...

u) 2, 22, 122, 622, 3122 ...   v) 0, 1, 1, 0, 1, 0, 1, 0, 0, 0, ...

1.1 Introduction

Sequences

A *sequence* is a list of real numbers a_1, a_2, a_3, ..., a_n or a_1, a_2, ..., a_k, ... that are arranged in a specific order. The individual numbers of the form a_k, $k \in \mathbb{N}$, are called the *terms* or *elements* of the sequence. The *index* k indicates the position of the term a_k within the sequence, generally denoted by (a_k). Unlike the elements of a set of numbers, some of the terms of a sequence (a_k) may have the same value or even all terms may coincide: $a_1 = a_2 = \ldots = a_k = \ldots$. This is the case for every *constant sequence*.

When the sequence (list) has a final term a_n, and thus a largest index $n \in \mathbb{N}$, the sequence is said to be *finite* or, occasionally, a sequence of length n. Otherwise, when the sequence does not end and, therefore, does not have a final term a_n, we speak of an *infinite* sequence. A familiar such sequence are the natural numbers $1, 2, 3, \ldots, k, \ldots$. If the general term $a_k \in \mathbb{R}$ is described by a term depending on the index k, we say that the sequence (a_k) is defined *explicitly*.

Examples of explicit definitions or representations:

- $a_k = 2k - 1$, $k \in \mathbb{N}$ • $a_k = k^2 - 3k + 14$, $k \in \mathbb{N}$ • $a_k = \cos(k \cdot \pi)$, $k \in \mathbb{N}$

An explicit representation of a sequence is basically a function $f : \mathbb{N} \longrightarrow \mathbb{R}$, which maps each index $k \in \mathbb{N}$ to its associated term $a_k \in \mathbb{R}$. For a finite sequence (a_k), the domain D_f is the finite subset $\{1, 2, 3, \ldots, n\}$ of $\mathbb{N}$.

If the general term $a_k \in \mathbb{R}$ is defined using one or more of its preceding terms and, perhaps, also k, the associated sequence (a_k) is defined *recursively*. To obtain the sequence, one or more of its initial terms are required.

Examples of recursive definitions or representations:

- $a_1 = 13$ and $a_k = 2a_{k-1} - 4$, $k > 1$ • $a_1 = 4$ and $a_{k+1} = 21 - 3a_k$, $k \in \mathbb{N}$
- $a_1 = 2$, $a_2 = 1$ and $a_{k+2} = a_{k+1} + a_k$, $k \in \mathbb{N}$ • $a_1 = 7$ and $a_{k+1} = a_k + k$, $k \in \mathbb{N}$

2. This exercise refers to the examples in the preceding box (Sequences).

    a) List the first six terms for each of the three explicitly defined sequences.

    b) List the first five terms for each of the four recursively defined sequences.

3. Find the explicit representation a_k for the given sequence, where $k \in \mathbb{N}$.

    a) 2, 4, 6, 8, 10, ...   b) 1, 3, 5, 7, 9, ...   c) 2, 4, 8, 16, ...   d) 1, 4, 9, 16, ...

    e) 7, 14, 21, 28, 35, ... f) 8, 15, 22, 29, 36, ... g) $-13, -3, 7, 17, 27, \ldots$ h) 3, 9, 27, 81, ...

4. Determine the explicit representation of the general term a_k for the given sequence, where $k \in \mathbb{N}$.

    a) $1, \frac{1}{2}, \frac{1}{3}, \frac{1}{4}, \ldots$ b) $\frac{1}{2}, \frac{2}{3}, \frac{3}{4}, \frac{4}{5}, \ldots$

    c) $\frac{-2}{3}, \frac{4}{5}, \frac{-8}{7}, \frac{16}{9}, \frac{-32}{11}, \ldots$ d) $\frac{4}{7}, \frac{12}{15}, \frac{20}{23}, \frac{28}{31}, \ldots$

5. Calculate the two terms a_{50} and a_{51} of the sequence defined by its first few terms.

    a) $\frac{1}{4}, \frac{3}{8}, \frac{5}{12}, \frac{7}{16}, \frac{9}{20}, \ldots$ b) $\frac{3}{4}, \frac{4}{7}, \frac{1}{2}, \frac{6}{13}, \frac{7}{16}, \frac{8}{19}, \ldots$

6. Find a recursive representation of the sequence defined by its first few terms.

    a) 3, 7, 11, 15, 19, ...   b) 6, 12, 24, 48, 96, ...

    c) 6, 13, 27, 55, 111, ...   d) 4, 11, 32, 95, 284, ...

7. For a sequence (a_k), the two initial terms and the recursion formula for calculating the other terms are known. List its first ten terms and determine a_{100}, a_{101}, a_{102} and a_{107}.

    a) $a_1 = 1, a_2 = 3, a_{k+2} = a_{k+1} - a_k, \; k \geq 1$ b) $a_1 = 2, a_2 = 1, a_{k+2} = \dfrac{a_{k+1}}{a_k}, \; k \geq 1$

8. For a sequence (a_k), the two initial terms a_1 and a_2 together with the following recursion formula are known: $a_{k+2} = 2a_{k+1} - a_k; \; k \in \mathbb{N}$. Using the given initial terms, calculate the five terms a_3, a_4, a_5, a_6 and a_7 and find the explicit representation of the general term a_k for $k \in \mathbb{N}$.

    a) $a_1 = 3, a_2 = 7$ b) $a_1 = 7, a_2 = 3$ c) $a_1 = 0, a_2 = 1$ d) $a_1 = -5, a_2 = -4$

9. The sequence (a_k) is defined by its initial terms a_1 and a_2 as well as the following recursion formula: $a_{k+2} = \frac{a_{k+1}^2}{a_k}$ $(k \in \mathbb{N})$. Using the given initial terms, calculate the five terms a_3, a_4, a_5, a_6 and a_7 and determine the explicit representation of the general term a_k of the sequence for $k \in \mathbb{N}$.

    a) $a_1 = 4, a_2 = 8$ b) $a_1 = 81, a_2 = 27$ c) $a_1 = 16, a_2 = -24$ d) $a_1 = -2, a_2 = -2$

10. Find a recursive representation of the sequence defined by the general term a_k for $k \in \mathbb{N}$.

    a) $a_k = 2k + 34$ b) $a_k = 1 - 2k$ c) $a_k = (-3)^k$ d) $a_k = k^2$

11. Give both an explicit and a recursive representation for the sequence having a periodic or a cyclic structure.

    a) 1, −1, 1, −1, 1, −1, 1, −1, 1, ...   b) −1, 1, −1, 1, −1, 1, −1, 1, −1, ...

    c) 1, 0, 1, 0, 1, 0, 1, 0, 1, ...   d) 0, 1, 0, −1, 0, 1, 0, −1, 0, 1, 0, ...

12. For a sequence (a_k), the first term and the recursion formula are given. List the first six terms of the sequence and use them to derive an explicit representation of the general term a_k of the sequence for $k \in \mathbb{N}$.

a) $a_1 = 1,\ a_{k+1} = \frac{a_k}{k+1}$

b) $a_1 = 1,\ a_{k+1} = a_k + \frac{1}{2^k}$

c) $a_1 = 1,\ a_{k+1} = a_k + (-1)^k \cdot \left(2k^2 + 2k + 1\right)$

d) $a_1 = 2,\ a_{k+1} = (-1)^{k+1} \cdot \frac{a_k}{2}$

13. The sequences (a_k), (b_k) and (c_k), where $k \in \mathbb{N}$, are given either recursively or explicitly. Determine the first five terms and compare these three sequences with one another.

a) $a_1 = 1,\ a_{k+1} = 10a_k + 1;\ \ b_1 = 1,\ b_{k+1} = b_k + 10^k;\ \ c_k = \frac{10^k - 1}{9}$

b) $a_1 = 1,\ a_{k+1} = a_k \cdot \frac{k}{k+1};\ \ b_1 = 1,\ b_{k+1} = b_k - \frac{1}{k(k+1)};\ \ c_k = \frac{1}{k}$

Series

By adding the first n terms of a given sequence (a_k), one obtains the *partial sum* s_n:

$$s_n = a_1 + a_2 + a_3 + \ldots + a_n = \sum_{k=1}^{n} a_k,\ n \in \mathbb{N}.$$

Additionally, we define that $s_1 = a_1$.

The sequence (s_n) associated with a sequence (a_k) is called the *series* or the *sequence of partial sums*. It can be defined recursively by: $s_1 = a_1$ and $s_{n+1} = s_n + a_{n+1}$ for $n \in \mathbb{N}$.

Calculation rules for the summation symbol:

$$\sum_{k=1}^{n}(a_k \pm b_k) = \sum_{k=1}^{n} a_k \pm \sum_{k=1}^{n} b_k \quad \text{and} \quad \sum_{k=1}^{n} c \cdot a_k = c \cdot \sum_{k=1}^{n} a_k$$

---

14. Calculate the first six terms of the sequence of partial sums (s_n) associated with the given sequence. What is s_{100}?

a) $1,\ 3,\ 5,\ 7,\ 9,\ 11,\ 13,\ \ldots$ b) $1,\ -2,\ 3,\ -4,\ 5,\ -6,\ 7,\ \ldots$ c) $1,\ 2,\ 4,\ 8,\ 16,\ 32,\ 64,\ \ldots$

15. Let the sequence (a_k) be given by the general term $a_k = 2k + 10$ for $(k \in \mathbb{N})$. Write down the associated partial sum without using the summation symbol and calculate its value.

a) $\displaystyle\sum_{k=1}^{4} a_k$

b) $\displaystyle\sum_{k=1}^{11} a_k$

c) $\displaystyle\sum_{k=1}^{14} a_k$

d) $\displaystyle\sum_{k=5}^{14} a_k$

e) $\displaystyle\sum_{k=1}^{n} a_k$

16. Rewrite the sum using the summation symbol $\sum$. You do not need to calculate its value.

a) $5 + 10 + 15 + 20 + 25 + 30 + 35$

b) $5 + 10 + 15 + 20 + \ldots + 250$

c) $25 + 30 + 35 + 40 + \ldots + 105$

d) $3 + 9 + 27 + 81 + 243$

e) $1 + 4 + 9 + 16 + \ldots + 400$

f) $0 + 3 + 6 + 9 + 12 + \ldots + 345$

17. For a sequence, its general term a_k is given and let s_n denote its nth partial sum.

a) Calculate s_3 and s_8 for $a_k = \sqrt{k+1} - \sqrt{k}$ and give s_n in its simplest form.

b) Calculate s_4 and s_9 for $a_k = \frac{1}{k^2} - \frac{1}{(k+1)^2}$ and give s_n in its simplest form.

c) Calculate s_{15} for $a_k = \frac{1}{\sqrt{k}} - \frac{1}{\sqrt{k+1}}$ and give s_n in its simplest form.

18. Let (a_k) denote the sequence of odd numbers, 1, 3, 5, 7, ...

    a) Find both a recursive and an explicit representation for the sequence (a_k), $k \in \mathbb{N}$.

    b) The series $a_1 + a_2 + a_3 + \ldots$ is associated with the sequence of partial sums (s_n) defined by $s_n = a_1 + a_2 + \ldots + a_{n-1} + a_n$ for $n \in \mathbb{N}$ with $s_1 = a_1$. Calculate the seven terms $s_2, s_3, s_4, \ldots, s_8$ of the sequence (s_n) and give a simple expression for the nth partial sum s_n.

    c) Find a simple geometric visualisation for your previous result.

    d) Determine the value of $a_{11} + a_{12} + a_{13} + \ldots + a_{1000}$ with as little effort as possible.

19. The sequence (a_k) is recursively defined by $a_1 = \frac{1}{2}$ and $a_{k+1} = a_k - \frac{2}{k(k+1)(k+2)}$, where $k \in \mathbb{N}$.

    a) Calculate the five terms a_2, a_3, a_4, a_5 and a_6 without using a calculator.

    b) Calculate the first five terms of the sequence of partial sums (s_n) associated with $s_n = a_1 + a_2 + \ldots + a_{n-1} + a_n$ (for $n \in \mathbb{N}$) and give a simple expression for s_n for arbitrary $n \in \mathbb{N}$.

    c) Find a formula for the explicit representation of the general term a_k with $k \in \mathbb{N}$ for the given sequence (a_k). *Hint:* $s_n = (a_1 + a_2 + \ldots + a_{n-1}) + a_n$.

20. Determine the partial sum $s_n = \sum_{k=1}^{n} a_k$ of the sequence with the general term $a_k = \ln\left(\frac{k+1}{k}\right)$.

1.2   Different Types of Sequences and Series

Arithmetic Sequences and Series

Arithmetic Sequence

A sequence (a_k) is called an *arithmetic sequence* or *arithmetic progression* (AP) if the common difference d of successive terms is constant and non-zero:

$$d = a_{k+1} - a_k, \ d \neq 0 \text{ is constant for all } k \in \mathbb{N}.$$

This yields the following recursion formula for the recursive representation or definition of an AP with the given initial term a_1:

$$a_{k+1} = a_k + d, \ k \in \mathbb{N}.$$

The explicit definition of an AP for given values of a_1 and d is

$$a_k = a_1 + (k-1)\,d, \ k \in \mathbb{N}.$$

Furthermore, this yields

$$d = \frac{a_k - a_1}{k-1} = \frac{a_k - a_m}{k-m}, \ k > m \geq 1 \ (k, m \in \mathbb{N}).$$

Apart from the first term a_1 (and, possibly, the final term) every term of an AP is the arithmetic mean of its two neighbouring terms:

$$a_k = \frac{a_{k-1} + a_{k+1}}{2}, \ k \geq 2.$$

21. Determine if the given sequence is an AP. If it is, give the recursive and explicit definitions for the sequence.

a) 1, 4, 7, 10, 13, ...

b) 1, 2, 3, 4, 5, 6, ...

c) −2, 3, 8, 13, 18, 25, ...

d) 12, 20, 28, 36, ...

e) 17, 14, 11, 9, 6, 3, 0, ...

f) 50, 40, 30, 20, 10, 0, −10, −20, ...

22. Calculate the fifth term of the given AP.

a) $a_1 = 6$, $a_{k+1} = a_k + 8$

b) 3, 7, 11, 15, ...

c) $a_k = 5 + (k - 1) \cdot 7$

d) $a_1 = 34$, $d = 5$

23. An AP starts with 2 and the 2775th term is 524'288.

a) What are the first three terms of the sequence?

b) Find the recursive and explicit definitions of the sequence.

24. Two terms of an AP are known. Give the explicit definition of the sequence and calculate the indicated term.

a) $a_3 = 5$, $a_5 = 6$, $a_{20} = ?$

b) $a_{10} = 12$, $a_{20} = 18$, $a_4 = ?$

25. Insert 24 numbers between the two numbers 800 and 1575 such that the resulting sequence is an AP. Find the recursive definition of this sequence.

26. How many terms of the AP with $a_1 = \frac{1}{7}$ and $a_3 = \frac{1}{11}$ are larger than 0?

27. Determine m such that the sequence m, $m^2 + 3$, $4m^2 - 2m$, ... is an AP.

28. The sum of the three terms of an AP of length 3 is 30 and the sum of the squares of the terms is 318. Determine the three terms of such an AP. *Hint:* Write the three terms as $m - d$, m, $m + d$.

Example: Find the sum of the first eleven terms of the AP 10, 17, 24, 31, 38, 45, 52, 59, 66, 73, 80, 87, 94, ..., that is the partial sum $s_{11} = 10 + 17 + 24 + 31 + ... + 80$.

To avoid the tedious work of adding up all numbers use the following trick:
Write down the sum twice: once as is and once in reversed order directly below. Then, add the two summands in each column:

$$
\begin{array}{r}
10 + 17 + 24 + 31 + \ldots + 66 + 73 + 80 = s_{11} \\
80 + 73 + 66 + 59 + \ldots + 24 + 17 + 10 = s_{11} \\
\hline
90 + 90 + 90 + 90 + \ldots + 90 + 90 + 90 = 2s_{11}
\end{array} \quad +
$$

On the left hand side, the number 90 appears eleven times, so $2s_{11} = 11 \cdot 90 = 990$ and

$$s_{11} = \frac{11 \cdot 90}{2} = \frac{11}{2} \cdot 90 = \frac{11}{2} \cdot (10 + 80) = 495.$$

---

29. Evaluate the sum $20 + 27 + 34 + 41 + \ldots + 1490$ using the method from the example in the grey box above.

Arithmetic Series

The sequence of partial sums (s_n) with $s_n = a_1 + a_2 + \ldots + a_n = \sum\limits_{k=1}^{n} a_k$ associated with an AP is called an *arithmetic series* (AS). An explicit representation of an AS is given by the two following formulas:

$$s_n = \tfrac{n}{2}\left(a_1 + a_n\right) \quad \text{and} \quad s_n = n \cdot a_1 + \frac{n(n-1)}{2} \cdot d, \; n \in \mathbb{N}.$$

30. *Proof of the arithmetic series formulas.* Verify the two summation formulas for arithmetic series using the method shown in the example in the previous grey box.

31. What number should replace the question mark?

   a) $1 + 2 + 3 + 4 + \ldots + 99 + 100 = ?$ b) $1 + 3 + 5 + 7 + \ldots + 97 + 99 = ?$

   c) $53 + 56 + 59 + \ldots + 335 = ?$ d) $2 + 4 + 6 + 8 + \ldots, \; s_{50} = ?$

   e) $a_1 = 0.5, \; d = 0.2, \; s_{45} = ?$ f) $a_2 = 48.8, \; a_{33} = 11.6, \; s_{50} = ?$

32. For an AP, the two terms $a_1 = 8$ and $a_{10} = 71$ are known.

   a) What are the first seven terms of the sequence?

   b) What is the term a_{50}?

   c) What is the sum of the terms from a_{21} to a_{50}?

33. Calculate the sum

   a) of all even numbers from 100 to 10'000.

   b) of all odd numbers from 999 to 9999.

   c) of all numbers that are divisible by seven from 77 to 7777.

34. An AP is given. Calculate the partial sums s_{111} and generally s_n.

   a) $7 + 9 + 11 + 13 + 15 + \ldots$ b) $12 + 14 + 16 + 18 + 20 + \ldots$

35. Calculate the sum.

   a) $\sum\limits_{i=1}^{20}(1000 - 4i)$ b) $\sum\limits_{i=0}^{12}\left(\tfrac{1}{2} + \tfrac{i}{4}\right)$ c) $\sum\limits_{i=10}^{50}\left(10 + \tfrac{3i}{2}\right)$ d) $\sum\limits_{i=0}^{24}3\left(-7 + \tfrac{5i}{9}\right)$

36. Rewrite the sum in terms of $\sum$ and evaluate the sum.

   a) $3 + 6 + 9 + 12 + 15 + 18 + 21$ b) $45 + 40 + 35 + 30 + 25 + 20 + 15 + 10$

   c) $7 + \tfrac{15}{2} + 8 + \tfrac{17}{2} + 9 + \tfrac{19}{2}$ d) $12 + 7 + 2 - 3 - \ldots - 48 - 53$

   e) $\tfrac{1}{3} + \tfrac{2}{3} + 1 + \ldots + 15$ f) $-31 - 23 - 15 - \ldots + 41 + 49$

37. The sum of the first, third and fifth term of an AP is 33. The product of the first three terms of the sequence is 231. Calculate a_1 and the difference d of the AP.

38. Sofia saws a roofing lath of 4 m into ten pieces. Each piece is 6 cm longer than the previous piece. There is no leftover. How long is the shortest piece?

39. Sabrina gets a loan of 120'000 Swiss francs, interest-free, from her parents, to be paid back in instalments as follows: She repays 6000 francs at the end of the first year and then 500 francs more than in the previous year for every following year. After how many years will Sabrina have repaid the loan? What is the last instalment that she has to pay?

40. A wagon starts rolling on a railway track that has a constant slope. It travels a distance of 0.3 m in the first second, 0.9 m in the second second and 1.5 m in the third second. In every following second it travels 0.6 m further than in the previous second. What distance will the wagon have travelled in the first 30 seconds? How many seconds does it take for the wagon to have travelled the first 120 m?

41. Ms M. Oney is starting a new job in a company where she may choose between the two salary schemes A and B.

- *Scheme A:* The annual salary for the first year is 120'000 francs and the annual wage rise is 8000 francs.

- *Scheme B:* The semiannual salary for the first semester is 60'000 francs and the semiannual wage rise is 2000 francs.

Which scheme is better in the long term? Calculate the overall earnings for a period of ten years from the start of employment (disregarding any possible interest).

Geometric Sequences and Series

Geometric Sequence

A sequence (a_k) with $a_k \neq 0$ is called a *geometric sequence* or *geometric progression* (GP) if the common ratio q of successive terms is constant and non-zero:

$$q = a_{k+1} : a_k = \frac{a_{k+1}}{a_k}, \; q \text{ is constant for all } k \in \mathbb{N}, \; q \neq 0, \; q \neq 1.$$

This yields the following recursion formula for the recursive representation or definition of a GP with a given initial term a_1:

$$a_{k+1} = a_k \cdot q, \; k \in \mathbb{N}.$$

The explicit definition of a GP for given values of a_1 and q is

$$a_k = a_1 \cdot q^{k-1}, \; k \in \mathbb{N}.$$

Except for the first term a_1 (and, possibly, the final term) every term of a GP consisting of positive terms is the geometric mean of its two neighbouring terms:

$$a_k = \sqrt{a_{k-1} \cdot a_{k+1}}, \; k \geq 2.$$

If the GP contains negative terms, then

$$|a_k| = \sqrt{a_{k-1} \cdot a_{k+1}}, \; k \geq 2.$$

42. Determine if the given sequence is a GP. If it is, give the recursive and explicit definitions of the sequence.

a) 1, 4, 16, 64, 256, ... b) 2, 3, 4.5, 6.75, 9, 13.5, ...

c) 2, 6, 18, 54, ... d) -2, 6, -18, 54, -189, ...

e) 12, 6, 3, 1.5, ... f) 10, -20, 40, -80, 160, ...

43. Given two terms of a GP (a_k) find the common ratio q and a_8.

a) $a_1 = 64$, $a_2 = 96$ b) $a_2 = 8$, $a_5 = 216$

c) $a_7 = 100$, $a_{10} = -12.5$ d) $a_4 = \frac{9}{2}$, $a_{26} = 9216$

44. Do the numbers given constitute the start of a GP? If they do, calculate its eighth term.

a) 1, 1.1, 1.21, 1.331, ... b) 0.1, 0.2, 0.4, ... c) 24, -18, 12, ...

d) 0.9, 0.99, 0.999, ... e) 0.9, 0.81, 0.729, ... f) 12, -18, 27, ...

45. Three numbers, of which the second is 17 greater than the first and the third 34 greater than the second, form a GP. What are the three numbers?

46. The sum of the terms of a GP of length 3 is 9 and their product is -216. Determine the three terms of this GP. *Hint:* Write the terms as $\frac{m}{q}$, m, mq.

47. How many terms of the GP 8, 9, ... are smaller than 10^{12}? In other words, find the largest index $k \in \mathbb{N}$ that satisfies $a_k < 10^{12}$.

48. How many terms of the GP 2022, 2021, ... are larger than 1291?

49. How many terms of the GP 1, 1.1, ... lie between 1000 and 10'000?

50. a) What annual percentage increase yields a doubling of the population of a country in ten years?

b) Approximately how many years does it take for the population of a country with a growth rate of 5 % to double?

51. The energy comsumption in Switzerland was $48.16 \cdot 10^9$ kWh in the year 1980 and $58.96 \cdot 10^9$ kWh in 1988.

a) Calculate the average annual increase in percent.

b) Calculate the energy comsumption in the year 2014, assuming that the average annual increase in percent is the same as it was from 1980 to 1988, and compare this value with the true energy comsumption of $69.63 \cdot 10^9$ kWh in the year 2014.

52. Insert two numbers u and v between 1 and 10 such that 1, u, v, 10 (in this order) is a GP. Which values do u and v take if they are rounded to the nearest natural number? Where do these numbers occur in everyday life?

Example: Find the sum $1 + 3 + 9 + 27 + \ldots + 19'683 + 59'049$, whose summands form a GP. The following approach can be used for evaluating this sum:

$$
\begin{aligned}
s &= 1 + 3 + 9 + 27 + \ldots + 59'049 & &|- \\
3s &= 3 + 9 + 27 + \ldots + 59'049 + 177'147 & &|+ \\
\hline
3s - s &= 177'147 - 1 & &|:2 \\
s &= 88'573
\end{aligned}
$$

53. Evaluate the sum $2 + 2 \cdot 0.5 + 2 \cdot 0.5^2 + 2 \cdot 0.5^3 + \ldots + 2 \cdot 0.5^{10}$ using the method from the above example in the grey box.

Geometric Series

The sequence of partial sums (s_n) with $s_n = a_1 + a_2 + \ldots + a_n = \sum_{k=1}^{n} a_k$ associated with a GP is called a *geometric series* (GS). An explicit formula for the GS is

$$
s_n = a_1 \frac{q^n - 1}{q - 1} = a_1 \frac{1 - q^n}{1 - q}, \ n \in \mathbb{N}, \ q \neq 1.
$$

54. *Proof of the geometric series formula.* Prove the sum formula for the geometric series using the method shown in the grey box.

55. Calculate using the geometric series formula.
   a) $32 + 48 + 72 + 108 + 162 + 243$ b) $2 - 6 + 18 - 54 + 162 - 486 + 1458 - 4374$
   c) $\frac{2}{3} + \frac{4}{3} + \ldots + \frac{4096}{3}$ d) $-3'188'646 + 1'062'882 \mp \ldots - \frac{2}{3} + \frac{2}{9}$

56. Rewrite the sum in terms of $\sum$ and evaluate the sum.
   a) $\frac{5}{4} + \frac{5}{2} + 5 + 10 + 20 + 40 + 80 + 160 + 320 + 640$
   b) $7680 + 3840 + 1920 + \ldots + 0.9375$ c) $-4 + 12 - 36 + \ldots + 708'588$
   d) $\frac{3}{2} + 1 + \ldots + \frac{256}{6561}$ e) $1 + \sqrt{3} + 3 + \ldots + 2187$

57. Evaluate the sum.

   a) $\displaystyle\sum_{i=1}^{11} (-4)^{i-1}$ b) $\displaystyle\sum_{i=1}^{8} 2 \cdot 3^{i-1}$ c) $\displaystyle\sum_{i=0}^{16} 4096 \cdot \left(\frac{3}{2}\right)^{i-1}$ d) $\displaystyle\sum_{i=6}^{13} \frac{(-4)^{i-1}}{1024}$

58. At least how many terms of the GP 15, 16, ... need to be added in order that the sum is greater than one billion?

59. Consider a GP consisting of ten positive terms with initial value 1 and final value 2. Calculate s_{10}.

60. *The Sierpinski triangle.* Let (A_k) denote the sequence given by the areas of the following figures.

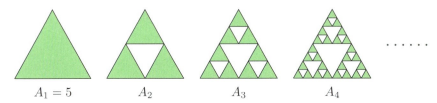

$A_1 = 5$ $\qquad\qquad$ A_2 $\qquad\qquad$ A_3 $\qquad\qquad$ A_4

Remark: The figures show the first four iterations of a famous *fractal* named after the Polish mathematician WACLAW FRANCISZEK SIERPINSKI (1882–1969).

a) Calculate A_2 and A_3.

b) Determine the explicit formula for A_k.

c) Which is the smallest $k \in \mathbb{N}$ such that A_k is less than 0.0001?

d) How many coloured triangles are there in the kth iterate of the Sierpinski triangle? Give an explicit formula.

e) Repeat part d), but count the white triangles instead.

   Hint: Derive a recursive formula first.

61. *The famous chessboard problem.* The Indian King Shihram is said to have asked Sissa, the inventor of chess, to wish for a reward. The latter asked for the total number of rice grains that would be placed on a chessboard if one were to place one grain on the first square, 2 on the second, 4 on the third, 8 on the fourth and so forth until the 64th square. How many grains does King Shihram owe the inventor?

1000 rice grains have a mass of 20 g and a volume of $25 \, \text{cm}^3$.

a) The annual, worldwide production of rice nowadays amounts to approximately 750 million tons. Convert the quantity of rice owed into tons and indicate how many of the current annual production of rice would be required for this.

b) Let this quantity of rice be equally distributed on the surface of the earth, assuming the earth is a sphere of radius 6370 km. How thick would this layer of rice be?

1.3   Infinite Geometric Sequences and Series

Infinite Geometric Series

For $|q| < 1$, the terms a_1, a_2, a_3, ... of a GP converge to 0. The GP is a so-called *null sequence* and we have: $a_k \to 0$ as $k \to \infty$. The corresponding series (s_n) converges to the limit s as $n \to \infty$:

$$s_n = a_1 \frac{1 - q^n}{1 - q} \quad \longrightarrow \quad s = \frac{a_1}{1 - q} \quad \text{as } n \to \infty$$

Remark: The number $s \in \mathbb{R}$ is called the *limit* or the *sum* of the series (s_n) if the terms s_n tend to s for increasing n.

A sequence having a limit is called *convergent*. Otherwise the sequence is called *divergent*.

A convergent series may also be expressed by using the summation symbol:

$$a_1 + a_2 + a_3 + \ldots + a_k + \ldots = \sum_{k=1}^{\infty} a_k.$$

Note: The notion of a limit is described more extensively in chapter 2.

62. Calculate the sum of the infinite GS.

a) $6 + 2 + \frac{2}{3} + \ldots$ b) $1 - \frac{1}{2} + \frac{1}{4} - \frac{1}{8} \pm \ldots$ c) $200 + 120 + 72 + \ldots$ d) $2\sqrt{3} + 2 + \frac{2}{3}\sqrt{3} + \ldots$

e) $\displaystyle\sum_{k=1}^{\infty} \frac{3}{10^k}$ f) $\displaystyle\sum_{k=0}^{\infty} \left(\frac{-7}{8}\right)^k$ g) $\displaystyle\sum_{i=10}^{\infty} 50'000 \cdot (0.1)^{i-1}$ h) $\displaystyle\sum_{k=0}^{\infty} 38 \left(\frac{2}{3}\right)^{3k}$

63. Consider the partial sum $s_n = 1 + \frac{9}{10} + \left(\frac{9}{10}\right)^2 + \left(\frac{9}{10}\right)^3 + \ldots + \left(\frac{9}{10}\right)^{n-1}$.

a) Find an explicit formula for s_n.

b) What is the sum of the first 100 summands?

c) Calculate the limit of s_n as $n \to \infty$.

64. The values $a_5 = 0.0972$ and $q = 0.3$ of a geometric series are known. Calculate the limit s of the infinite GS.

65. Two terms of a GP (a_k) with common ratio q are given. Calculate q and the limit s of the corresponding infinite GS.

a) $a_5 = 1296$, $a_8 = \frac{2187}{4}$ b) $a_3 = \frac{80}{3}$, $a_6 = \frac{-640}{81}$

66. Determine the common ratio q of the infinite GS if the following is known:

a) The limit s equals 6 times the first term of the GS.

b) The limit s equals 4.5 times the second term of the GS.

67. Determine the next three terms of the GP $4 \cdot \sqrt{2} - 4$, $2 \cdot \sqrt{2} - 4$, ... and, if it exists, calculate the limit s of the corresponding GS.

68. *Repeated decimals.* Repeated decimals can be converted into ordinary fractions using infinite geometric series.

Example: $0.\overline{3} = 0.33333\ldots = 0.3 + 0.03 + 0.003 + 0.0003 + \ldots = \frac{3}{10} \cdot \frac{1}{1-\frac{1}{10}} = \frac{1}{3}$

Use this method to convert the given repeated decimal into an ordinary fraction.

a) $0.\overline{4}$ b) $0.\overline{17}$ c) $0.\overline{9}$ d) $4.5\overline{135}$

69. For which values of x does the GP converge?

a) $3, 15x, 75x^2, \ldots$ b) $2, 2x - 4, 2x^2 - 8x + 8, \ldots$

70. Kubinski is building a tower in his mind using cubes with edges of lengths $8\,\text{cm}$, $4\,\text{cm}$, $2\,\text{cm}$, $1\,\text{cm}$, ... How tall can this tower be at most? What is the combined volume of all these cubes?

71. Finn has a ball of wool. He cuts off a metre of wool thread. Next, he cuts off a piece of wool thread, which is only $\frac{1}{3}$ times as long, from the ball of wool. Each successive piece he cuts off from the ball of wool is $\frac{1}{3}$ as long as the previous one. How large is the distance if Finn could cut off infinitely many pieces and lay them end to end?

72. Michèle draws a spiral consisting of semicircles whose radii form an infinite GP. The first semicircle has a radius of $r_1 = 5\,\text{cm}$, and the second of $r_2 = 4\,\text{cm}$. What is the total length of the spiral?

73. The spiral-shaped path shown at the left starts at the origin $(0 \mid 0)$ and consists of line segments (as depicted), whose lengths form a GP. How long is the path when continued infinitely and where is the final destination point Z of the spiral path located?

For exercise 73:   *For exercise 74:*

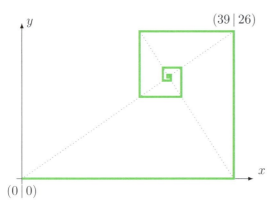

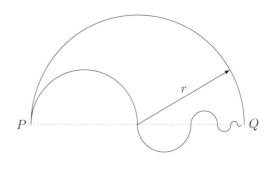

74. The serpentine path from P to Q is composed of infinitely many semicircular arcs whose radii form a GP with $q = \frac{1}{2}$. Which path from P to Q is shorter, the serpentine path or the semicircular path?

75. By what percentage must the terms of a infinite GS with initial term $a_1 = 3.21$ decrease at least from one term to the next in order that the sum of the series does not exceed $s = 4.56$?

76. *Geometric derivation of the formula* $s = \frac{a_1}{1-q}$. Let the points P, Q, R, S, ... lie on the right of the origin O on the positive x-axis such that the lengths of the line segments

$$a_1 = \overline{OP},\ a_2 = \overline{PQ},\ a_3 = \overline{QR},\ a_4 = \overline{RS},\ \ldots$$

form a decreasing GP, that is a GP with $0 < q < 1$. These line segments are now rotated upwards by $90°$ around their left endpoint. The rotated endpoints P', Q', R', S', ... then have coordinates

$$P'(0\,|\,a_1),\ Q'(a_1\,|\,a_2),\ R'(a_1 + a_2\,|\,a_3),\ S'(a_1 + a_2 + a_3\,|\,a_4),\ \ldots$$

Draw a suitable diagram and prove that all these points P', Q', R', S', ... lie on the same line g. This descending line g intersects the x-axis in a point $X(x_0\,|\,0)$. Show that the abscissa x_0 coincides with the cumulative length $s = a_1 + a_2 + a_3 + \ldots$ of the infinite GS of these line segments.

77. A path consisting of individual line segments $a_1, a_2, a_3, \ldots$ is inscribed in a square of side length ℓ (see the left-hand figure shown below). Show by algebraic as well as geometric arguments that the infinite path has length: $L = \sum\limits_{k=1}^{\infty} a_k = 2\ell + \ell \cdot \sqrt{2} = \ell\left(2 + \sqrt{2}\right)$.

For exercise 77:

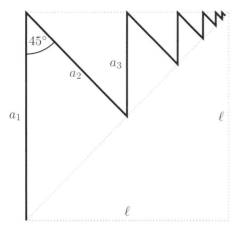

For exercise 78:

78. In the right-hand figure shown above, a circle of maximum size is inscribed into the outmost square. Then, a square of maximum size is inscribed in this circle. In this latter square a circle is inscribed again, and so forth. What percentage of the area of the outmost square do the infinitely many black areas cover?

1.4 Further Topics

Arithmetic Sequences of Higher Order

Difference Sequences and Initial Sequences

If a_1, a_2, ..., a_k, ... are the terms of a given sequence (a_k), the sequence which is defined by the difference of successive terms, $d_k = a_{k+1} - a_k$, is called the *first difference sequence* (1st DS) of (a_k) for $k \in \mathbb{N}$. One may also consider the difference sequence of a difference sequence, which successively yields the second, third or generally the kth difference sequence (kth DS). The sequence given initially will be called the *initial sequence* (IS).

Examples:

- IS: 2, 5, 10, 17, 26, 37, 50, ...

  1st DS: 3, 5, 7, 9, 11, 13, ...
  2nd DS: 2, 2, 2, 2, 2, ...
  3rd DS: 0, 0, 0, 0, 0, ...

- IS: 1, 2, 4, 8, 16, 32, 64, 128, ...

  1st DS: 1, 2, 4, 8, 16, 32, 64, ...

- IS: 1, −1, 1, −1, 1, −1, 1, −1, 1, ...

  1st DS: −2, 2, −2, 2, −2, 2, ...
  2nd DS: 4, −4, 4, −4, 4, ...

- IS: 1, $\frac{1}{2}$, $\frac{1}{3}$, $\frac{1}{4}$, $\frac{1}{5}$, $\frac{1}{6}$, ...

  1st DS: $\frac{-1}{2}$, $\frac{-1}{6}$, $\frac{-1}{12}$, $\frac{-1}{20}$, $\frac{-1}{30}$, ...
  2nd DS: $\frac{1}{3}$, $\frac{1}{12}$, $\frac{1}{30}$, $\frac{1}{60}$, ...

---

79. Determine as many terms as possible of the 1st DS using the initial sequence given by its first few terms. What might be the next two terms of the 1st DS?

  a) 10, 11, 15, 24, 40, 65, ...   b) 3, 10, 15, 18, 19, 18, 15, ...

  c) 4, 6, 9, 14, 21, 32, 45, ...   d) 1, 2, 0, 3, 4, 2, 5, 6, ...

80. Only the first few terms of some lengthy initial sequence are known. Using the obvious pattern or structure of its 1st DS, determine the next two terms of the initial sequence.

  a) 106, 107, 105, 108, 104, 109, 103, 110, ...   b) $\frac{11}{60}$, $\frac{41}{60}$, $\frac{61}{60}$, $\frac{91}{60}$, $\frac{37}{20}$, $\frac{47}{20}$, ...

  c) 13, 14, 25, 27, 49, 52, 85, 89, ...   d) 6, 7, 9, 10, 8, 9, 11, 12, 10, 11, 13, ...

81. Anna, Bryan, Carla and Daniel are assigned to «reconstruct» the initial sequence using its 1st DS given by 1, 2, 4, 8, 16, 32, 64, 128, ... They each arrive at a different initial sequence:

  Anna: 1, 2, 4, 8, 16, 32, 64, 128, ...   Bryan: 0, 1, 3, 7, 15, 31, 63, 127, ...

  Carla: 3, 4, 6, 10, 18, 34, 66, 130, ...   Daniel: −7, −6, −4, 0, 8, 24, 56, 120, ...

  Who made an error in «reconstructing»?

82. *Reconstruction of the initial sequence from its difference sequence.* Different initial sequences can have the same 1st DS, or put differently: For each 1st DS, there are multiple initial sequences. What statement can be made concerning two initial sequences whose 1st DS are identical?

83. The infinite sequence 2, 1, 4, 3, 6, 5, 8, 7, 10, ... with general term $a_k = k - (-1)^k$ for $k \in \mathbb{N}$ is given.

  a) Determine the 1st DS, 2nd DS, 3rd DS and 4th DS for this initial sequence step by step.

  b) What do you suspect for the 6th DS and 7th DS?

  c) How might one explicitly give the terms d_k of the nth DS for $n \geq 2$? (No proof required.)

Arithmetic Sequences of Higher Order

A sequence (a_k) is called an *arithmetic sequence of order n* or *arithmetic progression of order n* (APn) if its nth DS is constant and non-zero. The AP introduced earlier thus is an AP1, that is an arithmetic sequence of order 1. An APn may be described explicitly using a polynomial function of degree n. In the case of an AP this simply is a linear function. The general term a_k of an APn is: $a_k = c_n k^n + c_{n-1} k^{n-1} + \ldots + c_2 k^2 + c_1 k + c_0$ with $c_n \neq 0$.

84. Consider the sequence given by the general term $a_k = k^2 - k + 3$.

   a) Calculate a few of its terms and show that it is an AP2.

   b) What is the general term d_k of its 1st DS?

85. Consider the sequence given by the general term $a_k = k^2 - 2k + 2$. What is the general term d_k of its 1st DS?

86. The first few terms of an AP2 are given. Establish an explicit formula for the general term a_k for $k \in \mathbb{N}$ and use it to calculate a_{101}.

   a) 1, 3, 7, 13, 21, ...   b) 10, 7, 2, −5, −14, ...   c) 3, 3, 4, 6, 9, ...

   d) 12, 11, 7, 0, −10, ...   e) 2, 0, 0, 2, 6, ...   f) 3, 3, 1, −3, −9, ...

87. The first few terms of an AP3 are given. Deduce an explicit formula for the general term a_k for $k \in \mathbb{N}$ and use it to calculate a_{22}.

   a) 0, 6, 24, 60, 120, ...   b) 2, 9, 28, 65, 126, ...

   c) −8, −2, 0, 0, 0, ...   d) 1, 0, −1, −8, −27, ...

88. a) Let the general term of the 2nd DS be given by $e_k = (-1)^k$, the initial term of the 1st DS be $d_1 = 3$ and that of the initial sequence be $a_1 = 4$. List the first eight terms of the initial sequence.

   b) Let the general term of the 2nd DS be given by $e_k = 6k + 6$, the initial term of the 1st DS be $d_1 = 7$ and that of the initial sequence be $a_1 = 1$. List the first six terms of the initial sequence.

89. Let the general term of the 2nd DS be given by $e_k = 3k + 12$, the initial term of the 1st DS be $d_1 = 5$ and that of the initial sequence be $a_1 = -10$. List the first six terms of the initial sequence.

90. If (s_n) is the series associated with the sequence (a_k), i.e. $s_n = \sum_{k=1}^{n} a_k$, then the question arises whether, conversely, (a_k) is the 1st DS of (s_n). What is the correct answer?

91. a) Consider an arbitrary GP. What can be said about its 1st DS?

   b) Let the 1st DS be a GP. Is the associated initial sequence also a GP?

92. Is it true that the third difference sequence of an AP4 is an AP1? Justify your answer.

93. Show that the initial sequence must be an AP5 if its third difference sequence is an AP2.

94. What can you say about the fifth difference sequence of an AP2? Justify your answer.

95. How does the order of an APn change, if every term a_k of the sequence
   a) is increased by a fixed number $m > 0$?
   b) is decreased by a fixed number $m > 0$?
   c) is multiplied by a fixed number $m > 0$?

96. A function f has the property that the function values $f(x_k)$ for $k \in \mathbb{N}$ are an APn if the arguments x_k for $k \in \mathbb{N}$ are an APn for some $n \in \mathbb{N}$. What can be said about the function f?

97. Consider n circles lying in a plane such that each circle intersects any other circle in two points and no three or more circles intersect in a common point. Into how many regions do the circles partition the plane? Find a recursive and an explicit formula for the number of regions.

98. Let F_k be the number of points in the kth figure.
   a) Calculate F_4, F_5 and F_6.
   b) Find a recursive formula for F_k.
   c) Find an explicit definition of the sequence (F_k).
   d) Calculate F_{100}.

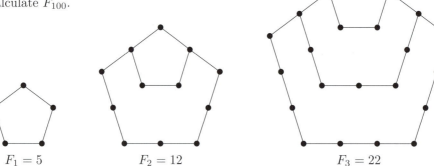

$F_1 = 5$ $F_2 = 12$ $F_3 = 22$

The Koch Curve and the Koch Snowflake

The Koch curve is a *fractal* first described in 1904 by the Swedish mathematician HELGE VON KOCH (1870–1924). The Koch curve is defined iteratively as follows:

1) First, start with a line segment of given length.
2) Then, remove the middle third of the line segment and place an equilateral triangle there upon.
3) In the next step, repeat the previous procedure on each of the four line segments obtained by the previous step.
4) Repeat the previous step infinitely many times.

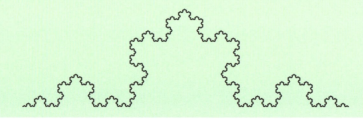

99. In our example, the initial segment is three units long: $a_0 = 3$. The sequence (a_k) denotes the length of the Koch curve after the kth iteration, where $k \in \mathbb{N}$.

$k = 0$ $\qquad\qquad$ $k = 1$ $\qquad\qquad$ $k = 2$ $\qquad\qquad$ $k = 3$

a) What are the values of a_1 and a_2?

b) Determine an explicit formula for a_k, where $k \in \mathbb{N}$.

c) To which value a does the length of the Koch curve tend? Calculate the limit of a_n as $n \to \infty$.

100. Let us start with an equilateral triangle of side s. On each side, we then apply the Koch curve construction described in the previous box. The resulting figure is called the Koch snowflake.

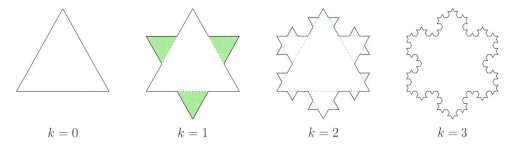

$k = 0$ $\qquad\qquad$ $k = 1$ $\qquad\qquad$ $k = 2$ $\qquad\qquad$ $k = 3$

In the following, u_k denotes the circumference of the snowflake in the kth iteration. A_k is the area of one of the newly formed triangles, where $k \in \mathbb{N}_0$.

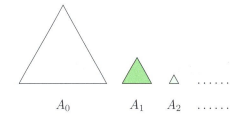

A_0 $\qquad$ A_1 $\quad$ A_2 $\quad$ ......

a) Determine the general term u_k. What is the limit of u_k as k increases infinitely?

b) Find an explicit formula for A_k.

   Note: The formulas for the remaining questions may be given as a function of A_0 – thus yielding a more compact solution.

c) What is the increase of area in the kth snowflake compared to the previous, $(k-1)$st snowflake?

d) Find an explicit formula for the complete area F_k of the kth snowflake.

e) Which value F does the sequence of areas (F_k) approach for increasing k?

Applications in Financial Mathematics

Basic Formula for Pension Calculation

The capital K after n years for an annual deposit R and an interest rate of $p\,\%$ is calculated as follows:

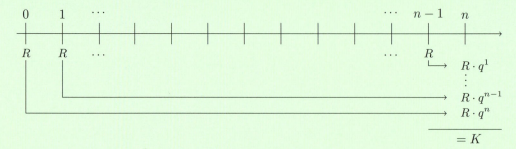

GS with $q = 1 + \frac{p}{100}$ $\Rightarrow$ $K = R \cdot q + R \cdot q^2 + \ldots + R \cdot q^{n-1} + R \cdot q^n = R \cdot q \cdot \dfrac{q^n - 1}{q - 1}$

Remark: The interest rates in the following exercises are chosen for illustrative effect and do not correspond to current reality.

101. Using a fixed annual interest rate of $5\,\%$, calculate the amount that a capital of 1000 francs

   a) will attain in 10 years.

   b) had 10 years ago.

102. A capital of 3500 francs earns interest at an annual rate of $1.2\,\%$ compounded quarterly.

   a) What is the capital after 6 years?

   b) What would the capital be after 6 years if the interest is compounded monthly?

103. Assume that you deposit 2000 francs into a bank account at the beginning of every year. How much money do you have in your account at the end of the sixth year, i.e. just before your seventh payment? The bank pays an annual interest rate of $5\,\%$.

104. a) A godfather deposits 500 francs into a savings account on the day of birth of his godchild and, subsequently, an amount of 200 francs on each birthday. He does so until the 18th birthday. What amount does the godchild have at her disposal if the annual interest rate is $4\,\%$?

   b) What amount does a godmother have to deposit into a savings account on the day of birth of her godchild and, subsequently, yearly until the 18th birthday, if her godchild is to have an amount of 10'000 francs at his disposal on his 20th birthday? The bank pays an annual interest of $4.5\,\%$.

105. At the end of a demanding work year, Mr Frei decides to start saving money in order to take a sabbatical in seven years' time. He calculates that he will need to have saved 50'000 francs for the expenses arising in Switzerland (rent, insurances, health insurance, ...). What amount does he need to deposit into an account at the beginning of each of the seven years in order to finance his sabbatical?

106. Amana wants to repay a debt of 100'000 francs in six instalments of 20'000 francs each. She pays the instalments at the end of each year. How much is the amount still owed at the end of the seventh year if the interest rate is 7 %?

107. How many times can an annual pension of 10'000 francs be received from a capital of 100'000 francs if the interest rate is 5 %? The pension is paid at the end of each year.

108. Zoé buys a television at an online shop for 2000 francs. She has to make a down payment of 200 francs and then pay six monthly instalments. The vendor charges an annual percentage rate of 16 %. How much is one instalment?

109. What capital is needed in order to receive an annual pension of 24'000 francs at the beginning of each year for 15 years? The annual interest rate is 4.5 %.

The Fibonacci Numbers

LEONARDO OF PISA, better known as FIBONACCI (Fibonacci means «son of Bonacci») was an Italian mathematician who lived approximately 1180–1250. The so-called Fibonacci numbers are credited to him.

110. FIBONACCI published his book «Liber abaci» (The Book of Calculation) in 1202. The famous problem about the population of rabbits was posed in this book:

We assume that the rabbits live indefinitely and that every month a pair of rabbits gives birth to another pair. A newborn pair gives birth for the first time only after two months. Let F_k be the number of pairs that exist at time k (in months) assuming that at time $k = 0$ we have one newborn pair.

a) Calculate the Fibonacci numbers F_0, F_1, ..., F_{10}.

b) Find a recursive formula for the sequence (F_k), by which F_{k+2} is calculated by means of F_k and F_{k+1}.

111. «Wer A sagt, muss auch B sagen.» (Translation: «In for a penny, in for a pound.») Using the two letters A and B, «words» consisting of k letters are to be formed. In doing so, the rule is that a letter B must follow each letter A. Moreover, a «word» is not allowed to end with an A. Let W_k be the number of possible «words» of length k. Find a recursive formula for W_k.

Example: $W_4 = 5$, «words»: $BBBB$, $BBAB$, $BABB$, $ABBB$, $ABAB$.

112. A rectangle of length k and width 2 is to be covered by dominoes of length 2 and width 1. Let M_k denote the number of possible patterns. Give a recursive formula for M_k. *Example:* $M_5 = 8$

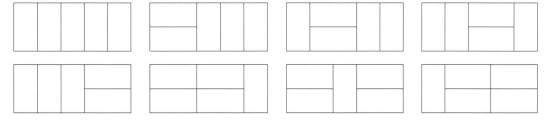

113. A staircase consists of $k \in \mathbb{N}$ steps. Andri can manage one or two steps in one stride. Let T_k be the number of possible sequences of strides that Andri can use to climb the stairs. Give a recursive formula for T_k.

114. Explain why the sequences (W_k), (M_k) and (T_k) from the exercises 111, 112 and 113 coincide term by term.

115. List the first ten terms of the sequence given by the recursive formula.

a) $a_1 = 1$, $a_2 = 2$, $a_{k+2} = a_{k+1} + a_k$, $k \in \mathbb{N}$ b) $a_1 = 2$, $a_2 = 1$, $a_{k+2} = a_{k+1} + a_k$, $k \in \mathbb{N}$

c) $a_1 = 3$, $a_2 = 4$, $a_{k+2} = a_{k+1} - a_k$, $k \in \mathbb{N}$ d) $a_1 = 2$, $a_2 = -1$, $a_{k+2} = a_{k+1} + a_k$, $k \in \mathbb{N}$

116. Let (F_n) denote the sequence of the Fibonacci numbers (see exercise 110). We define two sequences (a_n) and (b_n) as follows: $a_1 = 1$, $a_{k+1} = \frac{1}{a_k + 1}$ $(k \geq 1)$; $b_{k+1} = \frac{F_k}{F_{k+1}}$ $(k \geq 0)$.

a) Calculate the first six terms of both sequences (a_n) and (b_n).

b) Proof that the two sequences (a_n) and (b_n) are identical.

c) One can prove that a_n has a limit g as $n \to \infty$ with $g \in \mathbb{R}^+$. For large values of n we therefore have: $a_{n+1} \approx a_n \approx g$. Calculate g.

117. JACQUES PHILIPPE MARIE BINET (1786–1856) published a formula for the explicit calculation of the terms 1, 1, 2, 3, 5, 8, 13, ... of the Fibonacci sequence in 1843, however it should be noted that LEONHARD EULER (1707–1783), DANIEL BERNOULLI (1700–1782) and ABRAHAM DE MOIVRE (1667–1754) knew about this formula already a good hundred years earlier. The formula for the kth Fibonacci number published by BINET is as follows:

$$F_k = \frac{1}{\sqrt{5}} \cdot \left[\left(\frac{1 + \sqrt{5}}{2} \right)^{k+1} - \left(\frac{1 - \sqrt{5}}{2} \right)^{k+1} \right], \ k \geq 0$$

a) Verify this formula – preferably without using a calculator – for $k = 0$, $k = 1$ and $k = 2$.

b) Calculate F_{31}, F_{32}, F_{33} and show numerically that: $F_{31} + F_{32} = F_{33}$.

c) Clever Circa tells her friend Exactus that she finds this formula too complicated. She simply calculates the term $T_k = \frac{1}{\sqrt{5}} \left(\frac{1+\sqrt{5}}{2} \right)^{k+1}$ and then rounds the result to the nearest natural number. What do you think of her method?

118. Let (u_k) with $k \in \mathbb{N}$ denote the Fibonacci sequence 1, 1, 2, 3, 5, 8, 13, ... in this exercise.

a) Determine the 1st DS of the Fibonacci sequence (u_k). What do you notice?

b) Using three successive Fibonacci numbers u_{k-1}, u_k, u_{k+1} one may construct the following term: $T(k) = u_{k-1} \cdot u_{k+1} - u_k^2$, $k > 1$. Calculate the value of $T(k)$ for $k = 4, 5, 6, 10$ and 13. What do you notice? Describe your discovery as precisely as possible.

c) Consider the partial sums $s_n = u_1 + u_2 + \ldots + u_n$ and and calculate the particular values s_4, s_7, s_8 and s_{11}. What is your conjecture on how s_n may be described by a simple formula? *Hint:* Increase the calculated sums by 1.

d) Evaluate the sum $U(n) = \sum\limits_{k=1}^{n} u_{2k-1}$ for $n = 2, 3, 4, 8$ and 11. Which simple formula holds for this sum?

Mathematical Induction

119. List and check the next two rows of the array of equations. Make a conjecture for the nth row.

a)
$$2 \qquad\qquad = 1 \cdot 2$$
$$2 + 4 \qquad = 2 \cdot 3$$
$$2 + 4 + 6 \qquad = 3 \cdot 4$$
$$2 + 4 + 6 + 8 = 4 \cdot 5$$

b)
$$1 \qquad\qquad = \tfrac{1 \cdot 2}{2}$$
$$1 + 2 \qquad = \tfrac{2 \cdot 3}{2}$$
$$1 + 2 + 3 \qquad = \tfrac{3 \cdot 4}{2}$$
$$1 + 2 + 3 + 4 = \tfrac{4 \cdot 5}{2}$$

c)
$$1 \qquad\qquad\quad = 2 - 1$$
$$1 + \tfrac{1}{2} \qquad\quad = 2 - \tfrac{1}{2}$$
$$1 + \tfrac{1}{2} + \tfrac{1}{4} \qquad = 2 - \tfrac{1}{4}$$
$$1 + \tfrac{1}{2} + \tfrac{1}{4} + \tfrac{1}{8} = 2 - \tfrac{1}{8}$$

Mathematical Induction

Principle of Mathematical Induction: If M is a set of natural numbers with the property that the number 1 is in M and for each natural number k in M, the successive number $(k + 1)$ is also in M then the set M consists of all natural numbers, i.e. $M = \mathbb{N}$.

Proof by Mathematical Induction: To establish the validity of a statement for all natural numbers the following statements have to be verified:

1. *Basis step:* The statement is true for the number 1.

2. *Inductive step:* If the statement holds for a natural number k, then it also holds for the natural number $(k + 1)$.

The validity of the statement for all natural numbers follows from the basis and inductive steps due to the principle of mathematical induction, i.e. the *inference rule of mathematical induction*.

Abridged Version: The statement $A(n)$ is to be proven for all $n \in \mathbb{N}$:

1. Basis step: $A(1)$ is true.

2. Inductive step: $A(1) \Rightarrow A(1+1) = A(2)$ is true, $A(2) \Rightarrow A(2+1) = A(3)$ is true, $A(3) \Rightarrow A(3+1) = A(4)$ is true, $A(4) \Rightarrow A(4+1) = A(5)$ is true, ...

Example: Proof of the statement $A(n)$: $s_n = 1 + 2 + 3 + 4 + \ldots + n = \frac{n \cdot (n+1)}{2}$, $n \in \mathbb{N}$

1. Basis step: $s_1 = 1 = \frac{1 \cdot 2}{2} \Rightarrow A(1)$ is true.

2. Inductive step: $s_{k+1} = 1 + 2 + 3 + \ldots + k + (k + 1) = s_k + (k + 1) = \frac{k \cdot (k+1)}{2} + \frac{2(k+1)}{2}$
$$= \frac{(k+1)(k+2)}{2}, \text{ i.e. } A(k) \Rightarrow A(k + 1)$$

The inference rule of mathematical induction now yields that the statement $A(n)$ is true for all $n \in \mathbb{N}$.

Remarks:

- Occasionally it is simpler to write the inductive step using $(k - 1)$ and k instead of k and $(k + 1)$.

- The proof is commonly simply called «proof by induction».

- Verifying only the inductive step does not prove the statement.

- The truth value of $A(k)$ is irrelevant in the inductive step. It is important that the following is shown: If $A(k)$ holds, then $A(k + 1)$ also holds. The truth of $A(k + 1)$ is derived from the assumed truth of $A(k)$.

120. Prove the formulas for the nth row in the exercises 119 a) and 119 c) by mathematical induction.

121. Make a conjecture for the nth row of the array of equations and prove it by induction.

a) 2 $= 3^1 - 1$
 $2 + 6$ $= 3^2 - 1$
 $2 + 6 + 18$ $= 3^3 - 1$
 $2 + 6 + 18 + 54$ $= 3^4 - 1$

b) $1 \cdot 2$ $= (1 \cdot 2 \cdot 3) : 3$
 $1 \cdot 2 + 2 \cdot 3$ $= (2 \cdot 3 \cdot 4) : 3$
 $1 \cdot 2 + 2 \cdot 3 + 3 \cdot 4$ $= (3 \cdot 4 \cdot 5) : 3$
 $1 \cdot 2 + 2 \cdot 3 + 3 \cdot 4 + 4 \cdot 5$ $= (4 \cdot 5 \cdot 6) : 3$

122. Find a formula for s_n with $n \in \mathbb{N}$ and prove it by induction. Proceed analogously as in the preceeding exercise: Set up an array of equations and use it to derive a formula for the nth row.

a) $s_n = 1 + 3 + 5 + \ldots + (2n - 1)$

b) $s_n = \frac{1}{1 \cdot 2} + \frac{1}{2 \cdot 3} + \frac{1}{3 \cdot 4} + \ldots + \frac{1}{n \cdot (n+1)}$

c) $s_n = \frac{1}{1 \cdot 3} + \frac{1}{3 \cdot 5} + \frac{1}{5 \cdot 7} + \ldots + \frac{1}{(2n-1)(2n+1)}$

d) $s_n = 1 + \frac{1}{3} + \frac{1}{6} + \frac{1}{10} + \ldots + \frac{2}{n(n+1)}$

e) $s_n = 1 \cdot 1! + 2 \cdot 2! + 3 \cdot 3! + 4 \cdot 4! + \ldots + n \cdot n!$

f) $s_n = 1^3 + 2^3 + 3^3 + 4^3 + \ldots + n^3$

123. Find a simple formula for the general term a_k of the sequence given recursively by considering its first few terms and prove the formula by induction.

a) $a_1 = 2$, $a_{k+1} = a_k + 2k + 1$; $k \in \mathbb{N}$

b) $a_1 = \frac{1}{2}$, $a_{k+1} = a_k + \frac{1}{(k+1)(k+2)}$; $k \in \mathbb{N}$

124. Determine a simple formula for the product $P_n = \left(1 - \frac{1}{2}\right)\left(1 - \frac{1}{3}\right)\left(1 - \frac{1}{4}\right)\ldots\left(1 - \frac{1}{n}\right)$, where $n \geq 2$, and prove it by induction.

125. For an AS and GS we have the sum formulas $s_n = n \cdot a_1 + \frac{n(n-1)}{2} \cdot d$ and $s_n = a_1 \frac{1 - q^n}{1 - q}$, respectively; see p. 6 and p. 9. Prove these two sum formulas by induction.

126. Prove the statement by induction.

a) $a_n = n^2 + n$ is divisible by 2 for all $n \in \mathbb{N}$.

b) $a_n = 7^n - 1$ is divisible by 6 for all $n \in \mathbb{N}$.

c) $a_n = n^2 - n + 3$ is an odd number for all $n \in \mathbb{N}$.

d) $a_n = n^3 + 3n^2 + 2n$ is divisible by 6 for all $n \in \mathbb{N}$.

127. Prove the statement by induction.

a) The inequality $2^n \geq n + 1$ is valid for all $n \in \mathbb{N}$.

b) The inequality $2^n > 2n + 1$ is valid for all natural numbers $n \geq 3$.

128. For which $n \in \mathbb{N}$ is $2^n > n^2$? Prove your conjecture by induction.

129. Prove by induction:

$$\left(1 + x\right)\left(1 + x^2\right)\left(1 + x^4\right)\cdots\left(1 + x^{(2^n)}\right) = \frac{1 - x^{\left(2^{n+1}\right)}}{1 - x}, \quad x \neq 1, \ n \in \mathbb{N}_0$$

130. *Necessity of the Basis Step for Induction.* The statement $A(n)$ is given. Show that, while the inductive step from n to $n+1$ of a proof by induction may hold for all $n \in \mathbb{N}$, the statement may not hold for any $n \in \mathbb{N}$.

a) $A(n)$: $\frac{1}{2} + \frac{1}{4} + \frac{1}{8} + \ldots + \frac{1}{2^n} = 2 - \frac{1}{2^n}$ b) $A(n)$: $12^n + 1$ is divisible by 11.

131. For which values of $n \in \mathbb{N}$ is the statement $A(n)$ true, if

   a) $A(1)$ is true and the truth of $A(n)$ implies the truth of $A(n+2)$?

   b) $A(1)$ is true and the truth of $A(n)$ implies the truth of $A(2n)$?

   c) $A(1)$ and $A(2)$ are true and if $A(n)$ and $A(n+1)$ are true, then also $A(n+2)$ is true?

   d) $A(1)$ is true and the implication $A(n) \Rightarrow A(n+1)$ holds for all $n \geq 4$?

> For **132–134**: Find a formula for the number A_n, where $n \in \mathbb{N}$, and prove it by induction.

132. A_n denotes the maximum number of points in which n circles in the plane intersect one another.

133. A_n denotes the maximum number of finite regions that are bounded by n lines in the plane.

134. A_n denotes the number of regions that are enclosed by n lines in general in the plane. (In general means that no two lines are parallel and no three lines have a common intersection point.)

135. What is the error in the following «proof by induction» that n arbitrary points always lie on the same line? ($n \in \mathbb{N}$)

   1. Basis step: $n = 1$: A single point always lies on a line.

   2. Inductive step: Assumption: n arbitrary points always lie on the same line.

      Claim: $(n+1)$ arbitrary points also lie on the same line.

      Proof: Of the $(n+1)$ points $P_1, P_2, \ldots, P_n, P_{n+1}$ the first n points $P_1, P_2, \ldots, P_n$ lie on a line g by assumption. On the other hand, the n points $P_2, P_3, \ldots, P_{n+1}$ also lie on a line by assumption so that P_{n+1} lies on the same line as for example P_2 and P_3, which is g. Therefore, all $(n+1)$ points $P_1, P_2, \ldots, P_n, P_{n+1}$ lie on line g.

1.5 Miscellaneous Exercises

136. Is the sequence arithmetic, geometric or neither? If it is an AP or a GP, give the explicit and the recursive definitions for the sequence.

a) $3, 5, 9, 15, 23, \ldots$ b) $40, 30, 22.5, 16.875, \ldots$

c) $3, 7, 11, 15, \ldots$ d) $5, -15, 45, -135, \ldots$

137. Give an explicit as well as a recursive definition of the sequence $3, 33, 333, 3333, \ldots$.

138. Determine a_3, a_4, a_5, a_6 using $a_1 = 1$, $a_2 = m$ and the recursion formula $a_{k+2} = 2a_{k+1} - a_k$. What type of sequence is (a_k)? Give an explicit formula for the general term a_k with $k \in \mathbb{N}$.

139. The sequence (a_k) is given by its first two terms a_1 and a_2 and the recursion formula $a_{k+2} = \frac{a_k a_{k+1}}{2a_k - a_{k+1}}$ ($k \in \mathbb{N}$). Calculate the five terms a_3 through a_7 for the given initial terms. Give an explicit representation of the general term a_k ($k \in \mathbb{N}$).

a) $a_1 = \frac{1}{3}$, $a_2 = \frac{1}{4}$ b) $a_1 = \frac{1}{6}$, $a_2 = \frac{1}{5}$ c) $a_1 = \frac{1}{9}$, $a_2 = \frac{1}{11}$ d) $a_1 = \frac{1}{2}$, $a_2 = -\frac{1}{3}$

140. The natural numbers 4, 40 and 121 are terms of an AP. What is the largest possible common difference d that such a sequence can have?

141. We denote the arithmetic and geometric mean of two different real numbers $s > t > 0$ by $a = \frac{s+t}{2}$ and $g = \sqrt{st}$, respectively. Is the given sequence an AP, a GP or neither?

a) s, a, t b) s, g, t c) t^2, g^2, s^2 d) st, as, s^2

e) $s+t, a, 0$ f) $1, g, st$ g) a^5, a^4, a^3 h) $10g, 20g, 30g$

i) $s, 2a, 3t$ j) $s-t, s, 2a$ k) $\frac{t}{s}, g^2, s^3 t$ l) $s-t, 4a^2-2g^2, s+t$

142. a) The AP (a_k) with common difference d is given. What type of sequence is the sequence (b_k) with the general term $b_k = 10^{a_k}$?

   b) What type of sequence is the sequence (b_k) with the general term $b_k = \ln(a_k)$, if a_k is the general term of a GP with common ratio q?

143. Prove that the three numbers 1, 2 and 10 cannot be terms of a geometric sequence.

144. A house of cards with eight levels has the shape of a triangle. How many cards are needed for this house?

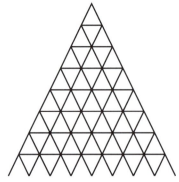

145. A possible strategy in roulette is to always bet on red and to double the bet money whenever a non-red outcome (a loss) occurs.

    a) Show that with this strategy one wins the initial bet, at the first appearance of red. *Remark:* In case of a win one receives a 1-to-1 payout.

    b) How much money should one have at one's disposal to be able to lose 20 consecutive times with an initial bet of 5 francs?

146. The following *fractal* is constructed iteratively. The first square has sides of length $1\,\mathrm{m}$. Trisect three of the four sides of the square and attach a square of side $\frac{1}{3}$ the original side on these sides. This process is repeated indefinitely. The first three iterations are depicted below.

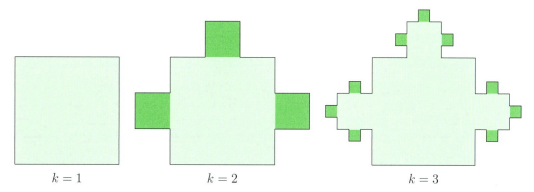

$k = 1$ $k = 2$ $k = 3$

    a) Determine if the sequence of areas converges and, if so, calculate the total area of the fractal.

    b) Determine if the sequence of circumferences converges and, if so, calculate the total circumference of the fractal.

147. The letter «Z» and the zigzag line are inscribed in two congruent squares, respectively. The infinitely many line segments of the zigzag line run alternately horizontally and diagonally and their lengths form a GP with common ratio $q = \frac{\sqrt{2}}{2}$, as can be seen easily; the whole upper side of the square is followed by half a diagonal, which is followed by half a side, and so forth. Show without calculation that the total length of the zigzag line is exactly the total length of the line segments yielding the letter «Z».

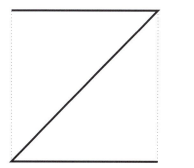 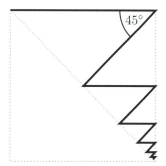

148. A container contains 20 litres of alcohol. Ten times in succession, 1 litre is removed from the container and replaced by 1 litre of water. How many litres of alcohol are contained in the alcohol-and-water mixture after these ten removals?

149. Two villages U and V are $119\,\text{km}$ apart. Alina walks from U to V covering a distance of $20\,\text{km}$ on the first day, $18\,\text{km}$ on the second, $16\,\text{km}$ on the third and on each further day, $2\,\text{km}$ less than on the previous day. Two days later, her friend Bianca sets off from V and hikes towards Alina. She covers a distance of $10\,\text{km}$ on her first day, $13\,\text{km}$ on her second and $16\,\text{km}$ on her third day, thus each day, walking $3\,\text{km}$ farther than the previous day. Where between U and V will Alina and Bianca meet?

150. A power function $y = p(x) = ax^r$ ($a \neq 0$, $r \in \mathbb{R}$) is given. Do the function values $p(x_k)$ form a GP if the arguments x_k ($k \in \mathbb{N}$) are a GP?

151. Based on measurements during the year 2014, it was estimated that a total of 2 million tons of a certain contaminant were emitted into the atmosphere. Because of the continuous growth of the economic output, an annual increase of $5\,\%$ in emissions is predicted. When will the emissions exceed the critical threshold of 8 million tons a year?

152. Gabriel and his mother have an agreement: If he succeeds in getting an average mark of at least 5 in the final report in the first year of school (primary), he will get 5 francs. At the end of each following year, the reward for an average mark of at least 5 increases by 10 francs. How much money does Gabriel get from his mother if he manages to obtain an average mark of at least 5 in all final reports (school years 1 to 12)?

153. Find three different numbers u, v and w whose sum is 3 such that they form an AP in the order u, v, w and a GP in the order v, w, u.

154. Prove that the quadratic equation $ax^2 + bx + c = 0$ ($a \neq 0$)
   a) always has a solution $x = -1$, if the three numbers a, $\frac{b}{2}$ and c form an AP.
   b) always has a double solution, if the three numbers a, $\frac{b}{2}$ and c form a GP.

155. List as many terms of the corresponding series $a_1 + a_2 + a_3 + a_4 + a_5 + \ldots$, as possible, for the given sequence of partial sums (s_n) ($n \in \mathbb{N}$).
   a) 5, 13, 24, 38, 55
   b) 2, 4, 8, 16, 32, 64
   c) 1, 1, 2, 2, 3, 3, 4, 4, 5
   d) 1, 0, 1, 0, 1, 0, 1
   e) 32, 16, 24, 20, 19
   f) 1, −1, 2, −2, 3, −3, 4

156. Is the infinite series $\frac{1}{1} + \frac{2}{2} + \frac{3}{4} + \frac{4}{8} + \frac{5}{16} + \frac{6}{32} + \ldots = \sum\limits_{k=1}^{\infty} \frac{k}{2^{k-1}}$ convergent or divergent? If it converges, determine its limit.

157. For $|x| < 1$, calculate the sum of the infinite series. *Hint:* Decompose the given series into two subseries.
   a) $2 + 5x + 2x^2 + 5x^3 + 2x^4 + 5x^5 + \ldots$
   b) $2 - 3x + 2x^2 - 3x^3 + 2x^4 - 3x^5 \pm \ldots$
   c) $x + x^2 - x^3 - x^4 + x^5 + x^6 - - + + \ldots$

158. The general term of a sequence (a_k) is $a_k = 0.5k^2 - 12k + 3$. What is the smallest term of the sequence?

159. Which is the largest term of the given sequence?
   a) $a_k = -3k^2 + 42k - 7$
   b) $b_k = -3k^2 + 32k$

160. For any arbitrary number $m \in \mathbb{N}$, we have:

$$\frac{m}{99} = \frac{m}{100 - 1} = \frac{\frac{m}{100}}{1 - \frac{1}{100}} = \frac{m}{100} + \frac{m}{100^2} + \frac{m}{100^3} + \frac{m}{100^4} + \ldots$$

For example, if $m = 72'485$ then $\frac{72'485}{99} = 724.85 + 7.25 + 0.07 + \ldots \approx 732.17$.

The exact answer is $\frac{72'485}{99} = 732.171717\ldots$, so 732.17 is a good approximation.

Use this method to calculate the quotient to two decimal places.

a) $6084 : 99$ b) $90'765 : 99$ c) $560'435 : 9999$

d) $3075 : 101$ e) $81'259 : 101$ f) $458'694 : 10'001$

For **161–163**: Some notations used are introduced in 1.4 «Further Topics».

- Mathematical induction is introduced on page 21.
- The notation AP2 is shorthand for an arithmetic sequence of order 2 (see the box *Arithmetic Sequences of Higher Order* on page 15).

161. Start with the function u defined by $u(x) = 2x - 1$. Then, recursively define $w_1 = u$, $w_2 = u \circ w_1 = u \circ u$, $w_3 = u \circ w_2 = u \circ u \circ u$, $\ldots$, $w_{n+1} = u \circ w_n = \underbrace{u \circ u \circ \ldots \circ u}_{(n+1) \text{ functions}}$, $n \in \mathbb{N}$.

a) Determine the function values of $w_2(x)$, $w_3(x)$ and $w_4(x)$.

b) Show that, for $n = 1, 2, 3, 4$, we have: $w_n(x) = 2^n x - (2^n - 1)$.

c) Does point $W(1 \,|\, 1)$ lie on the graph of the function w_n for all $n \in \mathbb{N}$?

d) Prove the formula for $w_n(x)$ given in b) for all $n \in \mathbb{N}$ by induction.

162. In this exercise, the formula for the maximum number of intersections A_n of n lines ($n \in \mathbb{N}$) shall be derived. Draw a sketch and count the maximum number of intersections A_n, where $n = 1, 2, \ldots, 6$. Examine the sequence $A_1, A_2, \ldots, A_6$ in order to develop a formula for the general term A_n. Finally, prove the formula obtained for A_n by induction.

163. Start with a finite or infinite sequence (a_k). Now, consider the function f, which maps each index k to its corresponding term: $f(k) = a_k$ with $k \in \mathbb{N}$.

a) What can be said about the graph of the function f, if (a_k)

  i) is an AP?   ii) is a GP?   iii) is an AP2?

b) Plot each f for the following concrete cases of sequences in separate coordinate systems while assuming that the given sequence has length $n = 8$.

  i) AP: 15, 13.25, 11.5, ...   ii) GP: 8, 4, 2, 1, ...   iii) AP2: 0, 3, 8, 15, 24, ...

1.6   Review Exercises

164. Calculate three more terms of the recursively defined sequence (a_k).

a) $a_1 = 9$, $a_{k+1} = 2a_k - 1$

b) $a_1 = -1$, $a_{k+1} = \frac{-a_k}{2}$

c) $a_1 = 2$, $a_{k+1} = a_k^2 - 1$

d) $a_1 = 2$, $a_2 = 3$, $a_{k+2} = 2a_{k+1} + a_k$

165. The sequence (a_k) is defined by its two initial terms a_1 and a_2 together with the recursion $a_{k+2} = 2a_{k+1} - a_k$ ($k \in \mathbb{N}$). List the terms a_3, a_4 and a_5 and give an explicit representation of the general term a_k.

a) $a_1 = 2$, $a_2 = 4$

b) $a_1 = 3$, $a_2 = 1$

166. How many four-digit natural numbers have a final digit 6?

167. Is the sequence arithmetic, geometric or neither? If it is an AP or a GP answer the following three questions:

- Explain why it is an AP or GP.
- Calculate the 15th term.
- Give the recursive and explicit definitions of the sequence.

a) $3, -6, 12, -24, 48, \ldots$

b) $2, -4, 6, -8, 10, \ldots$

c) $1, 2, 4, 8, 15, \ldots$

d) $765, 654, 543, 432, 321, \ldots$

e) $2, -6, 18, -54, 162, -486, \ldots$

f) $-1, 3, -5, 7, -9, 11, \ldots$

g) $1, 2, -4, -8, 16, 32, -64, -128, \ldots$

h) $-765, -654, -543, -432, -321, \ldots$

168. Define the sequence explicitly and recursively.

a) $1, 4, 7, 10, 13, \ldots$

b) $6, 13, 20, 27, 34, \ldots$

c) $2^2, 2^3, 2^4, 2^5, \ldots$

169. Determine the general term s_n of the sequence of partial sums (s_n) that corresponds to the

a) AS $37 + 30 + 23 + \ldots$

b) GS $16 + 12 + 9 + \ldots$

170. Write the sum of an AS or GS using the summation symbol $\sum$ and calculate it.

a) $15 + 18 + 21 + \ldots + 84$

b) $4 + 12 + 36 + \ldots + 972$

c) $15'360 - 23'040 + 34'560 \mp \ldots - 1'328'602.5$

d) $17 + 11 + 5 - 1 - \ldots - 31 - 37$

171. Calculate the indicated quantities using the information given for an AP.

a) Given: $a_1 = 9$, $a_2 = 11$. Find: a_{48} and s_{48}.

b) Given: $a_1 = 34$, $d = -6$. Find: All indices $i \in \mathbb{N}$ with $a_i = -80$.

c) Given: $a_8 = -2$, $a_{24} = 12$. Find: a_1 and the sum $\sum\limits_{k=8}^{24} a_k$.

172. a) How many terms of the AP $10, 18, \ldots$ are less than 1000?

b) How many terms of the sequence with $u_1 = 4$, $u_{k+1} = u_k + 1.25$ are less than 10^4?

173. Which AS of length 3 with common difference $d = 12$ has a sum of 345?

174. An AP starts with 3, ends with 37 and has a sum of 400. How many terms does it contain?

175. An AP with common difference $d = 3$ starts with 5 and ends with 302. What is its length?

176. An AP 975, 957, ... consists of positive terms. How many terms can this AP include at most and which value does the final and smallest term a_n have?

177. For a GP, calculate the indicated quantities using the given information.

    a) Given: $a_4 = 27$, $q = 0.3$. Find: a_1 and s_9.

    b) Given: $a_1 = 3840$, $q = \frac{3}{2}$. Find: All indices $i \in \mathbb{N}$ with $a_i = 98'415$.

    c) Given: $a_{11} = 162$, $a_{15} = 1458$. Find: q and a_1.

    d) Given: $a_1 = 0.2$, $q = 2$. Find: All indices $i \in \mathbb{N}$ with $s_i = 26'214.2$.

178. a) How many terms of the GP 1000, 999, ... are greater than 1?

    b) How many terms of the sequence (u_n) with $u_1 = 1024$, $u_{k+1} = 0.5 \cdot u_k$ are greater than 0.1?

    c) At least how many terms of the sequence with $a_k = 4 \cdot 3^{k-1}$ need to be added in order that the sum exceeds 10^4?

179. There are infinitely many GPs of length 3 with the property that the product $a_1 \cdot a_2 \cdot a_3$ takes the value 27'000. However, for all possible GPs with this property the middle term a_2 always takes the same value. What is a_2?

180. For a GP the values $a_7 = \frac{729}{32}$ and $q = \frac{3}{4}$ are known. Calculate the limit s of the infinite GS.

181. Calculate the limit of the infinite GS $\sqrt{12} + \sqrt{6} + \sqrt{3} + \dots$

182. Consider a square of side length 100 cm, in which more squares are inscribed (see figure). Let Q_1, Q_2, Q_3, ... denote the areas of the inscribed squares, whose side lengths form a GP.

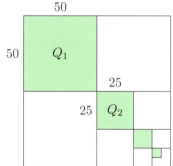

    a) Calculate Q_1, Q_2 and Q_3.

    b) Determine the explicit formula for Q_n.

    c) From which $n \in \mathbb{N}$ on is the area Q_n less than 10^{-34} cm^2?

    d) Calculate the sum of the areas of the first eight squares.

    e) What is the total area of all squares?

183. Calculate the limit of the infinite GS.

    a) $18.9 + 12.6 + 8.4 + \dots$ b) $18 - 6 + 2 \mp \dots$ c) $\sum_{k=1}^{\infty} \frac{3}{4^k}$ d) $\sum_{k=0}^{\infty} 6 \cdot \left(\frac{-2}{5} \right)^k$

184. For which values of $m \in \mathbb{R}$ ($m \neq 0$) does the infinite GS $4m + 6m^2 + 9m^3 + \dots$ converge?

2   Limits

1. *Infinite geometric series.* Which number do the partial sums of
$\frac{1}{2} + \frac{1}{4} + \frac{1}{8} + \frac{1}{16} + \ldots$ approach?

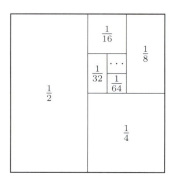

$$\frac{1}{2} + \frac{1}{4} + \frac{1}{8} = \frac{7}{8} = 0.875$$

$$\frac{1}{2} + \frac{1}{4} + \frac{1}{8} + \ldots + \frac{1}{64} = \frac{63}{64} = 0.984375$$

$$\frac{1}{2} + \frac{1}{4} + \frac{1}{8} + \ldots + \frac{1}{512} = \frac{511}{512} = 0.998046875$$

$$\frac{1}{2} + \frac{1}{4} + \frac{1}{8} + \ldots + \frac{1}{65'536} = \frac{65'535}{65'536} = 0.9999847412109375$$

$\vdots$

2. *Development of the temperature of a body over time*

   a) Lena takes an apple out of the refrigerator and places it on the kitchen table. The apple slowly gets warmer. What is the highest temperature the apple will reach?

   b) Louis places his hot cup of coffee on the kitchen table. The coffee and the cup slowly cool down. What is the lowest temperature the coffee and the cup will reach?

2.1   Limits of Sequences

For **3–5**: Examine whether the sequence approaches a certain number, i.e. if the sequence has a so-called limit. If so, find the limit value. In some cases, the formulas for infinite geometric sequences and series are useful.

3. a) 0.3, 0.33, 0.333, 0.3333, $\ldots$ b) 1, $1 + \frac{1}{2}$, $1 + \frac{1}{2} + \frac{1}{4}$, $1 + \frac{1}{2} + \frac{1}{4} + \frac{1}{8}$, $\ldots$

   c) 1, $1 - \frac{1}{2}$, $1 - \frac{1}{2} + \frac{1}{4}$, $1 - \frac{1}{2} + \frac{1}{4} - \frac{1}{8}$, $\ldots$ d) $3 + \frac{1}{2}$, $3 - \frac{1}{4}$, $3 + \frac{1}{8}$, $\ldots$

4. a) $\frac{1}{10}$, $\frac{1}{100}$, $\frac{1}{1000}$, $\frac{1}{10'000}$, $\ldots$ b) 1.9, 1.99, 1.999, 1.9999, $\ldots$

   c) $\frac{2}{3}$, $\left(\frac{2}{3}\right)^2$, $\left(\frac{2}{3}\right)^3$, $\left(\frac{2}{3}\right)^4$, $\ldots$ d) $\sqrt{2}$, $\sqrt[3]{3}$, $\sqrt[4]{4}$, $\sqrt[5]{5}$, $\ldots$

5. a) 0, 0.1, 0.12, 0.123, 0.1234, $\ldots$ b) 0, 0.10, 0.10100, 0.101001000, $\ldots$

   c) 1, 2, 1, 2, 1, 2, $\ldots$ d) 2, 2, 2, 2, 2, 2, $\ldots$

Limit of a Sequence

A number $a \in \mathbb{R}$ is called the *limit* of the sequence (a_k) if the terms a_k come arbitrarily close to the value a and remain arbitrarily close for increasing k where $k \in \mathbb{N}$.

Notation: $a_k \longrightarrow a$ for $k \longrightarrow \infty$ or $\lim\limits_{k \to \infty} a_k = a$

Example: $a_k = \frac{1}{k}$. The value of the fraction $\frac{1}{k}$ becomes ever smaller and remains arbitrarily close to the number 0 for increasing k. The sequence (a_k) with $a_k = \frac{1}{k}$ has the limit 0; in short $\lim\limits_{k\to\infty} \frac{1}{k} = 0$.

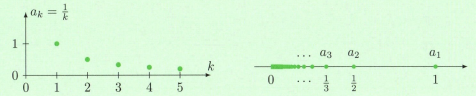

A sequence that has a limit is called *convergent*; if no limit exists, the sequence is said to be *divergent*. A sequence has at most one limit.

A sequence (a_k) is called *properly divergent* for $k \to \infty$ if the terms a_k either tend to ∞ or to $-\infty$.

A sequence (a_k) is called *oscillating* for $k \to \infty$ if the terms of the sequence a_k neither tend to a fixed number nor to ∞ or $-\infty$ for $k \to \infty$.

Examples for divergence:

- $\lim\limits_{k\to\infty} \sqrt{k} = \infty$; properly divergent

- $\lim\limits_{k\to\infty} (-1)^k$ does not exist; oscillating

- $\lim\limits_{k\to\infty} (-2)^k$ does not exist; oscillating

For **6–8**: What is the limit, if there is any, of the sequence (a_k)? List a few terms of the sequence if necessary and justify your guess.

6. a) $a_k = \dfrac{2k+1}{k} = 2 + \dfrac{1}{k}$
     b) $a_k = \dfrac{4k-3}{k} = 4 - \dfrac{3}{k}$
     c) $a_k = \dfrac{3k+1}{k}$

    d) $a_k = 5 + (-1)\cdot\dfrac{1}{k}$
     e) $a_k = 6 + (-1)^k \cdot \dfrac{1}{k}$
     f) $a_k = \dfrac{2 + (-1)^k}{k}$

7. a) $a_k = \dfrac{5}{k+5}$
     b) $a_k = \dfrac{7}{2k-1}$
     c) $a_k = \dfrac{k-10}{10}$

8. a) $a_k = \left(\dfrac{4}{5}\right)^k$
     b) $a_k = 3 + \left(\dfrac{-7}{8}\right)^k$
     c) $a_k = 2 + \left(\dfrac{4}{3}\right)^k$

9. Freshly brewed tea in a teapot has a temperature of $94\,^{\circ}\text{C}$. The process of cooling depends on the properties of the teapot (material and surface area) as well as on the temperature difference between the tea and its surroundings. The ambient temperature is $24\,^{\circ}\text{C}$ and the temperature difference between the surroundings and the tea diminishes by $\frac{1}{3}$ of its value every minute.

a) Determine an explicit formula for the sequence (a_k) that describes the time dependent temperature difference. Here, a_k denotes the temperature difference in $^{\circ}\text{C}$ at time k with $k \in \mathbb{N}_0$ given in minutes.

b) How do the terms of the sequence (a_k) behave with increasing index? What temperature does the tea in the teapot gradually approach?

For **10–12**: Formal definition of the limit of a sequence: A number $a \in \mathbb{R}$ is called the *limit* of the sequence (a_k) if, for every strip of width ε above and below the limit a, no matter how small, there exists an index n from which onwards all the terms of the sequence actually lie within this strip (see figure).

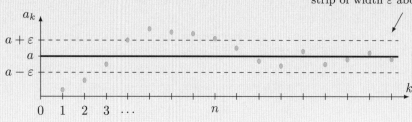

10. Determine the limit a of the sequence (a_k) and calculate the smallest index n such that the distances of all successive a_k to a are less than 0.01.

a) $a_k = \dfrac{1}{k(k+1)}$
b) $a_k = 2^{-k^2}$

11. Which is the smallest index n such that all successive a_k are closer to the limit of (a_k) than 10^{-3} or 10^{-10}, respectively?

a) $a_k = 1 + \dfrac{1}{k}$
b) $a_k = 7 - \dfrac{1}{\sqrt{k}}$
c) $a_k = \dfrac{k+1}{2k}$

12. Which is the smallest index n such that all successive $a_k = \frac{5k-1}{k}$ are closer to the limit of (a_k) than 0.02?

13. Calculate the smallest index n such that all successive terms of the properly divergent sequence (a_k) are greater than or equal to 10^6.

a) $a_k = 10k^5$
b) $a_k = \lg(100k)$
c) $a_k = 2\sqrt{5k-1}$

14. a) Write down the general term a_k for two different sequences (a_k) that both have the limit
$$\lim_{k\to\infty} a_k = 2.$$
b) Give a sequence (a_k) such that $0 < a_k < 1$ and $\lim\limits_{k\to\infty} a_k = \frac{1}{2}$.

15. Yes or no?

a) Does the sequence (a_k) with $a_k = (-1)^k$ converge?
b) Is the constant sequence (a_k) with $a_k = c \in \mathbb{R}$ convergent?
c) Are there convergent arithmetic sequences?
d) Are there convergent geometric sequences?
e) Are there sequences that are neither convergent nor divergent?

16. Is the statement true? What do you think?

a) $\lim\limits_{k\to\infty} \frac{1}{k-1000}$ does not exist as the fraction is not defined for $k = 1000$.

b) $\lim\limits_{k\to\infty} \sqrt{1 - \frac{1000}{k}}$ does not exist as the square root is not defined for $k = 1, 2, 3, \dots, 999$.

Properties of Limits

If (a_k) and (b_k) are convergent sequences with $\lim\limits_{k\to\infty} a_k = a$ and $\lim\limits_{k\to\infty} b_k = b$, then:

1) $\lim\limits_{k\to\infty} (a_k \pm b_k) = \lim\limits_{k\to\infty} a_k \pm \lim\limits_{k\to\infty} b_k = a \pm b$ 2) $\lim\limits_{k\to\infty} (a_k \cdot b_k) = \lim\limits_{k\to\infty} a_k \cdot \lim\limits_{n\to\infty} b_k = a \cdot b$

3) $\lim\limits_{k\to\infty} \dfrac{a_k}{b_k} = \dfrac{\lim\limits_{k\to\infty} a_k}{\lim\limits_{k\to\infty} b_k} = \dfrac{a}{b}$, if $b \neq 0$

Examples:

- $\lim\limits_{k\to\infty} \dfrac{4k^2 - 3}{2k^2 + 3k} = \lim\limits_{k\to\infty} \dfrac{k^2\left(4 - \frac{3}{k^2}\right)}{k^2\left(2 + \frac{3}{k}\right)} = \dfrac{\lim\limits_{k\to\infty}\left(4 - \frac{3}{k^2}\right)}{\lim\limits_{k\to\infty}\left(2 + \frac{3}{k}\right)} = \dfrac{\lim\limits_{k\to\infty} 4 - \lim\limits_{k\to\infty} \frac{3}{k^2}}{\lim\limits_{k\to\infty} 2 + \lim\limits_{k\to\infty} \frac{3}{k}} = \dfrac{4 - 0}{2 + 0} = 2$

- $\lim\limits_{k\to\infty} (\ln(k+1) - \ln(k)) = \lim\limits_{k\to\infty} \ln\left(\frac{k+1}{k}\right) = \lim\limits_{k\to\infty} \ln\left(1 + \frac{1}{k}\right) = \ln\left(\lim\limits_{k\to\infty}\left(1 + \frac{1}{k}\right)\right) = \ln(1) = 0$

Four important limits:

i) $\lim\limits_{k\to\infty} \sqrt[k]{a} = 1, \ a > 0$ ii) $\lim\limits_{k\to\infty} \sqrt[k]{k} = 1$

iii) $\lim\limits_{k\to\infty} \dfrac{k^m}{a^k} = 0, \ a > 1, m \in \mathbb{R}$ iv) $\lim\limits_{k\to\infty} \dfrac{\log_a(k)}{k^m} = 0, \ a > 1, m > 0$

17. The sequence (a_k) is given by the general term a_k. Calculate $\lim\limits_{k\to\infty} a_k$.

a) $a_k = \dfrac{2k^2 - 4k}{k^2 + 3k - 5}$

b) $a_k = \dfrac{3k - 21}{4k^2}$

c) $a_k = \dfrac{-k^2}{5k^2 + k}$

d) $a_k = \dfrac{(-1)^k \cdot 8}{k^2 + 1}$

e) $a_k = \dfrac{4k + 3 \cdot (-1)^k}{2k}$

f) $a_k = \dfrac{5\sqrt{k} - 3}{\sqrt{k}}$

g) $a_k = \dfrac{2 + 7\sqrt{k}}{4\sqrt{k} + 41}$

h) $a_k = \dfrac{(2k - 1)^3}{(k - 1)^3}$

i) $a_k = \dfrac{(2k - 1)^4}{1 - k^4}$

18. Determine the limit if it exists. Otherwise note the type of divergence.

a) $\lim\limits_{k\to\infty} \left(\dfrac{1 - 4k^3}{5k^3 + 4k - 2}\right)$

b) $\lim\limits_{k\to\infty} \left(\dfrac{2k^2 - 3k + 4}{5 - 6k}\right)$

c) $\lim\limits_{k\to\infty} \left(\dfrac{2k + (-1)^k \cdot 5k}{k}\right)$

19. For a sequence, the general term a_k is known. Calculate $\lim\limits_{k\to\infty} a_k$.

a) $a_k = \dfrac{1}{3^k + 3^{-k}}$

b) $a_k = e^{\frac{1}{k}}$

c) $a_k = \dfrac{k^3}{3^k}$

20. Calculate the limit if it exists. Otherwise note the type of divergence.

a) $\lim\limits_{k\to\infty} \sqrt[k]{5}$

b) $\lim\limits_{k\to\infty} \dfrac{4k^5}{3^k}$

c) $\lim\limits_{k\to\infty} \dfrac{\ln(k)}{7k^3}$

d) $\lim\limits_{k\to\infty} \dfrac{5^k - 2}{3k^2}$

e) $\lim\limits_{k\to\infty} \dfrac{4^k}{\ln(k)}$

f) $\lim\limits_{k\to\infty} \ln\left(\dfrac{k + 1}{k^2}\right)$

21. Does the sequence (a_k) converge? Justify your answer and determine the limit or note the type of divergence, respectively.

a) $a_k = (-1)^k + 5$

b) $a_k = \left(5 + (-1)^k\right) \cdot \frac{2}{k^2}$

c) $a_k = \cos\left(\frac{1}{k}\right)$

d) $a_{k+1} = a_k + 4, \quad a_1 = 2$

e) $a_{k+1} = a_k \cdot \frac{1}{4}, \quad a_1 = 100$

f) $a_{k+1} = a_k \cdot 1.1, \quad a_1 = -11$

22. The sequence (a_k) is given by the general term a_k. Determine the limit of the sequence.

a) $a_k = e^{1-k}$

b) $a_k = e^{-2k} \cdot \cos\left(\frac{1}{k}\right)$

c) $a_k = \dfrac{1000}{1 + 999 \cdot e^{-k/2}}$

23. Determine $\lim\limits_{k \to \infty} \frac{4 \cdot k^m}{7 \cdot k^\ell}$ for $m, \ell \in \mathbb{N}$. Distinguish the cases $m > \ell$, $m = \ell$ and $m < \ell$.

24. Calculate the limit depending on the value of the parameter $a \in \mathbb{R}$.

a) $\lim\limits_{k \to \infty} \dfrac{a^2 k}{a^2 k - 1}$

b) $\lim\limits_{k \to \infty} \dfrac{ak}{a^2 k + 2}$

25. The sequence (a_k) is given by $a_k = \frac{2k-1}{k+3}$. Calculate $\lim\limits_{k \to \infty} a_k$ and then construct a sequence (b_k) with $\lim\limits_{k \to \infty} b_k = 4$ using the sequence (a_k) by applying only one elementary operation.

26. Euler's number $e = 2.718281828\ldots$ can be expressed as a limit: $e = \lim\limits_{n \to \infty} \left(1 + \frac{1}{n}\right)^n$. Use this to determine the following limits. *Hint:* The substitution $m = \frac{n}{2}$ is useful for d).

a) $\lim\limits_{n \to \infty} \left(1 + \frac{1}{n}\right)^{2n}$

b) $\lim\limits_{n \to \infty} \left(1 + \frac{1}{n}\right)^{-n}$

c) $\lim\limits_{n \to \infty} \left(1 + \frac{1}{3n}\right)^{3n}$

d) $\lim\limits_{n \to \infty} \left(1 + \frac{2}{n}\right)^n$

Limit of an Infinite Series

If the sequence of partial sums (s_n), with $s_n = a_1 + a_2 + \ldots + a_n = \sum\limits_{k=1}^{n} a_k$, associated with the sequence (a_k) has a limit s, the associated *infinite series* $\sum\limits_{k=1}^{\infty} a_k$ is called convergent. We have:

$$s = \lim_{n \to \infty} s_n = \lim_{n \to \infty} \sum_{k=1}^{n} a_k = \sum_{k=1}^{\infty} a_k.$$

---

27. Calculate.

a) $\sum\limits_{k=1}^{\infty} \left(\frac{3}{4}\right)^{k-1}$

b) $\sum\limits_{k=1}^{\infty} \dfrac{3^k - 1}{6^k}$

c) $\lim\limits_{n \to \infty} \sum\limits_{k=1}^{n} \dfrac{1}{\left(\sqrt{2}\right)^k}$

d) $\lim\limits_{n \to \infty} \sum\limits_{k=1}^{n} (-1)^{k+1} \cdot \dfrac{3}{2^k}$

28. Calculate the limit $\lim\limits_{n \to \infty} \left(\frac{1}{n^2} + \frac{2}{n^2} + \frac{3}{n^2} + \ldots + \frac{n-1}{n^2} + \frac{n}{n^2}\right)$.
Hint: Use $1 + 2 + 3 + \ldots + n = \frac{n(n+1)}{2}$.

29. *The harmonic series.* From $\lim\limits_{k \to \infty} a_k = 0$ one may not conclude that $\sum\limits_{k=1}^{\infty} a_k$ exists.

Example: $a_k = \frac{1}{k}$. Although $\lim\limits_{k \to \infty} a_k = 0$, the harmonic series $\sum\limits_{k=1}^{\infty} a_k$ diverges, as

$$\sum_{k=1}^{\infty} \frac{1}{k} = 1 + \frac{1}{2} + \left(\frac{1}{3} + \frac{1}{4}\right) + \left(\frac{1}{5} + \frac{1}{6} + \frac{1}{7} + \frac{1}{8}\right) + (\ldots) + \ldots$$

$$> 1 + \frac{1}{2} + \left(\frac{1}{4} + \frac{1}{4}\right) + \left(\frac{1}{8} + \frac{1}{8} + \frac{1}{8} + \frac{1}{8}\right) + (\ldots) + \ldots = 1 + \frac{1}{2} + \frac{1}{2} + \frac{1}{2} + \ldots = \infty.$$

a) Does the series $\frac{1}{100} + \frac{1}{101} + \frac{1}{102} + \ldots$ converge?

b) Does the series $\frac{1}{11} + \frac{1}{111} + \frac{1}{1111} + \ldots$ converge?

2.2   Limits of Functions

Limit of a Function for $x \to \infty$ and $x \to -\infty$

The number $b \in \mathbb{R}$ is called the *limit* of the function f for $x \to \infty$ or $x \to -\infty$, if the function values $f(x)$ come arbitrarily close to the value b and stay arbitrarily close as x on the x-axis tends to the far right or left, respectively.

Notation: $\lim\limits_{x \to \infty} f(x) = b$ or $\lim\limits_{x \to -\infty} f(x) = b$, respectively

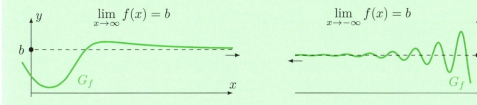

Remark: The following statements also hold for the case $x \to -\infty$.

- The limit of a function for $x \to \infty$ is unique. A function that has a limit for $x \to \infty$ is called *convergent* for $x \to \infty$. If no limit exists, the function is called *divergent* for $x \to \infty$.

- $x \to \infty$ signifies that x tends to ∞ (x becomes arbitrarily large), but never actually equals ∞.

- A function is called *properly divergent* for $x \to \infty$ if $f(x)$ either tends to ∞ or $-\infty$ for $x \to \infty$.

- A function is called *oscillating* for $x \to \infty$ if $f(x)$ neither tends to a fixed value nor to ∞ or $-\infty$ for $x \to \infty$.

Examples of divergence:

- $\lim\limits_{x \to \infty} x^3 = \infty$; proper divergence
- $\lim\limits_{x \to -\infty} x^2 = \infty$; proper divergence
- $\lim\limits_{x \to \infty} \ln(x) = \infty$; proper divergence
- $\lim\limits_{x \to \infty} \sin(x)$ does not exist; oscillating

---

30. Determine both limits.

a) $\lim\limits_{x \to \infty} \frac{1000}{x}$; $\lim\limits_{x \to -\infty} \frac{1000}{x}$
b) $\lim\limits_{x \to \infty} \frac{x}{x^3}$; $\lim\limits_{x \to -\infty} \frac{x}{x^3}$
c) $\lim\limits_{x \to \infty} \frac{1}{\sqrt{x}}$; $\lim\limits_{x \to -\infty} \frac{1}{\sqrt{-x}}$

31. Determine the limit if it exists. Otherwise note the type of divergence.

a) $\lim\limits_{x \to \infty} \frac{1}{x^2 + 1}$
b) $\lim\limits_{x \to -\infty} (2^x + 1)$
c) $\lim\limits_{x \to \infty} \cos(x)$
d) $\lim\limits_{x \to \infty} \frac{x^2 + 1}{x}$

32. True or false? Justify.

a) If $\lim\limits_{x \to \infty} f(x) = 3$, then the function f never has the function value 3.

b) If $\lim\limits_{x \to \infty} f(x) = 3$, then for continuously increasing values of x, the function value $f(x)$ comes arbitrarily close to the limit 3 and remains arbitrarily close.

c) If $\lim\limits_{x \to \infty} f(x) = 3$, then for continuously increasing values of x, the function value $f(x)$ continuously comes closer and closer to the limit 3.

Limit Laws

If both limits $\lim\limits_{x\to\infty} f(x) = u$ and $\lim\limits_{x\to\infty} g(x) = v$ exist, then:

1) $\lim\limits_{x\to\infty} (f(x) \pm g(x)) = \lim\limits_{x\to\infty} f(x) \pm \lim\limits_{x\to\infty} g(x) = u \pm v$

2) $\lim\limits_{x\to\infty} (f(x) \cdot g(x)) = \lim\limits_{x\to\infty} f(x) \cdot \lim\limits_{x\to\infty} g(x) = u \cdot v$

3) $\lim\limits_{x\to\infty} \dfrac{f(x)}{g(x)} = \dfrac{\lim\limits_{x\to\infty} f(x)}{\lim\limits_{x\to\infty} g(x)} = \dfrac{u}{v}$, if $v \neq 0$

Remark: The laws are also valid for the case $x \to -\infty$.

Examples:

- $\lim\limits_{x\to\infty} \dfrac{3x^2+3}{4x^2+3x} = \lim\limits_{x\to\infty} \dfrac{x^2\left(3+\frac{3}{x^2}\right)}{x^2\left(4+\frac{3}{x}\right)} = \dfrac{\lim\limits_{x\to\infty}\left(3+\frac{3}{x^2}\right)}{\lim\limits_{x\to\infty}\left(4+\frac{3}{x}\right)} = \dfrac{\lim\limits_{x\to\infty} 3 + \lim\limits_{x\to\infty}\frac{3}{x^2}}{\lim\limits_{x\to\infty} 4 + \lim\limits_{x\to\infty}\frac{3}{x}} = \dfrac{3+0}{4+0} = \dfrac{3}{4}$

- $\lim\limits_{x\to\infty} \dfrac{\sqrt{x^4+1}}{3x^2+x} = \lim\limits_{x\to\infty} \dfrac{x^2\sqrt{1+\frac{1}{x^4}}}{x^2\left(3+\frac{1}{x}\right)} = \dfrac{\lim\limits_{x\to\infty}\sqrt{1+\frac{1}{x^4}}}{\lim\limits_{x\to\infty}\left(3+\frac{1}{x}\right)} = \dfrac{1}{3}$

- $\lim\limits_{x\to-\infty} \dfrac{\sqrt{1-x}-1}{x} = \lim\limits_{x\to-\infty} \left(\dfrac{\sqrt{1-x}-1}{x} \cdot \dfrac{\sqrt{1-x}+1}{\sqrt{1-x}+1}\right) = \lim\limits_{x\to-\infty} \dfrac{1-x-1}{x(\sqrt{1-x}+1)} = \lim\limits_{x\to-\infty} \dfrac{-1}{\sqrt{1-x}+1} = 0$

Two important limits:

i) $\lim\limits_{x\to\infty} \dfrac{x^k}{a^x} = 0, \; a > 1, \, k \in \mathbb{R}$ \qquad\qquad ii) $\lim\limits_{x\to\infty} \dfrac{\log_a(x)}{x^k} = 0, \; a > 1, \, k > 0$

33. Determine the limit for $x \to \infty$ if it exists.

a) $\dfrac{x}{3x-1}$ \qquad b) $\dfrac{1-x^2}{x}$ \qquad c) $\dfrac{3x^7-4x^5}{2x^7+7x^6}$ \qquad d) $\dfrac{3-x^3}{2x^3+x^2}$

e) $\dfrac{6+3\sqrt{x}}{\sqrt{x}}$ \qquad f) $\dfrac{2x^2}{\sqrt{x}}$ \qquad g) $\dfrac{\sqrt{x}-x^2}{\sqrt{x}+x^2}$ \qquad h) $\dfrac{x^4+2\sqrt{x}}{3x^4-\sqrt{x}}$

34. Determine the limit.

a) $\lim\limits_{x\to\infty} \dfrac{\sqrt{2x^2+1}}{x-1}$ \qquad b) $\lim\limits_{x\to\infty} \dfrac{2x-3}{\sqrt{x^2+1}}$ \qquad c) $\lim\limits_{x\to\infty} \dfrac{\sqrt[4]{x}+1}{\sqrt[3]{x}+1}$ \qquad d) $\lim\limits_{x\to\infty} \dfrac{5x-\sqrt{x+1}}{x+1}$

35. Determine the limit of $\sqrt{2x+1} - \sqrt{2x}$ for $x \to \infty$ using a clever expansion.

36. What is the value of $\lim\limits_{x\to\infty} f(x)$?

a) $f(x) = \dfrac{2+3\cdot 2^x}{2^x \cdot 4 - 1}$ \qquad\qquad b) $f(x) = \dfrac{5^{-x}+3}{8-2\cdot 5^{-x}}$

37. Calculate.

a) $\lim\limits_{x\to\infty} \dfrac{x^3+\sin(x)}{x^3}$ \qquad b) $\lim\limits_{x\to-\infty} \dfrac{x^2-\sin(x)}{x^3}$ \qquad c) $\lim\limits_{x\to\infty} \dfrac{\cos(x)-3x}{1-2x}$

38. Determine both limits and compare them.

a) $\lim\limits_{x\to\infty} \dfrac{2x^6 - 3x^5 + 7x^3 + 4x - 8}{5x^6 + 8x^5 - 9x^4 + 11x^2}$; $\lim\limits_{x\to-\infty} \dfrac{2x^6 - 3x^5 + 7x^3 + 4x - 8}{5x^6 + 8x^5 - 9x^4 + 11x^2}$

b) $\lim\limits_{x\to\infty} \dfrac{10x^8 - 4x^7 + 12}{x^9 + 3x^6 + 4}$; $\lim\limits_{x\to-\infty} \dfrac{10x^8 - 4x^7 + 12}{x^9 + 3x^6 + 4}$

c) $\lim\limits_{x\to\infty} \dfrac{\sqrt{x+1}}{x-1}$; $\lim\limits_{x\to-\infty} \dfrac{\sqrt{x+1}}{x-1}$

d) $\lim\limits_{x\to\infty} \dfrac{x^9 + x^4}{e^x}$; $\lim\limits_{x\to-\infty} \dfrac{x^9 + x^4}{e^x}$

39. Determine the limit if it exists.

a) $\lim\limits_{x\to\infty} \left(\dfrac{2.46x^2 - 4.68x}{1.23x^2 + 4.56} \right)^2$

b) $\lim\limits_{x\to\infty} \sqrt{\dfrac{8x^2 - 5}{2x^2 + 3x}}$

c) $\lim\limits_{x\to\infty} \left(\dfrac{12 - 3x}{6x + 54} \right)^{-7}$

d) $\lim\limits_{x\to\infty} \left(\dfrac{x^2 - 3x + 5}{8x^2 - 13} \right)^{\frac{1}{3}}$

40. Is $f(x)$ convergent or divergent for $x \to \infty$ or $x \to -\infty$? If convergent, determine the limit.

a) $f(x) = 3^{-x}$ b) $f(x) = e^{-x}$ c) $f(x) = 3^{-x-2}$ d) $f(x) = 3^{-5x^2-2}$

e) $f(x) = 3^x$ f) $f(x) = e^x$ g) $f(x) = 3^{x-10}$ h) $f(x) = e^{x^2-1}$

41. Is $f(x)$ convergent or divergent for $x \to \infty$? If convergent, determine the limit.

a) $f(x) = \dfrac{3^x}{x^5 + 1}$ b) $f(x) = \dfrac{\lg(x)}{\sqrt{x}}$ c) $f(x) = \dfrac{x^3 - 1}{x \cdot \ln(x)}$ d) $f(x) = \dfrac{\left(\frac{1}{2}\right)^x}{\frac{1}{x^2}}$

42. Determine the limit for $x \to \infty$ and $x \to -\infty$.

a) $\dfrac{x^4 + x^2 + 6}{9 - x^4}$ b) $\dfrac{\sqrt{x^2 - 9}}{x}$ c) $\dfrac{\sqrt[3]{8x^6 + 5}}{3x^2 - 4x}$ d) $\dfrac{\lg(x)}{x^5 + 7}$

43. Determine the parameters a, b and c fulfilling: $\lim\limits_{x\to\infty} \dfrac{ax^b + 4x^5 + cx^3 - 1}{2x^3 - 11x^2 + 5} = 3.$

44. The function $f(x) = \dfrac{4x^4 - 6x^3 + 7}{2x^m + 5x + 1}$ is given.

Determine the exponent $m \in \mathbb{N}_0$ resulting in the following limits, or explain, why no m exists that yields this limit.

a) $\lim\limits_{x\to\infty} f(x) = 2$ b) $\lim\limits_{x\to\infty} f(x) = 0$ c) $\lim\limits_{x\to-\infty} f(x) = -2$

d) $\lim\limits_{x\to-\infty} f(x) = -\infty$ e) $\lim\limits_{x\to-\infty} f(x) = \infty$

Limit of a Function at a Point $a \in \mathbb{R}$

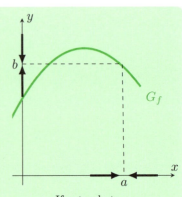

The number b is called the *limit* of the function f at the point a if the function values $f(x)$ come arbitrarily close to the value b and remain arbitrarily close when x approaches the value a.

... then $f(x)$ tends to b

Notation: $f(x) \longrightarrow b$ for $x \longrightarrow a$ or $\lim\limits_{x \to a} f(x) = b$

Remarks:

- $x \to a$ signifies that x tends to a. x can be larger or smaller If x tends to a, ...
  than a, but never equals the value a. So $x \neq a$ holds.
- The limit of f at the point a can exist even when f is not defined at the point $x = a$. (What is important is that f is defined in a neighbourhood of a.)
- The limit of f at the point a can differ from the value of f at a.
- A function is called *properly divergent* for $x \to a$ if the function values $f(x)$ tend either to ∞ or to $-\infty$ for $x \to a$.

  A function is called *oscillating* for $x \to a$ if the function values $f(x)$ neither tend to a fixed value nor to ∞ or $-\infty$ for $x \to a$.

  Examples of divergence:

  - Properly divergent: $\lim\limits_{x \to 0} \frac{1}{x^2}$ • Oscillating: $\lim\limits_{x \to 0} \frac{1}{x^3}$, $\lim\limits_{x \to 0} \sin\left(\frac{1}{x^2}\right)$

45. Determine the limit of the function f at a.

a) $f(x) = 2x - 6$; $a = 3$

b) $f(x) = 3x^2 + 5x - 1$; $a = 1$

c) $f(x) = \frac{x-2}{x}$; $a = 2$

d) $f(x) = x$; $a = -3$

e) $f(x) = x$; $a \in \mathbb{R}$ arbitrary

f) $f(x) = 3$; $a = 1$

46. Calculate $\lim\limits_{x \to a} f(x)$ and note the type of divergence when divergent.

a) $f(x) = \dfrac{2(x-1)}{x-1}$; $a = 1$ b) $f(x) = \dfrac{3x(x-2)}{x-2}$; $a = 2$ c) $f(x) = \dfrac{4x}{x+4}$; $a = -4$

d) $f(x) = \dfrac{(x-3)^3}{x-3}$; $a = 3$ e) $f(x) = \dfrac{x+1}{x}$; $a = 0$ f) $f(x) = \dfrac{x(x-7)}{(x-7)^2}$; $a = 7$

Calculation of Simple Limits

The limit laws for $x \to \pm\infty$ are also valid for $x \to a$ with $a \in \mathbb{R}$.

Example: Calculation of limits using algebraic manipulations (factorisation)

The function f is given by $f(x) = \frac{x^2-4}{x-2}$. Determine $\lim\limits_{x \to 2} f(x)$.

$$\lim_{x \to 2} \frac{x^2 - 4}{x - 2} = \lim_{x \to 2} \frac{(x-2)(x+2)}{x-2} \overset{x \neq 2}{=} \lim_{x \to 2} (x+2) = 4$$

> *Example: Calculating limits by use of the h-method*
>
> The function f is given by $f(x) = \frac{x^3-1}{x-1}$. Determine $\lim\limits_{x \to 1} f(x)$.
>
> Rewrite $x \to 1$ as $x = 1 + h$ with $h \to 0$ and insert it into the formula of f. Then determine $\lim\limits_{h \to 0} f(1+h)$ using algebraic manipulations.
>
> $$\lim_{x \to 1} \frac{x^3-1}{x-1} = \lim_{h \to 0} \frac{(1+h)^3-1}{(1+h)-1} = \lim_{h \to 0} \frac{h(3+3h+h^2)}{h} \overset{h \neq 0}{=} \lim_{h \to 0} (3+3h+h^2) = 3$$

47. Determine the limit.

a) $\lim\limits_{x \to 2} \dfrac{(2-x)(2+x)}{2-x}$

b) $\lim\limits_{x \to 1} \dfrac{10(x-1)(x+1)}{1-x}$

c) $\lim\limits_{x \to 3} \dfrac{8(x-3)}{x^2-9}$

d) $\lim\limits_{x \to -2} \dfrac{x^2+3x+2}{x^2+2x}$

48. Determine both limits and if divergent note the type of divergence.

a) $\lim\limits_{x \to 4} \dfrac{x-4}{x^2-16}$; $\lim\limits_{x \to -4} \dfrac{x-4}{x^2-16}$

b) $\lim\limits_{x \to 5} \dfrac{5+x}{25-x^2}$; $\lim\limits_{x \to -5} \dfrac{5+x}{25-x^2}$

c) $\lim\limits_{x \to 7} \dfrac{49-x^2}{7+x}$; $\lim\limits_{x \to -7} \dfrac{49-x^2}{7+x}$

d) $\lim\limits_{x \to 1} \dfrac{13x^2-13}{x-1}$; $\lim\limits_{x \to -1} \dfrac{13x^2-13}{x-1}$

49. Determine both limits and if divergent note the type of divergence.

a) $\lim\limits_{x \to 1} \dfrac{x^2+4x-5}{x^2-1}$

b) $\lim\limits_{x \to 1} \dfrac{x^2+4x-5}{x^2-2x+1}$

c) $\lim\limits_{x \to -9} \dfrac{x^2+7x-18}{x^2-81}$

d) $\lim\limits_{x \to -2} \dfrac{12x^2+24x}{x^2+10x+16}$

e) $\lim\limits_{x \to 2} \dfrac{3x^2-6x}{6-5x+x^2}$

f) $\lim\limits_{x \to 4} \dfrac{2x^2-14x+24}{12x-7x^2+x^3}$

50. Calculate.

a) $\lim\limits_{h \to 0} \dfrac{(1+h)^2-1}{h}$

b) $\lim\limits_{h \to 0} \dfrac{(2+h)^3-8}{h}$

c) $\lim\limits_{h \to 0} \dfrac{\frac{1}{(2-h)^2}-\frac{1}{4}}{h}$

51. Calculate the limit once *without using* the h-method and once *by using* the h-method.

a) $\lim\limits_{x \to 1} \dfrac{x^3-x^2+x-1}{x^2-1}$

b) $\lim\limits_{x \to 1} \dfrac{x^3+x^2-x-1}{x^2-1}$

52. Calculate.

a) $\lim\limits_{x \to 1} \dfrac{(x+1)^2-4}{x-1}$

b) $\lim\limits_{x \to 3} \dfrac{(x+3)^2-36}{x-3}$

c) $\lim\limits_{x \to 2} \dfrac{(2+x)^2-16}{2-x}$

d) $\lim\limits_{x \to p} \dfrac{(p+x)^2-4p^2}{p-x}$

53. Determine the limit by cleverly expanding the fraction.

a) $\lim\limits_{x \to 0} \dfrac{\sqrt{1-x}-1}{x}$

b) $\lim\limits_{x \to 0} \dfrac{\sqrt{x^2+9}-3}{x^2}$

c) $\lim\limits_{x \to -4} \dfrac{\sqrt{x^2+9}-5}{x+4}$

54. Are both equations true for $x \in \mathbb{R}$? Explain.

$$\frac{x^2+x-12}{x-3} = x+4; \qquad \lim_{x \to 3} \frac{x^2+x-12}{x-3} = \lim_{x \to 3} (x+4)$$

For **55–57**: Determine the limit or note the type of divergence.

55. a) $\lim\limits_{x \to 1} \dfrac{x^2 + 1}{x - 2}$
b) $\lim\limits_{x \to 0} \dfrac{x^2 - 1}{x - 2}$
c) $\lim\limits_{x \to 2} \dfrac{x^2 - 4}{x - 2}$

d) $\lim\limits_{x \to -4} \dfrac{x - 2}{x^2 - 4}$
e) $\lim\limits_{x \to 3} \dfrac{x^2 - 4x + 3}{x^2 - 9}$
f) $\lim\limits_{x \to 1} \dfrac{x^2 - 4x + 3}{x^2 - 9}$

56. a) $\lim\limits_{x \to 1} \dfrac{1 - \sqrt{x}}{1 - x}$
b) $\lim\limits_{x \to 6} \dfrac{x^2 - 36}{x^4 - 1296}$
c) $\lim\limits_{x \to 0} \dfrac{x^2 - 4x + 4}{x^5 + 2x^4 + x^3}$

d) $\lim\limits_{x \to -1} \dfrac{x^2 - 4x + 4}{x^5 + 2x^4 + x^3}$
e) $\lim\limits_{x \to 0} \dfrac{\frac{1}{2+x} - \frac{1}{2}}{x}$
f) $\lim\limits_{x \to 0} \dfrac{(4 + x)^3 - 64}{x}$

57. a) $\lim\limits_{x \to 0} \dfrac{(x + 2)^4 - 2^4}{2x}$
b) $\lim\limits_{x \to 0} \dfrac{(x + 5)^4 - 5^4}{5x}$
c) $\lim\limits_{x \to 0} \dfrac{(x + m)^4 - m^4}{mx}$

58. Calculate both limits and compare the two.

a) $\lim\limits_{x \to 0} \dfrac{\sqrt{x + 2} - \sqrt{2}}{x}$, $\quad \lim\limits_{x \to 2} \dfrac{\sqrt{x} - \sqrt{2}}{x - 2}$
b) $\lim\limits_{x \to 0} \dfrac{\sqrt{x + m} - \sqrt{m}}{x}$, $\quad \lim\limits_{x \to m} \dfrac{\sqrt{x} - \sqrt{m}}{x - m}$

59. In this exercise, it is assumed that the limits exist.

a) The function u fulfils $\lim\limits_{x \to 3} \frac{u(x)-2}{x-1} = 4$. Find the limit $\lim\limits_{x \to 3} u(x)$.

b) The two limits $\lim\limits_{x \to 5} \frac{f(x)-4}{x-5} = 3$ and $\lim\limits_{x \to 5} \frac{g(x)-4}{x-5} = 6$ are given. Using them, calculate the limits $\lim\limits_{x \to 5} f(x)$ and $\lim\limits_{x \to 5} g(x)$.

c) The two limits $\lim\limits_{x \to 0} u(x) = 3$ and $\lim\limits_{x \to 0} v(x) = -4$ are given. Using them, calculate the limit $\lim\limits_{x \to 0} \frac{2 \cdot u(x) - v(x)}{u(x) + 5}$.

60. *When the calculator fails.* The function f with $f(x) = \frac{\sqrt{x^2 + 16} - 4}{x^2}$ is given for $x \neq 0$.

a) Calculate $\lim\limits_{x \to 0} f(x)$ using algebraic manipulations.

b) The table on the right was computed using a calculator. (As $x^2 \geq 0$, the table only lists positive values of x.)

   Verify the table using your calculator. Based on the table, what value do you suspect the limit $\lim\limits_{x \to 0} f(x)$ to equal?

c) Comment on your results in a) and b).

d) Which transformed term from a) yields correct approximations for the limit using the values of x in the table and computing them by calculator as in b)?

x	$f(x)$
1	0.123105...
0.5	0.124515...
0.1	0.124980...
0.01	0.124999...
0.001	0.125000...
0.0001	0.125000...
10^{-5}	0.125000...
10^{-6}	0.100000...
10^{-7}	0.000000...
10^{-8}	0.000000...

61. *Geometric calculation of limits.* $\lim\limits_{x \to 0} \frac{\sin(x)}{x}$ is to be calculated using the figure on the right.

a) Explain the following steps:

$$F_{\Delta OAB} \leq F_{\text{sector } AOB} \leq F_{\Delta OAC}$$

$$\overset{(1)}{\Rightarrow} \frac{\sin(x)}{2} \leq \frac{x}{2} \leq \frac{\tan(x)}{2} \qquad \overset{(2)}{\Rightarrow} \frac{\sin(x)}{2} \leq \frac{x}{2} \leq \frac{\sin(x)}{2\cos(x)}$$

$$\overset{(3)}{\Rightarrow} 1 \leq \frac{x}{\sin(x)} \leq \frac{1}{\cos(x)} \qquad \overset{(4)}{\Rightarrow} 1 \geq \frac{\sin(x)}{x} \geq \cos(x)$$

$$\overset{(5)}{\Rightarrow} \lim_{x \to 0} \frac{\sin(x)}{x} = 1$$

b) Determine the following limits: (i) $\lim\limits_{x \to 0} \frac{\sin^2(x)}{x}$ (ii) $\lim\limits_{x \to 0} \frac{\sin(2x)}{x}$ (iii) $\lim\limits_{x \to 0} \frac{\tan(x)}{x}$

2.3 Further Topics

Further Properties of Sequences and Theorems about Convergent Sequences and Series

> **Further Properties of Sequences**
>
> A sequence with limit 0 is called a *null sequence*. Accordingly, the limit of an arbitrary sequence (a_k) can be reformulated: (a_k) has the limit a $\Leftrightarrow$ $(a_k - a)$ is a null sequence
>
> A non-constant sequence (a_k) is called *monotonically increasing* (or *decreasing*) if, for all $k \in \mathbb{N}$, we have: $a_{k+1} \geq a_k$ (or $a_{k+1} \leq a_k$, respectively).
>
> A sequence (a_k) is called *strictly increasing* (or *decreasing*) if, for all $k \in \mathbb{N}$, we have: $a_{k+1} > a_k$ (or $a_{k+1} < a_k$, respectively).
>
> A sequence (a_k) is called *alternating* if its terms are alternately positive or negative, i.e. if, for all $k \in \mathbb{N}$, we have: $a_{k+1} \cdot a_k < 0$.
>
> A sequence (a_k) is called *bounded from below and above* if there exist two numbers m and M such that, for all terms a_k, we have: $m \leq a_k \leq M$ for all $k \in \mathbb{N}$.

62. Is the sequence defined by the general term a_k ($k \in \mathbb{N}$) a null sequence?

a) $a_k = (-0.99)^k$ b) $a_k = \frac{\sqrt{k}}{\sqrt[3]{k}}$ c) $a_k = \frac{k^2}{2^k}$ d) $a_k = \frac{k^{-3}}{\left(\frac{1}{3}\right)^k}$ e) $a_k = \frac{\ln(k)}{k}$

63. Is the sequence defined by the general term a_k ($k \in \mathbb{N}$) a null sequence?

a) $a_k = \frac{2 \cdot 3^k - 3 \cdot 2^k}{6^k}$ b) $a_k = \frac{1}{2} - \sum_{i=1}^{k}\left(\frac{1}{2^i} - \frac{1}{2^{i+1}}\right)$

64. Check whether the sequence given by the general term a_k, $k \in \mathbb{N}$, is monotonically increasing or monotonically decreasing. *Hint:* For c) and d), examine the difference $a_{k+1} - a_k$.

a) $a_k = \frac{3}{k+1}$ b) $a_k = \frac{1}{2k+1} - \frac{1}{2k-1}$

c) $a_k = \frac{3k}{2k-1}$ d) $a_k = \frac{2k}{2k+1}$

65. Examine the sequence with the general term a_k for monotonicity and boundedness.

    a) $a_k = \frac{1}{3k+4}$ b) $a_k = \frac{k}{k^2+1}$; *Hint*: Examine $a_k - a_{k+1}$.

66. Examine the sequence with the general term a_k for monotonicity and boundedness.

    a) $a_k = \frac{2^k}{2^k+2^{-k}}$ b) $a_k = \frac{k}{\sqrt{k^2+1}}$

67. Examine the monotonicity of the recursively defined sequence.

    a) $a_1 = 1$, $a_{k+1} = a_k + 2(3 - 4k)$; $k \in \mathbb{N}$ b) $a_1 = 4$, $a_{k+1} = a_k + \frac{1}{k+1} - \frac{1}{k}$; $k \in \mathbb{N}$

68. Justify or disprove by way of a counterexample.

    a) Every stricly increasing sequence is divergent.

    b) An alternating sequence cannot converge.

    c) If the sequence (a_k) is monotonically decreasing, then $(-a_k)$ is monotonically increasing.

    d) A convergent alternating sequence is always a null sequence.

69. Justify or disprove by way of a counterexample.

    a) If the sequence (a_k) with $a_k \neq 0$ diverges, then the sequence $\left(\frac{1}{a_k}\right)$ is a null sequence.

    b) If the sequence (a_k) diverges, then the sequence (a_k^2) also diverges.

    c) If the sequence (a_k) with $a_k > 0$ is monotonically increasing, then the sequence $\left(\frac{1}{a_k}\right)$ is monotonically decreasing.

70. Give an explicit description of a sequence (a_k) that has the property described.

    a) The sequence converges to 5 from above.

    b) The sequence converges and its terms oscillate around the value -1.

71. Justify or disprove by way of a counterexample.

    a) Every monotonic sequence is convergent.   b) Every convergent sequence is monotonic.

    c) Every bounded sequence is convergent.   d) Every divergent sequence is not bounded.

    e) Every null sequence is monotonically decreasing.

    f) Every sequence without an upper bound is monotonically increasing.

    g) The limit of a sequence is both an upper and lower bound.

    h) Every convergent sequence has terms that are not equal to its limit.

Theorems about Convergent Sequences and Series

- Every convergent sequence is bounded.
- Every monotonically increasing sequence bounded from above is convergent.
- Every monotonically decreasing sequence bounded from below is convergent.
- If $\lim\limits_{n\to\infty} \sum\limits_{k=1}^{n} a_k$ exists, then $\lim\limits_{k\to\infty} a_k = 0$ holds.

  In other words: If $\lim\limits_{k\to\infty} a_k \neq 0$ holds, then $\lim\limits_{n\to\infty} \sum\limits_{k=1}^{n} a_k$ does not exist.

72. Show that the sequence is monotonic and bounded and use this information to determine its limit.

a) $a_1 = 2$, $a_{k+1} = \dfrac{a_k + 6}{2}$, $k \in \mathbb{N}$ $\qquad\qquad$ b) $a_1 = 2$, $a_{k+1} = \dfrac{1 + a_k}{2}$, $k \in \mathbb{N}$

73. Justify the theorem from the green box above: If $\lim\limits_{n \to \infty} \sum\limits_{k=1}^{n} a_k$ exists, then $\lim\limits_{k \to \infty} a_k = 0$.

74. Does the given infinite series converge?

a) $\displaystyle\sum_{k=1}^{\infty} \dfrac{k^2}{5k^2 + 4}$ $\qquad\qquad\qquad$ b) $\dfrac{1}{3} + \dfrac{2}{5} + \dfrac{3}{7} + \dfrac{4}{9} + \ldots$

c) $\dfrac{1}{1 + \frac{1}{2}} + \dfrac{1}{1 + \frac{1}{4}} + \dfrac{1}{1 + \frac{1}{8}} + \dfrac{1}{1 + \frac{1}{16}} + \ldots$ $\qquad$ d) $\dfrac{1 \cdot 2}{3 \cdot 4} + \dfrac{3 \cdot 4}{5 \cdot 6} + \dfrac{5 \cdot 6}{7 \cdot 8} + \dfrac{7 \cdot 8}{9 \cdot 10} + \ldots$

Right-Handed and Left-Handed Limits of a Function at a Point $a \in \mathbb{R}$

$\lim\limits_{x \to a^-} f(x)$ is called the *left-handed* limit and $\lim\limits_{x \to a^+} f(x)$ is called the *right-handed* limit. Here, right-handed means that x tends to a on the x-axis from the right and left-handed means that x tends to a on the x-axis from the left.

We also write: $\lim\limits_{x \to a^-} f(x) = \lim\limits_{x \uparrow a} f(x)$; $\quad \lim\limits_{x \to a^+} f(x) = \lim\limits_{x \downarrow a} f(x)$

It holds: $\lim\limits_{x \to a} f(x) = b \in \mathbb{R} \;\Leftrightarrow\; \lim\limits_{x \to a^-} f(x) = \lim\limits_{x \to a^+} f(x) = b$

In words: The function f has the limit b at a point a if and only if the left-handed and right-handed limits coincide and are equal to b.

Remark: For points a at the boundary of an interval the right-handed or left-handed limit at point a suffices, respectively.

Example: The graph G_f of the function is given. Filled dots denote points of the curve, while unfilled dots are not part of the curve, i.e. gaps.

We have: $\lim\limits_{x \to 2^-} f(x) = 3$, $\lim\limits_{x \to 2^+} f(x) = 1$,
i.e., $\lim\limits_{x \to 2} f(x)$ does not exist.

We have: $\lim\limits_{x \to 5^-} f(x) = 2$, $\lim\limits_{x \to 5^+} f(x) = 2$,
i.e., $\lim\limits_{x \to 5} f(x) = 2$, while $f(5) = 1.5 \neq 2$.

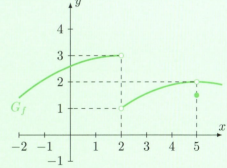

Remark: The theorems for calculating with limits for $x \to \pm\infty$ are also valid for right-handed and left-handed limits.

75. Do both limits coincide if they exist at all?

a) $\lim\limits_{x\uparrow 4} \dfrac{x^2 - 16}{x - 4}$, $\lim\limits_{x\downarrow 4} \dfrac{x^2 - 16}{x - 4}$ b) $\lim\limits_{x\uparrow -1} \dfrac{1 + x}{1 - x^2}$, $\lim\limits_{x\downarrow -1} \dfrac{1 + x}{1 - x^2}$ c) $\lim\limits_{x\uparrow 1} \dfrac{1 - x}{1 - x^2}$, $\lim\limits_{x\downarrow 1} \dfrac{1 - x}{1 - x^2}$

76. $f(x) = \begin{cases} 3 - x, & x < 2 \\ \frac{x}{2} + 1, & x \geq 2 \end{cases}$ is given.

a) Determine $\lim\limits_{x\to 4^+} f(x)$ and $\lim\limits_{x\to 4^-} f(x)$. Does $\lim\limits_{x\to 4} f(x)$ exist?

b) Determine $\lim\limits_{x\to 2^+} f(x)$ and $\lim\limits_{x\to 2^-} f(x)$. Does $\lim\limits_{x\to 2} f(x)$ exist?

77. Do both limits coincide if they exist at all?

a) $\lim\limits_{x\to 0^-} \dfrac{x}{|x|}$, $\lim\limits_{x\to 0^+} \dfrac{x}{|x|}$ b) $\lim\limits_{x\to 0^-} 2^{\frac{1}{x}}$, $\lim\limits_{x\to 0^+} 2^{\frac{1}{x}}$

78. Determine the right-handed and left-handed limits at the points where the function is not defined.

a) $f(x) = \dfrac{2}{x}$ b) $f(x) = \dfrac{x^2 - 9}{x - 3}$ c) $f(x) = 5^{\frac{1}{x}}$

79. Calculate $\lim\limits_{x\to a^-} f(x)$ and $\lim\limits_{x\to a^+} f(x)$ and compare them.

a) $f(x) = \dfrac{x^2 + 1}{x - 2}$, $a = 2$ b) $f(x) = \dfrac{x^2 + 2x + 1}{x + 1}$, $a = -1$

c) $f(x) = e^{-\frac{1}{x}}$, $a = 0$ d) $f(x) = e^{-\frac{1}{x^2}}$, $a = 0$

e) $f(x) = \dfrac{|1 + x| - 1}{x}$, $a = 0$ f) $f(x) = \dfrac{|x| - 1}{x^2 - 1}$, $a = -1$

80. True or false? Pay attention to the domain of the function.

a) $\lim\limits_{x\to 1^-} \sqrt{x} = 1$ b) $\lim\limits_{x\to 0^+} \sqrt{x} = 0$ c) $\lim\limits_{x\to 0} \sqrt{x} = 0$

d) $\lim\limits_{x\to 0} \sqrt[3]{x} = 0$ e) $\lim\limits_{x\to 3} \sqrt[4]{7(x - 3)} = 0$ f) $\lim\limits_{x\to \pi} \sin\left(\dfrac{1}{x - \pi}\right) = 1$

81. The function is not defined at x_0. Does the limit at x_0 exist?

a) $f(x) = \dfrac{x^2 - 1}{x - 1}$, $x_0 = 1$ b) $f(x) = \dfrac{x + 2}{x^2 - 4}$, $x_0 = -2$

c) $f(x) = \dfrac{1}{1 - e^{\frac{1}{x}}}$, $x_0 = 0$ d) $f(x) = \dfrac{1}{1 - e^{-\frac{1}{x^2}}}$, $x_0 = 0$

e) $f(x) = x\sqrt{1 + \dfrac{1}{x^2}}$, $x_0 = 0$ f) $f(x) = \dfrac{\sin(x)}{\cos(x)}$, $x_0 = \frac{\pi}{2}$

82. Is the function f divergent or convergent for $x \to a$? If convergent determine the limit. Pay attention to the domain of the function.

a) $f(x) = \ln(x + 2)$, $a = 0$ b) $f(x) = \ln(5x)$, $a = 0$ c) $f(x) = \ln(x^3 - 1)$, $a = 1$

83. Calculate the limit if it exists.

a) $\lim\limits_{x\to -2} \dfrac{2 - |x|}{x + 2}$ b) $\lim\limits_{u\downarrow 4} \dfrac{4 - u}{|4 - u|}$ c) $\lim\limits_{x\to 0^-} \left(\dfrac{1}{x} - \dfrac{1}{|x|}\right)$ d) $\lim\limits_{x\to 0^+} \left(\dfrac{1}{x} - \dfrac{1}{|x|}\right)$

2.4   Miscellaneous Exercises

For Chapter 2.1: Limits of Sequences

84. Is the sequence $\frac{1}{2}$, $\frac{1}{4}$, $\frac{3}{4}$, $\frac{1}{8}$, $\frac{3}{8}$, $\frac{5}{8}$, $\frac{7}{8}$, $\frac{1}{16}$, ... convergent or divergent?

85. Determine $\lim\limits_{n\to\infty} \sqrt[n]{a^n + b^n}$ for $0 < a < b$.

86. The partial sums $s_n = \sum\limits_{k=1}^{n} a_k = \frac{n-1}{n+1}$ of a sequence (a_k) are given.
Determine a_n as well as $\lim\limits_{n\to\infty} s_n$.

87. The sequence (a_k) is described explicitly by the general term $a_k = \frac{1}{k} - \frac{1}{k+1}$ $(k \in \mathbb{N})$.
   a) Calculate the four terms a_5, a_7, a_{15} and a_{99} without using a calculator and the four terms a_{110}, a_{1110}, a_{11110} and a_{111110} using a calculator.
   b) Calculate the first five terms s_1, s_2, s_3, s_4 and s_5 of the sequence of partial sums (s_n) with $s_n = a_1 + a_2 + \ldots + a_{n-1} + a_n$ and describe s_n with the simplest possible expression.
   c) Determine both limits $\lim\limits_{k\to\infty} a_k$ and $\lim\limits_{n\to\infty} s_n$ if they exist.
   d) Show that the series $\sum\limits_{k=1}^{\infty} \frac{1}{(k+1)^2}$ converges. *Hint:* Compare the terms of this series with the terms of the series $\sum\limits_{k=1}^{\infty} \left(\frac{1}{k} - \frac{1}{k+1} \right)$ from the previous parts of the exercise.

88. Eric writes down: $\frac{1}{3} = \frac{1}{5} + \frac{2}{15} = \frac{1}{5} + \frac{1}{25} + \frac{7}{75} = \frac{1}{5} + \frac{1}{25} + \frac{1}{125} + \frac{32}{375} = \frac{1}{5} + \frac{1}{25} + \frac{1}{125} + \frac{1}{625} + \ldots$

Review Erics calculation using the formula for infinite geometric series. Comment on his calculation and explain your answer.

89. $\sum\limits_{n=1}^{\infty} \frac{1}{n^2} = \frac{\pi^2}{6}$ is given (Leonhard Euler, 1735).
   a) What is the limit if one only takes the summands with even natural numbers n?
   b) What is the limit if one only takes the summands with odd numbers $n \in \mathbb{N}$?

90. The Fibonacci sequence (f_n) is defined recursively: $f_1 = 1$, $f_2 = 1$, $f_{n+1} = f_n + f_{n-1}$, $n \geq 2$. Fibonacci used this sequence to describe the growth of a population of rabbits (see first exercise of the topic «The Fibonacci numbers» on page 19).
   a) Determine the first twelve terms of the Fibonacci sequence.
   b) Determine the first ten terms of the sequence of ratios (q_n) with $q_n = \frac{f_{n+1}}{f_n}$, $n \geq 1$.
   c) List the first ten terms of the sequence (q_n) as decimal numbers. What do you suspect?
   d) Determine a recursion formula for q_n.
   e) Determine $\lim\limits_{n\to\infty} q_n$ assuming that the sequence (q_n) converges to a limit $q > 0$.
   f) Show that the sequence (q_n) indeed has a limit. To this end, examine the difference of consecutive terms.
   g) Does the limit q from part e) depend on the choice of the values of f_1 and f_2?
   h) Write the limit q as an infinite continued fraction and list its first four fractional approximations.

For Chapter 2.2: Limits of Functions

91. True or false? Justify.

a) $\lim\limits_{x\to a} f(x) = G$ means: If x_2 is closer to a than x_1 is, then $f(x_2)$ is closer to G than $f(x_1)$ is.

b) Whether $\lim\limits_{x\to a} f(x)$ exists, depends on how or whether $f(a)$ is defined, respectively.

c) $\lim\limits_{x\to 1} \dfrac{x^2 + 6x - 7}{x^2 + 5x - 6} = \dfrac{\lim\limits_{x\to 1}(x^2 + 6x - 7)}{\lim\limits_{x\to 1}(x^2 + 5x - 6)}$

d) $\lim\limits_{x\to 1} \dfrac{x - 3}{x^2 + 2x - 4} = \dfrac{\lim\limits_{x\to 1}(x - 3)}{\lim\limits_{x\to 1}(x^2 + 2x - 4)}$

e) If $\lim\limits_{x\to a} f(x) = \infty$ and $\lim\limits_{x\to a} g(x) = \infty$, then $\lim\limits_{x\to a}(f(x) - g(x)) = 0$.

f) If $\lim\limits_{x\to a} f(x) = \infty$ and $\lim\limits_{x\to a} g(x) = \infty$, then $\lim\limits_{x\to a} \frac{f(x)}{g(x)} = 1$.

g) If $\lim\limits_{x\to\infty} (f(x) - g(x)) = 0$, then $\lim\limits_{x\to\infty} \frac{f(x)}{g(x)} = 1$.

92. In this exercise, it is assumed that the limits exist.

a) It is known of the function f that it fulfils. $\lim\limits_{x\to -2} \frac{f(x)}{x^2} - 1$. Using this, calculate both limits $\lim\limits_{x\to -2} f(x)$ and $\lim\limits_{x\to -2} \frac{f(x)}{x}$.

b) The function g has the property that $\lim\limits_{x\to 0} \frac{g(x)}{x^2} = 1$. What values of $\lim\limits_{x\to 0} g(x)$ and $\lim\limits_{x\to 0} \frac{g(x)}{x}$ result from this?

93. Determine $\lim\limits_{x\to 0} f(x)$ if

a) $\lim\limits_{x\to 0} \dfrac{4 - f(x)}{x} = 1$

b) $\lim\limits_{x\to -4} \left(x \cdot \lim\limits_{x\to 0} f(x) \right) = 2$

94. *Interesting graph progression*: Solve this exercise if possible without using a calculator.

Consider: $f_1(x) = 2^{\frac{1}{x}}$, $f_2(x) = \dfrac{1}{1 + 2^{\frac{1}{x}}}$, $f_3(x) = 2^{\frac{-1}{x^2}}$.

All three functions f_1, f_2 and f_3 have their only singularity at the same point x_0. First determine this common singularity. Then, examine and characterise the behaviour of f_1, f_2 and f_3 in the neighbourhood of this point x_0. Finally, sketch by hand the general course of the graphs of f_1, f_2 and f_3 in the neighbourhood of x_0 – each in its own coordinate system – and also draw in any asymptotes.

95. The thin lens equation $\frac{1}{b} + \frac{1}{g} = \frac{1}{f}$ arises from the field of optics. When an object is imaged through a lens, the equation relates the image distance b, the object distance g and the focal length f. Calculate the image distance, when the object lies arbitrarily far from the lens, i.e. tends to infinity.

96. Two point masses with masses m_1 and m_2 lie on the x-axis at the points x_1 and x_2. The centre of mass of both point masses lies on the x-axis at the point $s = \frac{x_1 \cdot m_1 + x_2 \cdot m_2}{m_1 + m_2}$.

a) How does the centre of mass shift, when m_2 becomes arbitrarily large?

b) How does the centre of mass shift, when m_1 becomes arbitrarily large?

97. *Fixed-point iteration.* Let $x_1 = c$, $x_2 = f(x_1)$, $x_3 = f(x_2) = f(f(c))$, ..., $x_{n+1} = f(x_n)$, where f is a continuous function. If the limit $\lim\limits_{n\to\infty} x_n = L$ exists, then $f(L) = L$ by the limit laws, i.e. L is a solution of the fixed-point equation $x = f(x)$.

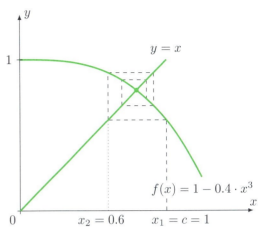

Example: Solve $2x^3 + 5x - 5 = 0$. By algebraic manipulations one arrives at the fixed-point equation $x = 1 - 0.4 \cdot x^3$. The intersection point of the line $y = x$ with the curve $f(x) = 1 - 0.4x^3$ yields the solution. The spiraling path leads to the intersection point. The x-coordinates of the vertices of this path are given by $x_1 = c = 1$, $x_{n+1} = 1 - 0.4 \cdot x_n^3$. This recursion thus yields: $x_1 = 1$, $x_2 = 0.6$, $x_3 = 0.9136, \ldots$, $x_{30} = 0.79720, \ldots$

a) Solve the equation $x = 0.5 - 0.1x^3$ to a precision of three decimal places using the fixed-point iteration. Start with $x_1 = 1$.

b) Find an approximate solution of the equation $x = 2 + e^{-x}$ with seven significant digits using the fixed-point iteration. Start with $x_1 = 2$.

c) Solve the equation $\cos(x) - x = 0$ to a precision of five decimal places using the fixed-point iteration. Choose a suitable starting value yourself.

2.5   Review Exercises

For Chapter 2.1: Limits of Sequences

98. Examine whether the sequence has a limit. If so, calculate the limit.

a) $\frac{10}{11}$, $\left(\frac{10}{11}\right)^2$, $\left(\frac{10}{11}\right)^3$, $\left(\frac{10}{11}\right)^4$, ...   b) $(-2) - \frac{1}{2}$, $(-2) + \frac{1}{3}$, $(-2) - \frac{1}{4}$, $(-2) + \frac{1}{5}$, ...

99. From which index n on are all $a_k = \frac{1}{k^2}$ closer to the limit of the sequence (a_k) than 10^{-6}?

100. a) Is the sequence $a_k = (-1)^k$ convergent?

b) Is the sequence $a_k = (-3)^k$ oscillating?

c) Give an example of a properly divergent sequence.

101. The sequence (a_k) is defined by the general term a_k. Determine the limit of the sequence for $k \to \infty$ if it exists. Otherwise note the type of divergence.

a) $a_k = 10 - \dfrac{3}{k}$ b) $a_k = \dfrac{6k - 5k^2}{4k^2}$ c) $a_k = \dfrac{1 - k^4}{2k^3}$

d) $a_k = \dfrac{3k^2}{k^2 + k + 1}$ e) $a_k = \dfrac{4k + 3k \cdot (-1)^k}{2k}$ f) $a_k = \dfrac{1 - 2\sqrt{k}}{4\sqrt{k} - 3}$

102. The sequence (a_k) is defined by the general term a_k. Determine the limit.

a) $a_k = \mathrm{e}^{-\frac{1}{k}}$

b) $a_k = \frac{k^{10}}{2^k}$

c) $a_k = \frac{\ln(k)}{\sqrt{k}}$

d) $a_k = \cos\left(-\frac{1}{k}\right)$

e) $a_k = \sqrt[k]{\frac{1}{k}}$

f) $a_k = \sqrt[k]{k} + \frac{1}{k}$

103. Does the sequence (a_k) converge? Justify your answer, and determine the limit or note type of divergence.

a) $a_k = a_{k-1} \cdot \frac{1}{3}, \quad a_1 = -3$

b) $a_{k+1} = a_k - 2, \quad a_1 = 1000$

104. Calculate the limit of the infinite series $\sum\limits_{k=1}^{\infty} \frac{1}{\left(\sqrt{5}\right)^k}$.

For Chapter 2.2: Limits of Functions

105. Determine both limits and compare them.

a) $\lim\limits_{x \to \infty} \frac{2x+3}{x+1}, \quad \lim\limits_{x \to -\infty} \frac{2x+3}{x+1}$

b) $\lim\limits_{x \to -\infty} \frac{7x}{x-7}, \quad \lim\limits_{x \to \infty} \frac{7x}{x-7}$

106. Determine the limit.

a) $\lim\limits_{x \to -\infty} \frac{x^2-1}{3x^2+x+1}$

b) $\lim\limits_{x \to \infty} \frac{x}{\sqrt{x^2+1}}$

c) $\lim\limits_{x \to \infty} \frac{\sqrt{x}+1}{\sqrt{x}+2}$

107. Is the function f divergent or convergent for $x \to \infty$? If convergent, determine the limit.

a) $f(x) = \frac{x}{\sqrt{2x}}$

b) $f(x) = \frac{x^7 - \cos(x)}{3x^5 - 1}$

c) $f(x) = \frac{-x^3 + 2x}{2x^3 + x^2}$

d) $f(x) = \frac{2^x + 1}{3 \cdot 2^x - 1}$

108. Is the function f divergent or convergent when $x \to \infty$? What happens when $x \to -\infty$? If convergent, determine the limit.

a) $f(x) = \mathrm{e}^{x-2}$

b) $f(x) = \left(\frac{1}{2}\right)^{-x-2}$

c) $f(x) = 3^{x^2}$

d) $f(x) = \ln\left(\frac{x-10}{x}\right)$

For **109–113**: Determine the limit if it exists.

109. a) $\lim\limits_{x \to 4} \frac{(x-4)(x+5)}{x-4}$

b) $\lim\limits_{x \to 2} \frac{(x-2)(x+7)}{x^2-4}$

110. a) $\lim\limits_{x \to 3} \frac{x^2-9}{x-3}$

b) $\lim\limits_{x \to -5} \frac{x^2-25}{x+5}$

111. a) $\lim\limits_{h \to 0} \frac{(x+h)^2 - x^2}{h}$

b) $\lim\limits_{x \to 0} \frac{(x+h)^2 - x^2}{h}$

112. a) $\lim\limits_{h \to 0} \frac{(3+h)^3 - 27}{h}$

b) $\lim\limits_{x \to 3} \frac{x^3-27}{x-3}$

113. a) $\lim\limits_{x \to 7} \frac{x^2-6x-7}{x^2-5x-14}$

b) $\lim\limits_{x \to \infty} \frac{x^2}{\ln(x)}$

3 Differential Calculus

3.1 Introduction

1. 📄 *What a position-time graph can tell us about speed.* From the following position-time graph, one can read the length of the path s that a point mass has travelled until the time t. After 5 seconds it has travelled 2 metres, for example. The following four parts of the exercise are to be solved in the given order using the diagram in a graphical and calculatory manner.

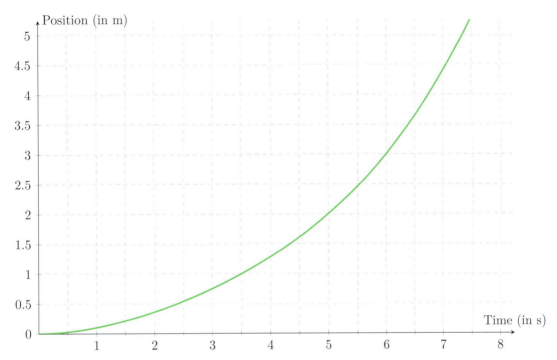

a) What is the average velocity v_1 with which the point mass travels in the first six seconds? In other words, find the average velocity during the time interval $[0; 6]$.
   Note: $v_{\text{average}} = \frac{\text{distance travelled}}{\text{elapsed time}}$.

b) Calculate the average velocities v_2 and v_3 with which the point mass travels during the time intervals $[0; 3]$ and $[3; 6]$, respectively.

c) With which average velocities v_4, v_5, v_6 and v_7 does the point mass travel during the successively diminishing time intervals $[1.25; 6]$, $[2.5; 6]$, $[3.5; 6]$ and $[5; 6]$? (Accuracy: one decimal place)

d) Now, try to determine the instantaneous velocity $v = v(6)$ of the point mass at time $t = 6$. In other words, find the velocity v with which the point mass travels at time $t = 6$.

2. *Interpreting algebraic expressions geometrically.* The displayed graph G_f describes the behaviour of a function f in the specified range.

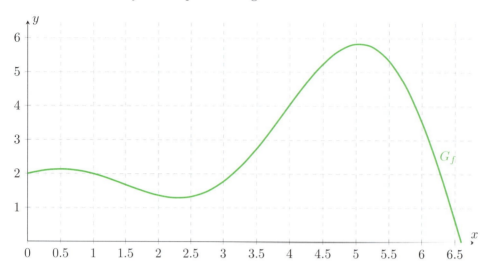

Which of the two expressions is greater? Or are both equal? Fill the signs $<$, $>$ or $=$ into the dots ... . You can also answer the question without determining the numerical values of the expressions.

a) $\dfrac{f(4) - f(1)}{4 - 1}$ ... 0

b) $\dfrac{f(2) - f(1)}{2 - 1}$ ... 0

c) $\lim\limits_{x \to 4} \dfrac{f(x) - f(4)}{x - 4}$ ... 0

d) $\lim\limits_{x \to 6} \dfrac{f(x) - f(6)}{x - 6}$ ... 0

e) $\dfrac{f(2) - f(1)}{2 - 1}$ ... $\dfrac{f(6) - f(5)}{6 - 5}$

f) $\lim\limits_{x \to 3} \dfrac{f(x) - f(3)}{x - 3}$ ... $\lim\limits_{x \to 4} \dfrac{f(x) - f(4)}{x - 4}$

g) $\lim\limits_{x \to 0.5} \dfrac{f(x) - f(0.5)}{x - 0.5}$ ... $\lim\limits_{x \to 5} \dfrac{f(x) - f(5)}{x - 5}$

3. 📄 *How fast does the rolling ball travel?* We consider a ball rolling down an inclined plane. The corresponding position function s is given by the equation $s(t) = 0.25t^2$ (s in metres and t in seconds).

a) Draw the graph of the function s for $0 \le t \le 5$ into a coordinate system.

b) With which average velocity v_{average} does the ball roll down the inclined plane during the time interval $[1; 3]$? *Note:* $v_{\text{average}} = \frac{\text{distance travelled}}{\text{elapsed time}}$.

c) How can the value v_{average}, calculated in b), also be interpreted by purely geometric means using the graph from a)?

d) How could the velocity of the rolling ball at time $t_0 = 1$, the so-called instantaneous velocity one second after the start, be determined? Do you have a clever idea of how this may be achieved in a calculatory or graphical manner?

e) Calculate the average velocities v_{average} for successively shorter time intervals $[t_0; t]$ and enter them into the table.

t_0	$t,\ (t > t_0)$	$s(t_0)$	$s(t)$	$v_{\text{average}} = \frac{\Delta s}{\Delta t} = \frac{s(t) - s(t_0)}{t - t_0}$
1	1.2			
1	1.1			
1	1.05			
1	1.001			

Can you guess what the instantaneous velocity v_{inst} at time $t_0 = 1$ is by inspecting the last column of the table?

f) Formulate the instantaneous velocity v_{inst} at time $t_0 = 1$ as the limit of the average velocities v_{average}. Calculate this limit and check if the result confirms your guess from the previous part e).

g) The value v_{inst} calculated in f) can also be interpreted purely geometrically using the graph from a). With which geometric quantity does v_{inst} coincide?

h) Determine a formula for calculating the instantaneous velocity $v(t_0)$ at any given time t_0. Use this to calculate the instantaneous velocities $v(t_0)$ for $t_0 = 0, 2, 3$ and 4.

Difference Quotient – Average Rate of Change

The expression $m_s = \frac{\Delta y}{\Delta x} = \frac{f(b) - f(a)}{b - a}$ is called the *difference quotient* or the *average rate of change* of the function f on the interval $[a; b]$.

Geometrically, m_s is the *slope of the secant s* through the points $(a \mid f(a))$ and $(b \mid f(b))$. This is a suitable measure for the *average rate of change* of the function f over the interval $[a; b]$.

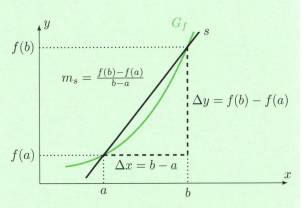

Instantaneous Rate of Change

The limit $\lim\limits_{\Delta x \to 0} \frac{\Delta y}{\Delta x}$ is the *instantaneous rate of change* of the function f at point a. A more thorough discussion follows on page 54.

4. We consider the graph of a function as the elevation profile of a road. x indicates the horizontal distance from the starting point A and y the altitude relative to the starting point A.

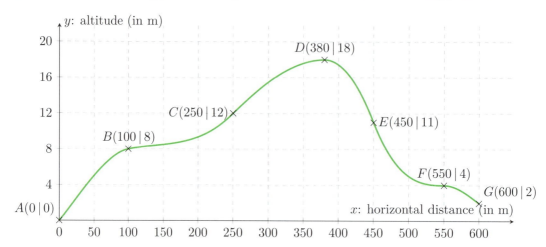

a) Calculate the slope of the secant through the given points.

   i) A and B ii) C and D iii) D and E iv) F and G

b) Which of these secants have the greatest or smallest slope?

c) Specify the interval in which the graph increases or decreases.

d) In which points is the slope of the graph zero?

e) What is the average slope if one follows the course of the graph from A to G? Can you use the average slope from A to G to say something about the course of the graph or shape of the elevation profile between the two positions A and G?

5. Give the domain of the function f and calculate its average rate of change on the intervals $I_1 = [-2; 0]$ and $I_2 = [2; 7]$.

  a) $f(x) = x^2 - 1$ b) $f(x) = -x^2 + x + 1$ c) $f(x) = 2\sqrt{x + 2}$ d) $f(x) = \frac{6}{x+3}$

6. The function T given by $T(t) = 75 \cdot 3^{-0.05t} + 20$ describes the temperature of a cup of tea as a function of time (T in degrees Celsius and t in minutes).

a) Calculate the average rate of cooling, that is the average change of temperature per minute, for the time intervals $[5; 40]$ and $[15; 50]$.

b) What do these values mean? Describe in words.

c) Draw the graph of the function T and the two secants belonging to the two intervals considered in a). How can you tell by means of these secants that the average cooling rate on the interval $[5; 40]$ is greater than on the interval $[15; 50]$?

7. The function $N = N(t)$ describes the bacterial growth at time t. We assume that the bacteria do not die during their reproduction. Describe the meaning of the following quantities.

  a) $N(t_1) - N(t_0)$ b) $\dfrac{N(t_1) - N(t_0)}{t_1 - t_0}$ c) $\displaystyle\lim_{t_1 \to t_0} \dfrac{N(t_1) - N(t_0)}{t_1 - t_0}$

8. According to experience, light loses $75\,\%$ of its luminosity per metre of depth in clean seawater. Let $L(x)$ denote the luminosity at a depth of x metres below the sea surface, where $L(0) = 1 = 100\,\%$ signifies full luminosity.

   a) Determine the equation of the function of luminosity $L = L(x)$.

   b) What is the meaning of the quantity $L(x_2) - L(x_1)$? Calculate $L(5) - L(2)$.

   c) Calculate the average rate of change in luminosity per metre between the depths of $2\,\mathrm{m}$ and $5\,\mathrm{m}$.

   d) What is the meaning of the term $\lim\limits_{\Delta x \to 0} \dfrac{L(5+\Delta x) - L(5)}{\Delta x}$?

9. The average inflation rate of a currency is defined as the loss of purchasing power per unit of time, most commonly per month or per year. Let $W = W(t)$ denote the purchasing power of the Swiss franc at time t.

   a) Express the average inflation rate $I_{[t_1;t_2]}$ of one franc for the time interval $[t_1;t_2]$ as a formula.

   b) Define and describe the instanteneous inflation rate $I(t_1)$ at time t_1 by a formula.

10. During a chemical reaction $A + B \to C$ two chemical substances A and B combine to form a new substance C. Let $M = M(t)$ denote the existing amount of substance C at time t. The average reaction rate $\nu_R = \nu_R(t)$ is defined as the change in the amount of substance per time unit, in this case it is the increase in the amount of substance C.

    a) Deduce a formula to calculate the average reaction rate ν_R (that is the average increase in the amount of substance C per time unit) for the time interval $[t_0; t_0 + \Delta t]$.

    b) How can the instantaneous reaction rate at time t_0 be determined from this?

11. The mass of a heavy gas that is contained in an upright cylindrical container with a cross-sectional area of Q is vertically
not evenly distributed, that is the density is not the same ev-
erywhere. Let $m = m(z)$ denote the mass of the gas that lies
between the upper rim of the cylinder and the depth z (see
sketch).

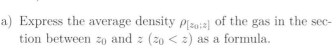

   a) Express the average density $\rho_{[z_0;z]}$ of the gas in the section between z_0 and z $(z_0 < z)$ as a formula.

   b) Define and describe the gas density ρ_{z_0} at a point z_0 by
a formula.

12. True or false? As the radius r_1 increases to r_2, the average rate of change in the area of the circle is

   a) smaller than $2\pi r_2$.   b) greater than $2\pi r_1$.   c) equal to $2\pi \cdot \dfrac{r_1 + r_2}{2}$.

13. A freely falling object obeys the law $s = \frac{1}{2}g\,t^2$, where g denotes the gravitational acceleration (distance fallen s in metres, time t in seconds, use $g = 10\,\text{m/s}^2$).

   a) Determine the average velocity of the object for the time interval from $t_0 = 5$ to $t_1 = 5 + \Delta t$, where $\Delta t = 1,\ 0.1,\ 0.05,\ 0.001$.

   b) What is the instantaneous velocity of the object at time $t_0 = 5$?

   c) Develop a formula for the instantaneous velocity of the object at an abitrary time $t_0 \geq 0$.

14. The displacement s of a particle at time t is described by the function $s(t) = 5t^2 + 2t + 1$ (t in seconds and s in metres).

   a) What distance has the particle travelled by the time $t = 5$?

   b) Determine the average velocity of the particle for the time interval $[3; 7]$.

   c) Can you give the instantaneous velocity of the particle at time $t = 9$?

Instantancous Rate of Change – Differential Quotient – Derivative

Let the function f be given by the equation $y = f(x)$. The *derivative* $f'(x_0)$ of this function f at the point x_0 is the limit of the difference quotient

$$f'(x_0) = \lim_{h \to 0} \frac{f(x_0 + h) - f(x_0)}{h}$$

if it exists. The act of determining this limit is called *differentiation* and the function f is called differentiable at point x_0 if the limit exists.

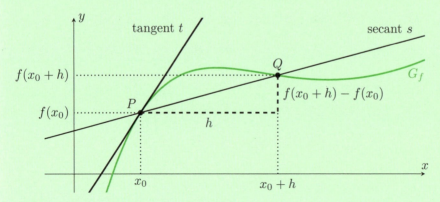

As the point Q on the curve approaches P, which may happen from the left or right side, the secant turns into the tangent t.

The tangent t to the curve with equation $y = f(x)$ at point $P(x_0 \,|\, y_0)$ is the straight line through P whose slope m_t coincides with the derivative $f'(x_0)$, that is: $m_t = f'(x_0)$.

The derivative $f'(x_0)$ is also called the *differential quotient* or *instantaneous rate of change* of the function f at point x_0.

The graph above illustrates that by letting h go to 0 the slope of the secant approaches the slope of the tangent at point $P(x_0 \,|\, f(x_0))$.

Further notations for derviatives: $y' = \lim\limits_{\Delta x \to 0} \dfrac{\Delta y}{\Delta x} = \dfrac{dy}{dx}$ and $f'(x) = \lim\limits_{\Delta x \to 0} \dfrac{\Delta f}{\Delta x} = \dfrac{d}{dx} f(x) = \dfrac{df}{dx}$

Remarks:

- Despite its name the differential quotient is a limit and not a quotient.
- It is possible that the tangent t at $P(x_0 \,|\, y_0)$ touches or intersects the curve at other points $x \neq x_0$.
- If the function f is differentiable at each point of its domain, the derivative f' is also a function of x on that domain.

15. The function f is defined by the equation $y = f(x) = x^2$.

  a) Calculate the three following difference quotients:

   i) $\dfrac{f(4+2) - f(4)}{2}$ ii) $\dfrac{f(4+h) - f(4)}{h}, \ h \neq 0$ iii) $\dfrac{f(x_0 + h) - f(x_0)}{h}, \ h \neq 0$

  b) Calculate the three following difference quotients:

   i) $\dfrac{f(6) - f(4)}{6 - 4}$ ii) $\dfrac{f(x_1) - f(4)}{x_1 - 4}, \ x_1 \neq 4$ iii) $\dfrac{f(x_1) - f(x_0)}{x_1 - x_0}, \ x_1 \neq x_0$

  c) Calculate the three following derivatives:

   i) $\lim\limits_{h \to 0} \dfrac{f(4+h) - f(4)}{h}$ ii) $\lim\limits_{x_1 \to 4} \dfrac{f(x_1) - f(4)}{x_1 - 4}$ iii) $\lim\limits_{h \to 0} \dfrac{f(x_0 + h) - f(x_0)}{h}$

   What is the meaning of these three limits for the function f? Also give a geometrical interpretation.

16. What is the instantaneous rate of change of the function $f(x) = \frac{1}{x}$ at the point a?

  a) $a = 1$ b) $a = 2$ c) $a = 5.5$

17. Determine the term $f'(\ldots)$ by calculating the difference quotient.

  a) $f'(2)$, if $f(x) = x^2$ b) $f'(5)$, if $f(x) = \frac{1}{x^2}$ c) $f'(-1)$, if $f(x) = x^3$

18. Determine the derivative of the function f at point a with the aid of the difference quotient.

  a) $f(x) = 7x - 4; \ a = 5$ b) $f(x) = x^2 + 1; \ a = -1$
  c) $f(x) = c = \text{const.}; \ a = 0$ d) $f(x) = mx + q; \ a = 3$

19. Calculate the derivative $\frac{d}{dx} f(x)$ of the function f given by $y = f(x) = \frac{x}{x+1}$ at point $a = 1$ by evaluating the limit of the difference quotient.

20. The limit given is the derivative of a function f at a point x_0. Give the simplest possible function f and the associated point x_0.

a) $\lim\limits_{h\to 0} \dfrac{\sqrt{49+h}-7}{h}$

b) $\lim\limits_{h\to 0} \dfrac{2^{5+h}-32}{h}$

c) $\lim\limits_{\Delta x\to 0} \dfrac{\lg(100+\Delta x)-2}{\Delta x}$

d) $\lim\limits_{\Delta x\to 0} \dfrac{\sin\left(\frac{5\pi}{6}+\Delta x\right)-\frac{1}{2}}{\Delta x}$

e) $\lim\limits_{x\to 3} \dfrac{x^4-81}{x-3}$

f) $\lim\limits_{x\to \frac{1}{5}} \dfrac{\frac{1}{x}-5}{x-\frac{1}{5}}$

21. Determine the derivative of the given function by means of the difference quotient.

a) $f(x)=x$

b) $f(x)=x^2$

c) $f(x)=x^3$

d) $f(x)=3x^2+4$

e) $f(x)=\frac{x}{2}$

f) $f(x)=2x^3-x^2+1$

g) $f(x)=\sqrt{x}$

h) $f(x)=ax^2+bx$

i) $f(x)=\frac{c}{x}$

22. The function f is given by a formula. Determine its *symmetric difference quotient* defined as $\frac{f(x+h)-f(x-h)}{2h}$ and its corresponding *symmetric differential quotient*.

a) $f(x)=x^2$

b) $f(x)=x^3$

c) $f(x)=x^{-1}=\frac{1}{x}$

d) $f(x)=x^2-x$

23. Given are the curve k defined by $y=\frac{x}{x+1}$ for $x>-1$ as well as the points $A\left(1\,\middle|\,\frac{1}{2}\right)$, $B\left(\frac{1}{2}\,\middle|\,\frac{1}{3}\right)$, $C\left(\frac{1}{3}\,\middle|\,\frac{1}{4}\right)$, $D\left(\frac{1}{4}\,\middle|\,\frac{1}{5}\right)$, $E\left(\frac{1}{5}\,\middle|\,\frac{1}{6}\right)$ and $O(0\,|\,0)$. Verify that the points A, B, ..., E and O lie on k and draw them into a suitable coordinate system. Sketch the curve k passing through these points up to the origin. Draw all of the secants AO, BO, ..., EO and calculate their slopes. By passing to the limit, find the slope m_O of the curve k at the origin O.

24. Consider a square of side a and the function F defined by the formula $F(a)=a^2$, which is the area of the square with side a. Now, the side a of the square is increased by Δa. Draw a sketch.

a) Determine the increase in area geometrically by use of the sketch as well as by calculation.

b) What is the average increase in area per length unit (difference quotient $\frac{\Delta F}{\Delta a}$)?

c) What is the instantaneous increase in area per length unit (differential quotient $\frac{\mathrm{d}F}{\mathrm{d}a}$)?

25. Consider a circle of radius r and the function $F=F(r)$ indicating the area of the circle as a function of r. Calculate the instantaneous increase $\frac{\mathrm{d}F}{\mathrm{d}r}$ of the area of the circle $F(r)$. What is the meaning of the term thus obtained?

26. Is the following statement accurate? «If the values of the function $y=w(x)$ are monotonically increasing for growing x in the interval $a\le x\le b$, then the corresponding derivative $y'=w'(x)$ is also monotonically increasing in the interval $a\le x\le b$.»

27. True or false? A cylindrical vase is filled with water at a steady pace. In the process, the water height in the vase increases continuously. The instantaneous increase of the quantity of water as a function of h is

i) constant.

ii) proportional to h.

iii) proportional to h^2.

3.2   Graphing Derivatives

28. Figures I to IV show the graphs of four functions f and figures a) to d) the graphs of their derivatives f'. Which graph of f matches with which graph of f'?

I
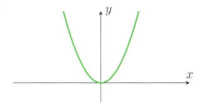

II
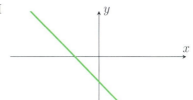

III
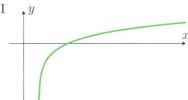

IV
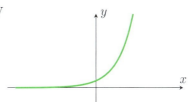

a)
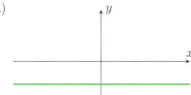

b)
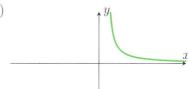

c)
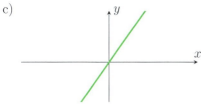

d)
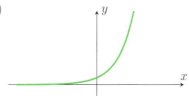

29. Sketch the graph of the derivative f' of the depicted function f.

a)
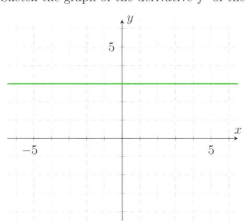

b)
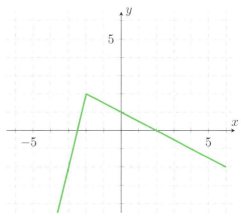

30. Sketch the graph of the derivative f' of the depicted function f.

a)

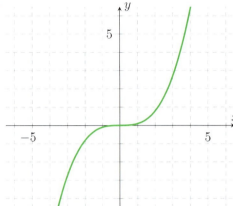

b)

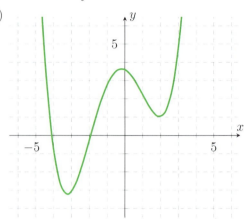

c)

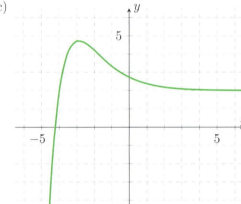

d)

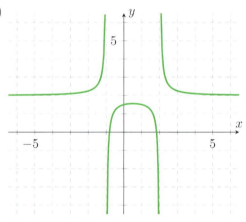

31. Sketch the graph of the derivative f' of the depicted function f.

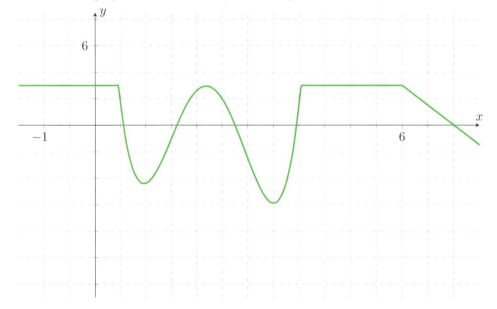

32. 📄 Consider the given graph of the derivative f'. Sketch the graph of a possible function f.

a)

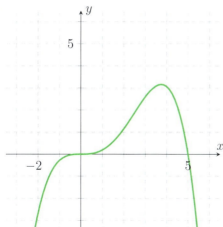

b)

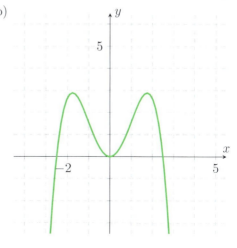

33. In this exercise, consider the two functions f and g (having coinciding domains $D_f = D_g = D$) with their graphs G_f and G_g. Furthermore, let $G_{f'}$ and $G_{g'}$ be the graphs of their derivatives. Are the following statements true?

a) If $G_{f'} = G_{g'}$ holds, that is the graphs $G_{f'}$ and $G_{g'}$ coincide, then $G_f = G_g$ also holds.

b) If reflecting the graph G_f through the y-axis yields the graph G_g, then the same holds for the graphs $G_{f'}$ and $G_{g'}$ of their derivatives.

c) If the functions f and g satisfy the relation $g(x) = -f(x)$ for all $x \in D$, then the graphs $G_{f'}$ and $G_{g'}$ of their derivatives are symmetrical about the x-axis.

d) If the graphs G_f and G_g are point reflections at the origin of each other, then the same holds for the graphs $G_{f'}$ and $G_{g'}$ of their derivatives.

e) If the graph G_g is one unit below G_f, then the same holds for the graphs $G_{g'}$ and $G_{f'}$.

34. Which of the following graphs belongs to the function f satisfying $f'(x) = 2x$ and $f(1) = 4$?

Graph 1

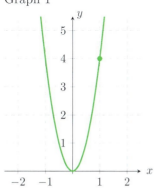

Graph 2

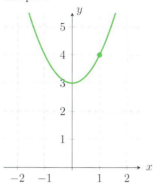

Graph 3

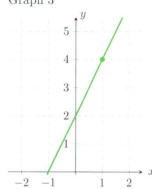

35. Which of the following graphs belongs to the function f given by $y = f(x)$, where $\frac{dy}{dx} = -x$ and $f(-1) = 1$?

Graph 1   Graph 2   Graph 3

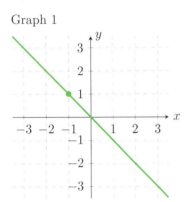 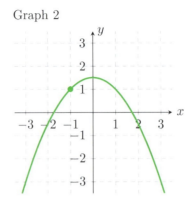 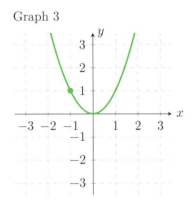

36. 📄 The graphs of the functions f and g are given in the figure:

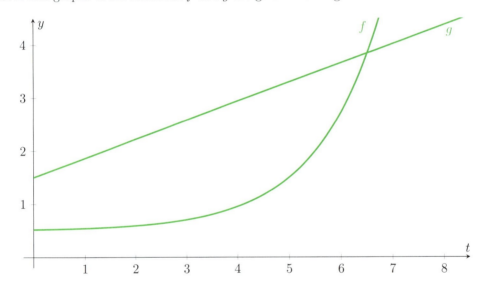

The corresponding terms $f(t)$ and $g(t)$ each give the position of a moving object at time t. When do the two objects move with exactly the same velocity during the time interval $[0;8]$? Do this exercise purely graphically by means of the above figure.

3.3 Differentiation Rules

Sum, Constant Factor and Power Rules

Sum rule: $(f(x) \pm g(x))' = f'(x) \pm g'(x)$ Shorthand notation: $(f \pm g)' = f' \pm g'$

Constant factor rule: $(c \cdot f(x))' = c \cdot f'(x)$ Shorthand notation: $(c \cdot f)' = c \cdot f'$

Power rule: $(x^n)' = n \cdot x^{n-1}$, $n \in \mathbb{R}$;   specifically: $\left(\frac{1}{x}\right)' = \frac{-1}{x^2}$ and $(\sqrt{x})' = \frac{1}{2\sqrt{x}}$

Constant term rule c: $(c)' = 0$

37. Apply the suitable differentiation rules to determine $f'(x)$.

a) $f(x) = 12$

b) $f(x) = -9x$

c) $f(x) = x^{16}$

d) $f(x) = 3x - 1$

e) $f(x) = x^3 + 11$

f) $f(x) = 7x^2 + 2x - 3$

g) $f(x) = -\frac{1}{10}x^5$

h) $f(x) = \frac{1}{2}x^3 - 2x - 7$

i) $f(x) = \frac{1}{12}x^4 - \frac{1}{9}x^3 - \frac{1}{11}$

j) $f(x) = \frac{1}{x^5}$

k) $f(x) = \sqrt[4]{x}$

l) $f(x) = -3\sqrt[5]{x}$

38. Apply the suitable differentiation rules to determine $g'(t)$.

a) $g(t) = -2t^2 + 3t - 6$

b) $g(t) = t^2 + 5t^3$

c) $g(t) = -3t^4 - 3t - 3t^{-2}$

d) $g(t) = at^4 + bt^2$

e) $g(t) = 2a\sqrt{t} + b$

f) $g(t) = 3x^2 t^3$

39. Apply the suitable differentiation rules to determine $\frac{dy}{dx}$.

a) $y = -4x^5 + 2x$

b) $y = \frac{9}{x^2}$

c) $y = \frac{1}{2}x^3 - 4$

d) $y = 1.25x^{\frac{4}{5}} - 1.5x^{\frac{2}{3}}$

e) $y = \sqrt[3]{x} + \sqrt[5]{x}$

f) $y = \sqrt{x} - 6\sqrt[3]{x}$

40. For each of the functions f and g the formulas for their derivatives are known: $\frac{d}{dx}f(x) = 2x + 3$ and $\frac{d}{dx}g(x) = 1 - 4x$. Use this to determine $\frac{dy}{dx}$ for the given y.

a) $y = 5f(x)$

b) $y = f(x) - g(x)$

c) $y = \frac{1}{2}f(x) + \frac{1}{4}g(x) - \frac{7}{4}$

41. Rewrite the formula of the function such that you can determine the derivative y' using the sum, constant factor or power rules.

a) $y = x(x^2 + 1)$

b) $y = x^3(x - 2)$

c) $y = (3x - 1)^2$

d) $y = \dfrac{4x^3 + 3x^2}{x}$

e) $y = \dfrac{2x^2 - 3x + 1}{x}$

f) $y = \dfrac{x^3 - 6}{x^2}$

42. Differentiate the function with respect to its argument. Interpret the function and its derivative geometrically.

a) $V(a) = a^3$

b) $O(h) = 2\pi \cdot r^2 + 2\pi \cdot r \cdot h$

c) $A(r) = \pi \cdot r^2$

d) $V(r) = \frac{4}{3}\pi \cdot r^3$

43. Let the function $h\colon x \mapsto h(x) = 2x - 1$ be given. What is the simplified formula for the requested derivative?

a) $\frac{d}{dx}(2h(x))$

b) $\frac{d}{dx}(h(x))^2$

c) $\frac{d}{dx}h(2x)$

d) $\frac{d}{dx}h(-x)$

e) $\frac{d}{dx}h(x^2)$

f) $\frac{d}{dx}\left(h(x) + x^2\right)$

44. Let the derivative f' be given. Find a possible formula for the function f.

a) $f'(x) = 2x$

b) $f'(x) = 8x^3 + 3$

c) $f'(x) = x^5 - x^3 + x$

45. During the first exam on the topic «Derivatives of simple functions», Leonie cheated off of her neighbour Gabriela, due to lack of time, and directly noted the result $y' = 3x^2 - 1$ for question 8 on her answer sheet. However, Leonie was annoyed after the exam, as she had not noticed that she and Gabriela had different exam variants to solve. Could it be that Leonie was very lucky in that her result was still correct although they had to differentiate different functions?

46. For which values of q does the graph of the function f with $f(x) = \frac{-3}{8}x^4 + qx^3 - x + 1$ have the slope $m = -1$ at point $P(2\,|\,y)$?

Product and Quotient Rules

Product rule: $(f(x) \cdot g(x))' = f'(x) \cdot g(x) + f(x) \cdot g'(x)$

Quotient rule: $\left(\dfrac{f(x)}{g(x)} \right)' = \dfrac{f'(x) \cdot g(x) - f(x) \cdot g'(x)}{(g(x))^2}$, $g(x) \neq 0$

Shorthand notation: $(f \cdot g)' = f' \cdot g + f \cdot g'$ or $\left(\dfrac{f}{g} \right)' = \dfrac{f' \cdot g - f \cdot g'}{g^2}$, respectively

47. Differentiate using the product rule. Pay attention to the argument.

a) $f(x) = x^2(1 - 3x^2)$
b) $g(z) = (z^2 - z + 1)(1 - z)$
c) $h(t) = (4 + 3t^2)(7 - 4t^3) + 1$
d) $f(x) = 3x^2\sqrt{x}$
e) $h(x) = (x^2 - 1)\sqrt{x}$
f) $g(u) = (2u + 1)\sqrt[5]{u^2}$
g) $f(z) = (z^3 - 2)(4z + \sqrt{z})$
h) $f(x) = (2x^2 + 5x + 3)^2$

48. Differentiate the function with respect to its argument.

a) $f(x) = x(a - x)$
b) $f(x) = a(1 - x)$
c) $f(x) = ax^3(1 - ax)$
d) $g(t) = \sqrt{t} \cdot (t + a)$
e) $g(t) = t^2\left(a + \frac{1}{t}\right)$
f) $g(t) = (t^2 + a^2)^2$

49. Let the function f be given by the formula $y = f(x) = (x^2 + 1)(3x - 4)$.

a) Determine the derivative of f first as follows: Expand the product, then simplify the expression, and finally differentiate it.

b) Determine f' also as follows: Apply the product rule first, then expand and simplify the expression.

50. The formula of a function f is represented in two ways; once as a simplified expression and once as a product. Show that no matter which representation of the formula one differentiates one arrives at the same result. That is, differentiating the first formula or the second formula by means of the product rule yields the same.

a) $f(x) = x^3 = x \cdot x^2$
b) $f(x) = x^4 = x^2 \cdot x^2$

51. Show by way of a simple counterexample that the statement $(f(x) \cdot g(x))' = f'(x) \cdot g'(x)$ does not hold in general.

52. Let $g(x) = k \cdot f(x)$. Show by use of the product rule that $g'(x) = k \cdot f'(x)$ holds.

53. Let f and g be two functions, then the following rule for the derivative of their product (shorthand notation) $(f \cdot g)' = f' \cdot g + f \cdot g'$ holds.

a) Deduce a corresponding rule for the derivative P' for a product of three factors: $P(x) = f(x) \cdot g(x) \cdot h(x)$; or in shorthand: $P = f \cdot g \cdot h$. The formula for P' must consist only of the terms f, g and h and their derivatives f', g' and h'.

Use the rule thus found to differentiate the function $P(x) = x^3(x + 1)(x^2 - 1)$. The formula does not need to be simplified.

b) Deduce a corresponding rule for the derivative of the special product P consisting of three identical factors: $P(x) = f^3(x) = (f(x))^3 = f(x) \cdot f(x) \cdot f(x)$.

Use the rule found to differentiate the function $P(x) = (2x - 1)^3$.

54. Use the product rule and the equation $x = \sqrt{x} \cdot \sqrt{x}$ for $x \geq 0$ to determine the derivative of $\sqrt{x}$.

55. Differentiate using the quotient rule.

a) $f(x) = \dfrac{x-1}{x+1}$ b) $f(x) = \dfrac{3x^2+1}{x}$ c) $f(x) = \dfrac{1-2x}{3x+4}$ d) $f(x) = \dfrac{x^2+1}{2x}$

e) $g(x) = \dfrac{x-1}{3x}$ f) $g(x) = \dfrac{3x}{1-x}$ g) $g(x) = \dfrac{1}{x+1}$ h) $g(x) = \dfrac{3}{1+x^2}$

i) $v(t) = \dfrac{3}{2t-1}$ j) $v(t) = \dfrac{t^2}{3t-1}$ k) $v(t) = \dfrac{t^3-1}{t^3+2}$ l) $v(t) = \dfrac{t}{1-3t^2}$

m) $w(t) = \dfrac{t^2-2}{t-1}$ n) $w(t) = \dfrac{1+4t^3}{2-4t^3}$ o) $w(t) = \dfrac{t+2}{\sqrt{t}}$ p) $w(t) = \dfrac{1}{1-t} - \dfrac{1}{1+t}$

56. Show that $(x^{-1})' = -x^{-2}$ by use of the quotient rule.

57. Differentiate the function f given by $f(x) = \frac{p(x)}{q(x)}$ (with $q(x) \neq 0$) without directly using the quotient rule. To do this, multiply both sides of the equation by $q(x)$, then differentiate both sides and, finally, solve for $f'(x)$.

58. Show by counterexample that the statement $\left(\frac{f(x)}{g(x)}\right)' = \frac{f'(x)}{g'(x)}$ does not hold in general.

Higher Order Derivatives I

The derivative of a differentiable function f is denoted by f'. In general, the function f' in turn can be differentiated; this derivative is called the second derivative of f and denoted by f''. Based on f'', one can establish the third derivative f''', and so on. We denote the nth derivative by $f^{(n)}$; $n \in \mathbb{N}$.

Example: $f(x) = x^2 + x$, $f'(x) = 2x + 1$, $f''(x) = 2$, $f'''(x) = 0$, $f^{(4)}(x) = 0$.

Further Notations: $y'' = \frac{\mathrm{d}^2 y}{\mathrm{d}x^2} = \frac{\mathrm{d}}{\mathrm{d}x}\left(\frac{\mathrm{d}y}{\mathrm{d}x}\right) = \frac{\mathrm{d}}{\mathrm{d}x}(y'(x))$, analoguously, the nth order derivative: $y^{(n)} = \frac{\mathrm{d}^n y}{\mathrm{d}x^n}$.

59. Determine the second derivative of the given function.

a) $f(x) = 3x^5 + 4x - 2$ b) $f(x) = \frac{1}{12}x^4 - \frac{1}{6}x^3 + \frac{1}{2}x^2$ c) $f(x) = (5x - 2)^2$

60. Determine $\frac{\mathrm{d}^2 y}{\mathrm{d}x^2} = \frac{\mathrm{d}}{\mathrm{d}x}\left(\frac{\mathrm{d}y}{\mathrm{d}x}\right)$.

a) $y = 4x^3 - 21$ b) $y = x^3 - 4x^2 + 5x - 1$

61. Determine the second derivative of the given function. Pay attention to the argument.

a) $x \mapsto 3x^6 - 4x + 5$ b) $x \to \frac{1}{24}x^4 + \frac{1}{18}x^3 - \frac{1}{8}x^2$

c) $x \mapsto (2x^3 - 3)(x^4 - 2x^3 + 7)$ d) $x \mapsto (ax - b)^3$

e) $x \mapsto 3xt^2 - tx^2$ f) $t \mapsto 3xt^2 - tx^2$

62. 📄 Sketch the graph of the second derivative f'' of the depicted function f.

a)
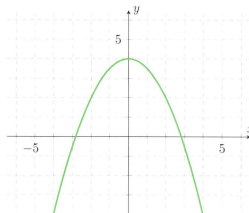

b)
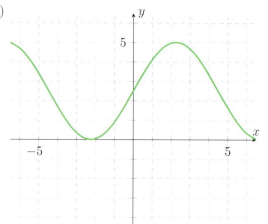

c)
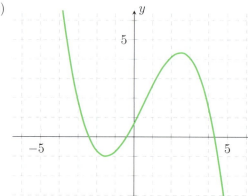

d)
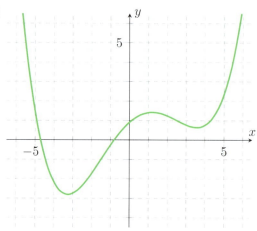

63. a) Determine the derivatives f', f'', $f^{(3)}$, $f^{(4)}$ and $f^{(5)}$ of the given function.

   i) $f(x) = x^3$ ii) $f(x) = x^4$ iii) $f(x) = x^5$ iv) $f(x) = x^6$

   b) What is the fifth derivative $f^{(5)}$ of the function $f(x) = x^n$, $n \in \mathbb{N}$ with $n \geq 5$?

   c) Can the power function $f(x) = x^n$ $(n \in \mathbb{N})$ be differentiated any number of times?

64. Give the simplest possible formula for the function $y = f(x)$ fulfilling the given equation.

   a) $y' = x^2$ b) $y'' = x^2$ c) $y''' = x^2$

65. Determine the first three derivatives y', y'' and y'''.

   a) $y = x^n$ b) $y = n \cdot x^{n-1}$ c) $y = \frac{x^n}{n!}$

66. Let $u(x) = k \cdot f(x)$, where k is an arbitrary constant. Show that the statement is true.

   a) $u''(x) = k \cdot f''(x)$ b) $u'''(x) = k \cdot f'''(x)$

67. Let $v(x) = f(x) \cdot g(x)$. Is the statement true? If not, how should it be corrected?

   a) $v''(x) = f''(x) \cdot g(x) + f'(x) \cdot g''(x)$

   b) $v'''(x) = f'''(x) \cdot g(x) + f''(x) \cdot g'(x) + f'(x) \cdot g''(x) + f(x) \cdot g'''(x)$

68. Differentiate the polynomial function p as many times as necessary until the result is 0 for the first time.

a) $p(x) = -2x^2 + 1$ 　　　　　 b) $p(x) = 9 - x + 5x^3$ 　　　　　 c) $p(x) = x^5$

d) How many times must a polynomial function of degree n be differentiated until the result is 0 for the first time?

Chain Rule

> Chain rule: $(f\,(g(x)))' = f'\,(g(x)) \cdot g'(x)$ 　　　　　 Shorthand notation: $(f \circ g)' = (f' \circ g) \cdot g'$
>
> Or with $y = f(u)$ and $u = g(x)$: $\dfrac{dy}{dx} = \dfrac{dy}{du} \cdot \dfrac{du}{dx}$
>
> *Mnemonic:* «outer derivative times inner derivative». A longer, more precise version of it is: «Derivative of outer function evaluated at inner function times derivative of inner function».

69. Determine the derivative y' using the chain rule.

a) $y = (4 - 3x)^2$ 　　　　　 b) $y = (x^2 + 1)^3$ 　　　　　 c) $y = (1 - x^2 + x^3)^4$

d) $y = 4 \cdot (3 - x^2)^{-1}$ 　　　　　 e) $y = (x^3 + 21)^{-4}$ 　　　　　 f) $y = \dfrac{10}{(10 - x^{10})^{10}}$

70. Differentiate the function w.

a) $w(x) = \sqrt{6x + 1}$ 　　　　　 b) $w(x) = 3 \cdot \sqrt{x^2 - 1}$ 　　　　　 c) $w(x) = (\sqrt{x} + 1)^4$

d) $w(t) = \dfrac{2}{\sqrt{2 - t^2}}$ 　　　　　 e) $w(t) = \sqrt{\dfrac{t - 1}{t + 1}}$ 　　　　　 f) $w(t) = \sqrt{1 + \sqrt{t}}$

71. Differentiate the function with respect to its argument.

a) $f(x) = \left(\dfrac{1 - x}{1 + x}\right)^2$ 　　　　　 b) $g(t) = \dfrac{(1 - t)^2}{1 + t^2}$ 　　　　　 c) $h(u) = \dfrac{3u^2 + 1}{(2u + 1)^2}$

72. Let the functions f and g fulfil the following relationship: $f(x) = g(ax + b)$; $a, b, x \in \mathbb{R}$. Show that this implies the following relationship for their derivatives f' and g': $f'(x) = a \cdot g'(ax + b)$.

73. Let the two functions f and g be defined by the formulas $f(x) = x^2 + 1$ and $g(x) = (x + 1)^2$. Determine the derivative of the indicated function.

a) $p(x) = f(g(x))$ 　　　 b) $q(x) = g(f(x))$ 　　　 c) $r(x) = f(f(x))$ 　　　 d) $s(x) = g(g(x))$

74. Differentiate the function with respect to its argument.

a) $f(x) = (3x^5 + 7x)^2$ 　　　　　 b) $g(x) = \dfrac{-2}{(1 - 2x^2)^3}$ 　　　　　 c) $f(t) = 2t^3(1 - 7t^2)^2$

d) $g(t) = \dfrac{3t}{(t - 1)^2}$ 　　　　　 e) $f(x) = \dfrac{x}{\sqrt{x + 1}}$ 　　　　　 f) $g(t) = \dfrac{(2t + 1)^2}{3t}$

75. Determine the value of the parameter $a \in \mathbb{R}$ such that the function f satisfies $f'(1) = 24$.

a) $f(x) = (2x - a)^3$ 　　　　　 b) $f(x) = \dfrac{-6}{\sqrt{ax}}$ 　　　　　 c) $f(x) = \dfrac{2a}{a - 4x}$

76. Let the function f be given. Determine the formula for P' using the chain rule.

    a) $P(x) = (f(x))^2$ b) $P(x) = (f(x))^3$ c) $P(x) = (f(x))^n$

77. For the square root function w defined by $w = w(x) = \sqrt{x}$, the equation $w^2 = x$ holds for $x > 0$. Determine w' by differentating both sides of the equation. What is the formula for the derivative $w' = w'(x) = (\sqrt{x})'$?

78. Let the three functions $u(x)$, $v(x)$ and $w(x)$ be given, where $x \in \mathbb{R}$. Deduce the formula of the derivative f' of the composite function f satisfying $f(x) = (u \circ v \circ w)(x) = u(v(w(x)))$.

    Note: The desired «chain rule» for f' must consist only of the functions u, v and w and their derivatives u', v' and w'.

79. Let the two functions u and v with $v(x) \neq 0$ be given.

    a) Determine the derivative f' of the function f defined by $f(x) = \frac{1}{v(x)}$ using the chain rule.

    b) Determine the derivative f' of the function f defined by $f(x) = \frac{u(x)}{v(x)} = u(x) \cdot \frac{1}{v(x)}$ using the product rule and the result of the previous part.

80. True or false? Give reasons. If false, how should the formula be corrected?

    a) $(f(-x))' = -f'(x)$ b) $\left(f\left(\frac{1}{x}\right)\right)' = \frac{1}{f'(x)}$

Trigonometric, Exponential and Logarithmic Functions

Derivatives of Elementary Functions

Trigonometric functions (x in radians):

$$\sin'(x) = (\sin(x))' = \cos(x) \qquad\qquad \cos'(x) = (\cos(x))' = -\sin(x)$$

$$\tan'(x) = (\tan(x))' = 1 + \tan^2(x) = \frac{1}{\cos^2(x)}$$

Exponential functions:

$$(e^x)' = e^x \qquad\qquad (e^{ax})' = ae^{ax} \qquad\qquad (b^x)' = b^x \cdot \ln(b)$$

Logarithmic functions:

$$\ln'(x) = (\ln(x))' = \frac{1}{x} \qquad\qquad\qquad \log_b'(x) = (\log_b(x))' = \frac{1}{x \cdot \ln(b)}$$

81. Determine $f'(x)$.

    a) $f(x) = \sin(x) - \cos(x)$ b) $f(x) = 3\sin(x) + 5\cos(x)$ c) $f(x) = \frac{1}{3}\sin(x) - 3$

    d) $f(x) = 1 - \cos(x)$ e) $f(x) = x + \cos(x)$ f) $f(x) = \frac{\sin(x)}{\pi}$

82. Determine the formula for the derivative.

a) $f(t) = -5\sin(t) + t^2$ b) $f(t) = -\sqrt{2} \cdot \sin(t)$ c) $f(t) = 2\sqrt{t} - \frac{1}{3}\cos(t)$

d) $g(t) = \sqrt{3} \cdot \sin(t)$ e) $g(t) = \frac{1}{t^2} - \pi \cdot \sin(t)$ f) $g(t) = \frac{-1}{\sqrt{2}} \cdot \cos(t) + \pi$

83. Determine the derivative y'.

a) $y = x \cdot \sin(x)$ b) $y = \cos(x) \cdot 2x^3$ c) $y = \frac{1}{x^2} \cdot \sin(x) + 3x$

d) $y = (x^2 - 3x + 1) \cdot \cos(x)$ e) $y = \sqrt{x} \cdot \cos(x) - 2$ f) $y = \sin(x) \cdot \cos(x)$

84. Determine the derivative y'.

a) $y = \dfrac{\sin(x)}{x}$ b) $y = \dfrac{2x}{\cos(x)}$ c) $y = \dfrac{-2\cos(x)}{\sqrt{x}}$

d) $y = \dfrac{\pi}{\sin(t)} + 2$ e) $y = \dfrac{t^3 + 2t}{\cos(t)}$ f) $y = \dfrac{\cos(t)}{\sin(t)} + t^2$

85. Prove the formula $(\tan(x))' = 1 + (\tan(x))^2 = \frac{1}{(\cos(x))^2}$ using $\tan(x) = \frac{\sin(x)}{\cos(x)}$.

86. Determine $\frac{\mathrm{d}}{\mathrm{d}t} f(t)$.

a) $f(t) = t - \tan(t)$ b) $f(t) = t \cdot \tan(t)$ c) $f(t) = \dfrac{1}{\tan(t)}$

87. Determine the derivative f' of the function f.

a) $f(x) = \sin(2x)$ b) $f(x) = \sin(x^2 + x)$ c) $f(x) = \cos(\pi \cdot x)$

d) $f(x) = \cos(x^2)$ e) $f(x) = \tan(x^2)$ f) $f(x) = \sin(\sqrt{x})$

g) $f(x) = \cos^2(x)$ h) $f(x) = \tan^2(x)$ i) $f(x) = \sqrt{\sin(x)}$

88. Differentiate the function with respect to its argument.

a) $f(x) = \dfrac{\sin(x)}{1 + \cos(x)}$ b) $f(t) = \dfrac{1 + \sin(t)}{1 - \cos(t)}$ c) $f(u) = \frac{1}{2} \cdot \cos^3(u)$

d) $f(x) = \dfrac{1}{x \cdot \cos(x)}$ e) $f(t) = t \cdot \sin(t^2)$ f) $f(u) = \cos\left(\dfrac{1 - u}{1 + u}\right)$

89. Differentiate the given function.

a) $f(x) = \mathrm{e}^{-2x}$ b) $g(x) = \mathrm{e} + \mathrm{e}^{x^2}$ c) $h(x) = \mathrm{e}^{\frac{1}{x}} + x^2$

d) $f(t) = 3 \cdot \mathrm{e}^{-\frac{t^2}{2}}$ e) $g(t) = t^2 \cdot \mathrm{e}^{1-t}$ f) $h(t) = \sqrt{\mathrm{e}^t}$

90. Determine $\frac{\mathrm{d}y}{\mathrm{d}x}$.

a) $y = 6^x$ b) $y = 5^{-x}$ c) $y = \frac{1}{2} \cdot 3^{2x} + 7$

d) $y = 3x \cdot 10^x$ e) $y = \dfrac{2^x}{x}$ f) $y = (2\mathrm{e})^x$

91. Determine y'.

a) $y = \ln(2 - 3x)$ b) $y = \lg(x + 1)$ c) $y = 3x^2 + \frac{1}{4} \cdot \mathrm{lb}(x^2)$

d) $y = \dfrac{t}{\ln(t)}$ e) $y = \sqrt{t} \cdot \ln(t)$ f) $y = \ln(2 + \sqrt{t})$

92. Differentiate the given function.

a) $f(x) = \dfrac{1}{1 + e^x}$

b) $g(x) = e^{-x} + \frac{x}{e^x}$

c) $h(x) = \ln\left(\frac{1}{x}\right) + \sqrt{2}$

d) $f(u) = \ln\left(\dfrac{u}{u-1}\right)$

e) $g(u) = \ln\left(\dfrac{\cos(u)}{e^{3u}}\right)$

f) $h(u) = 10^{\frac{2u}{u+1}}$

g) $f(x) = \dfrac{e^x}{x^2 - 4} + \pi^2$

h) $g(x) = \log_3(\cos(x))$

i) $h(x) = \sin(x) \cdot \ln(x)$

j) $f(u) = u^e \cdot e^{-u}$

k) $g(u) = \ln\left(\dfrac{\sqrt{u^2 + 1}}{u+1}\right)$

l) $h(u) = \sin(\ln(\cos(u)))$

93. Let $f(x) = a^x$, where $a > 0$. Show that the derivative fulfils $f'(x) = f'(0) \cdot f(x)$.

94. Use the relationship $e^{\ln(x)} = x > 0$ to deduce the formula for the derivative of the function $y = \ln(x)$.

95. Let the function f be given by the equation $y = f(x)$. Differentiate the function g given by the equation $g(x) = \ln(f(x))$ with respect to x.

96. Differentiate the function with respect to its argument.

a) $y(t) = A \cdot \sin(\omega \cdot t) + c$

b) $y(t) = \dfrac{A}{2\omega} t \cdot \sin(\omega \cdot t)$

c) $y(t) = a \cdot e^{-\lambda \cdot t}$

97. Differentiate the function $y = \ln(x) - \ln(2x) + \ln(3x) - \ln(4x)$.

98. Differentiate with respect to x.

a) $f(x) = x^x$

b) $g(x) = x^{\sin(x)}$

c) $h(x) = \sqrt[x]{x}$

d) $i(x) = x^{\sqrt{x}}$

e) $j(x) = x^{(e^x)}$

f) $k(x) = \log_x(2)$

Higher Order Derivatives II

99. Let the function f be given by the equation $f(x) = x \cdot e^x$. Answer all parts of this exercise in the given order.

i) Determine $f'(x)$ and $f''(x)$.

ii) What do you guess the formula for the fourth derivative $f^{(4)}(x)$ is? Verify or falsify your conjecture.

iii) What do you think is the formula for $f^{(n)}(x)$, that is the nth order derivative of f?

iv) Prove your conjecture by mathematical induction.

100. Let the function f be given by $f(x) = x^2 \cdot e^x$. Calculate the first four derivatives $f'(x)$, $f''(x)$, $f'''(x)$ and $f^{(4)}(x)$. What is the formula for $f^{(n)}(x)$, i.e. the nth order derivative of f?

101. The setting for this exercise is the sine function with the equation $y = \sin(x)$. Answer all parts of this exercise in the given order.

i) Determine the following higher order derivatives: $y^{(3)}$, $y^{(4)}$ and $y^{(5)}$.

ii) What are the 11th, the 101st and the 1001st derivatives of the sine function?

iii) Deduce the 1001st derivative of the cosine function from the 1001st derivative of $y = \sin(x)$.

102. Determine the first six derivatives of $y = x \cdot \sin(x)$. What conjecture arises from these for the 12th and 13th derivatives of $y = x \cdot \sin(x)$? What do you think is the 25th derivative of y?

First Applications

In physics, the time t is a commonly occurring argument of functions. In combination with the position $s = s(t)$, the velocity $v = v(t)$ or the acceleration $a = a(t)$, the derivatives with respect to t are occasionally indicated by dots instead of prime marks as follows: $v(t) = \frac{\mathrm{d}}{\mathrm{d}t}s(t) = \dot{s}(t)$ and $a(t) = \frac{\mathrm{d}}{\mathrm{d}t}v(t) = \dot{v}(t) = \ddot{s}(t)$.

103. In the case of straight-line trajectories, the position of the moving point mass is commonly described by the so-called *position function* $s(t)$, where $s(0)$ marks the initial position. Determine the velocity $v(t) = \dot{s}(t)$ of the point mass as well as its acceleration $a(t) = \ddot{s}(t)$ as a function of the time t for the given trajectory.

a) Vertically launched projectile: $s(t) = s_0 + v_0 \cdot t - \frac{g}{2} \cdot t^2$; $g \mathrel{\hat{=}}$ standard gravitational acceleration

b) Simple harmonic motion: $s(t) = C \cdot \cos(\omega_0\, t + \varphi)$

104. The course of a viral illness is modelled by a function V with $V(t) = \frac{10^6}{8}(6t^2 - t^3)$. Here, t denotes the time (in days) since the beginning of the infection and V the number of so-called virions in 1 millilitre of blood.

a) For what values of t is this model valid?

b) Calculate $V'(t)$. What is the meaning of the first derivative V'?

c) Draw the graphs G_V and $G_{V'}$ into a coordinate system.

d) Interpret the progression of both graphs. When is $V'(t) = 0$, $V'(t) < 0$ or $V'(t) > 0$? What does this signify?

105. The distance $s(t)$ of an object from a fixed point F as a function of time t is described by the following equation of motion: $s(t) = -0.25t^2 + 4t$; t in seconds, s in metres.

a) At which time is the object at point F?

b) What is the velocity of the object after $t_1 = 2\,\mathrm{s}$, $t_2 = 5\,\mathrm{s}$ and $t_3 = 13\,\mathrm{s}$? How should a negative sign be interpreted?

c) Calculate the acceleration $a = a(t)$ imparted to this object.

106. In a $30\,\mathrm{km/h}$ speed zone, Olev gets on his e-bike and rides off briskly, his velocity v (in km/h) in the first t seconds hereafter is as follows: $v = v(t) = \frac{30t^2}{t^2+2}$. *Note:* All results are to be rounded to at most one decimal place.

a) How fast is Olev travelling after $t = 1$, $t = 2$, $t = 3$ and $t = 7$ seconds?

b) Calculate the average change in velocity (per second) for the following three time intervals in seconds: $[0; 3]$, $[3; 6]$ and $[6; 9]$.

c) Deduce a formula for the acceleration $a = a(t)$, imparted to the e-bike at time t, and use it to calculate the accelerations $a(0.5)$, $a(1)$, $a(1.5)$ and $a(5)$ in $\mathrm{m/s^2}$.

d) Determine $\lim\limits_{t \to \infty} v(t)$.

107. The following graphs show the position $s = s(t)$, the velocity $v = v(t) = \dot{s}$ and the acceleration $a = a(t) = \ddot{s}$ of an object that travels along the y-axis as a function of time t. Which of the graphs A, B and C corresponds to which of the functions s, v and a?

For exercise 107:   *For exercise 108:*

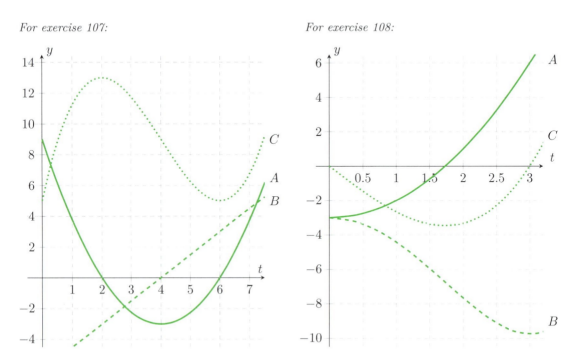

108. The plotted graphs above right show the position $s = s(t)$, the velocity $v = v(t)$ and the acceleration $a = a(t)$ of a vehicle travelling along the y-axis as functions of time t. Which of the graphs A, B and C corresponds to which of the functions s, v and a?

109. A mass m attached to a spring moves up and down. The function given by the formula $s(t) = 3 + \cos(\pi \cdot t)$ describes the distance of the mass from the ceiling to which the spring is attached (see figure).

a) What is the maximum (or minimum) displacement of the mass?

b) Determine the first and second derivatives $\dot{s}(t)$ and $\ddot{s}(t)$. What are the physical meanings of these two derivatives?

c) Sketch the graph of $\dot{s}(t)$ for a complete cycle.

d) Calculate $\dot{s}(0)$, $\dot{s}\left(\frac{1}{2}\right)$, $\dot{s}(1)$ as well as $\dot{s}\left(\frac{3}{2}\right)$ and interpret the numerical values.

110. The number of bacteria in a Petri dish is a function of time t (in days) and is denoted by N (in 1000 per cm^2). The growth rate of this bacterial culture is modelled by the function N as follows:

$$N(t) = \frac{100}{1 + 99 \cdot e^{-\frac{4}{5}t}}$$

a) How many bacteria per cm^2 are there at the initial time?

b) After approximately how many days are there 60'000 bacteria per cm^2?

c) What is the instantaneous daily increase in the middle of the third day (in 1000 per cm^2)?

111. In a simple electrical circuit the voltage U (in Volts, V), the current I (in Amperes, A) and the built-in resistance R (in Ohms, Ω) are related as follows: $U = R \cdot I$. In what follows, we assume that the applied voltage U continually decreases by $1\,\text{V/s}$, while the resistance continually increases by $4\,\Omega/\text{s}$. What is the rate of change in the current I if the circuit at the time in question is subject to a voltage of $30\,\text{V}$ and the resistance measures $40\,\Omega$? Is the current increasing or decreasing?

112. A beacon rotates uniformly and requires 4 seconds per revolution. The two strongly bundled light beams alternately hit a wall, which is 10 metres from the light source.

   a) Determine the velocity $v = v(t)$ of the beam spot moving across the wall depending on the displacement $s = s(t) \geq 0$.

   b) Calculate the velocity of the beam spot for both of the displacements $s_1 = 0\,\text{m}$ and $s_2 = 10\,\text{m}$.

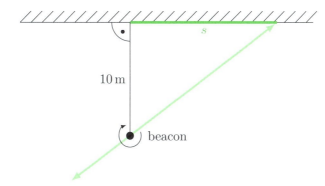

The acceleration $a = a(t)$ in everyday trajectories is not always constant. Three typical examples are: a lift ride, emergency braking with a vehicle or the start of a 100 metre sprint. Common to all examples is that either a phase of motion follows a phase of rest or vice versa, usually with a rapid change. In the phase of motion $a = a(t) \neq 0$ holds, while in the phase of rest $a = a(t) = 0$.

Thus, as the acceleration $a(t)$ is not always constant, we know that its derivative $a'(t)$ does not constantly equal zero, but changes with the time t and thus itself is a function of t. The kinematic function defined by $j(t) = a'(t)$ is called the *jerk* or the *jolt* and has the unit of m/s^3. The use of the letter j is naturally derived from the inital letter of the words jerk and jolt.

113. As is well known, the velocity $v = v(t)$ depends on the acceleration $a = a(t)$.

   a) How are the velocity function $v(t)$ and the jerk function $j(t)$ related?

   b) Determine the jerk $j(t)$ for $v(t) = \sin(t)$.

   c) Give the simplest possible velocity function $v(t)$ such that $j(t) = 1.2t$.

114. a) Since the distance $s = s(t)$ travelled during the time t depends on the acceleration $a = a(t)$, there also has to be a relationship between the two functions $s(t)$ and $j(t)$. What is this relationship?

b) Determine the jerk $j(t)$ for the situation of free fall, that is for $s(t) = \frac{1}{2}gt^2$.

c) For which simplest possible position function $s = s(t)$ is $j(t) = 3t$?

115. a) A ball is launched vertically upward with an initial velocity of v_0. What can be said of the jerk $j(t)$ in the phase of ascent, if $s(t)$ denotes the ball's increasing height above the ground?

b) What can be said of the velocity profile $v(t)$ if $j(t)$ is very strongly negative?

116. a) 🖺 Let the graph G_a of the acceleration function a be given. Sketch the course of the jerk function j into the same figure with a different colour.

For a)   *For b)*

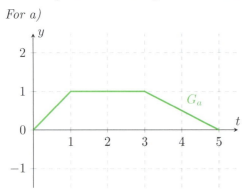   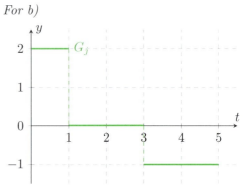

b) 🖺 Let the graph G_j of the jerk function j be given. Amend the figure with the course of the acceleration function a using a different colour such that $a(0) = 0$.

117. The diagrams for the exercises 107 and 108 on page 70 both depict the course of the three functions s, v and a. For each diagram, consider what you can say with certainty about the course of the jerk function.

A First Look at Differential Equations

A *differential equation*, abbreviated as *DE*, is an equation which describes a mathematical relationship between an unknown function $y = y(x)$, its first or higher order derivatives y', y'', ... and the argument x.

Examples:

- $y' = x + y$
- $y'' + y = 0$
- $y' = -y^2$
- $xy' - y = 0$
- $2yy' = 1$

If the function $y = y(x)$ fulfils a given DE, the function is called a *solution* of the DE. Generally, differential equations have infinitely many solutions.

Examples:

- $y = y(x) = 2x$ is a solution of the DE $xy' - y = 0$.
- $y = e^x$ is a solution of the DE $y' = y$; $y = 3.21 \cdot e^x$ is also a solution of this DE.
- $y = \cos(x)$ is a solution of the DE $y'' + y = 0$; $y = \sin(x)$ is also a solution of this DE.

Note: The topic of DEs will be revisited in chapter 4.7 on page 152.

Remark: Differential equations play a central role in the modelling of phenomena in the natural sciences, engineering and economics.

118. Let a differential equation and a function $y = y(x)$ be given. Verify that the given function is a solution of the DE.

a) $xy' - y = 0, \ y = 4x$

b) $xy' - y + 1 = 0, \ y = 1 - x$

c) $y' = \frac{2y}{x}, \ y = 2x^2$

d) $xy' - 2y + 4 = 0, \ y = x^2 + 2$

e) $y' = x - y + 1, \ y = x + 12e^{-x}$

f) $yy' - x = 0, \ y = \sqrt{x^2 - 1}$

119. Is the given function $y = y(x)$ a solution of the given DE?

a) $y'' - 9y = 0, \ y = 12 \cdot e^{3x}$

b) $y'' - 4y = 0, \ y = e^{2x} - e^{-2x}$

c) $y'' + 4y = 0, \ y = \sin(2x)$

d) $y'' + 16y = 0, \ y = 3 \cdot \sin(4x) - 5 \cdot \cos(4x)$

e) $2yy' + 1 = 0, \ y = \sqrt{x}$

f) $y' = (y-2)^2, \ y = 2 + \frac{1}{x+3}$

120. For which value $k \in \mathbb{R}$ is the function y given by its formula a solution of the DE?

a) $xy' - y + 12 = 0, \ y = k - x$

b) $xy' - 2y + 10 = 0, \ y = x^2 + k$

c) $y' = \frac{3}{2x+1}, \ y = k \cdot \ln(2x + 1); \ 2x + 1 > 0$

d) $y' = 1 - y, \ y = e^{kx} + 1$

e) $y' = x - y + 1, \ y = x + ke^{-x}$

f) $y'' + 9y = 0, \ y = \sin(kx); \ k \neq 0$

121. Determine the value of the constant $m \in \mathbb{R}$ in the differential equation such that the function y given by its formula is a solution of the DE.

a) $xy' - y + m = 0, \ y = 3x + 21$

b) $xy' - 2y + m = 0, \ y = \frac{1}{3}x^2 - 1$

c) $y' = \frac{m}{4x-2}, \ y = 3 \cdot \ln(2x - 1); \ 2x - 1 > 0$

d) $y' = y - 2x + m, \ y = x + e^{-x}$

e) $y' = m - y, \ y = e^{-x} + 1001$

f) $y'' + my = 0, \ y = \cos(-2x)$

122. For trajectories, the distance s travelled, the velocity v and the acceleration a are generally written as a function of time f, that is in the form $s = s(t)$, $v = v(t)$ and $a = a(t)$.

a) The physical dependence between $s(t)$ and $v(t)$, between $v(t)$ and $a(t)$, or between $s(t)$ and $a(t)$ can be stated as differential equations of the form $T(f, g) = 0$; here $T(f, g)$ denotes a formula, which only comprises the functions f and g and their derivatives. List three differential equations of the form $T(s, v) = 0$, $T(v, a) = 0$ and $T(s, a) = 0$ that are as simple as possible.

b) In the context of free fall $a = g$ is the gravitational acceleration. Show that $s = s(t) = \frac{1}{2}gt^2$ is a solution of the differential equation $s'' - g = 0$. This DE has more solutions. Which ones?

123. Show that the function s occurring in the solution of exercise 112 given by $s(t) = 10 \tan\left(\frac{\pi}{2}t\right)$ is a solution of the following differential equation: $\dot{s}(t) = \frac{\pi}{20}s^2(t) + 5\pi$.

3.4   Tangent, Normal and Intersecting Angle

Tangent and Normal

The tangent t and the normal n to the graph G_f of a function f given by $y = f(x)$ at the point a fulfil the following:

- slope of tangent t: $m_t = f'(a) = \tan(\varphi)$, $\varphi \mathrel{\hat{=}}$ angle of inclination of the tangent
- $n \perp t \;\Rightarrow\; m_t \cdot m_n = -1 \;\Rightarrow\;$ slope of normal n: $m_n = \frac{-1}{m_t} = \frac{-1}{f'(a)}$

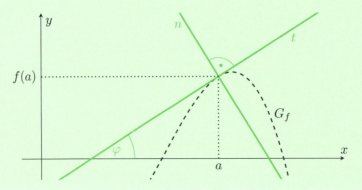

Intersecting Angle

If the curves k_1 and k_2 intersect at a point S, the intersection angle σ of the two curves is defined as the (acute) angle enclosed by the two tangents t_1 and t_2 to the curves at the point of intersection S.

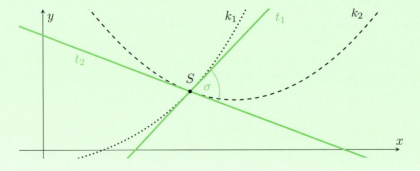

Polynomial Functions

124. Let the parabola be given by the equation $y = x^2 + 4$. Determine the equation of the tangent t to the parabola at the specified point P.

a) $P(3 \,|\, y_P)$ b) $P(-2 \,|\, y_P)$

125. Determine the tangent and the normal to the graph of the function f given by $f(x) = 2x^3 - 7x^2 + 3$ at the point $P(3 \,|\, -6)$.

126. Let the function f be given by $f(x) = 2x^3 - 14x + 9$. At which points of the graph G_f is the tangent parallel to the line $g: y = 10x - 8$?

127. Determine the value of the parameter k such that the tangent to the graph of the function $f(x) = 2x^3 + kx - 3$ at point $x = 2$ has the slope 4.

128. Let the function f be given by $f(x) = 1 - 3x + 12x^2 - 8x^3$. Determine the equation of the tangent t to the graph of f that is parallel to the tangent at the point $x = 1$.

129. At which point P of the graph of f defined by $f(x) = 2x^2 - 12x + 13$ is the tangent parallel to the x-axis?

130. Determine the equation of the horizontal tangents to the graph of the function f, where $f(x) = 2x^3 + 3x^2 - 12x + 1$.

131. Let the curve k be given by $k(x) = 2\left(\frac{x}{3} - 2\right)^3$. The tangent at point P intersects the line $g(x) = \frac{-1}{2}x + 7$ at a right angle. Determine P and the equation of the tangent.

132. Let the curve k be given by the equation $y = k(x) = x^3 - 4x^2 + 1$.
 a) Verify that the point $K(1 \,|\, {-2})$ is on the curve k and determine the equation of the tangent t_K to the curve at K.
 b) At which point $E \in k$ of the curve is its tangent t_E to the curve parallel to the x-axis?
 c) At which point $P \in k$ of the curve is its tangent t_P to the curve parallel to the line $g\colon y = 3x - 12$?

133. The chord s of the parabola $y = x^2 - 2x + 3$ connects the two points $P(1 \,|\, y_P)$ and $Q(3 \,|\, y_Q)$ of the curve. Determine the equation of the tangent t to the parabola that is parallel to s.

134. Let the two functions $f(x) = x^3 - x - 2$ and $g(x) = x^2 - 3$ be given. Show that the graphs G_f and G_g touch each other at the point $x_0 = 1$.

135. Verify that the two curves defined by the equations $y = k(x) = \frac{1}{3}x^3 - 4x + \frac{28}{3}$ and $y = p(x) = \frac{-1}{2}x^2 + 2x + 2$ are tangent to each other.

136. For which value of a does the graph G_f of f with $f(x) = \frac{1}{2}x^2 + ax + 4.5$ touch the x-axis?

137. At which points of the graph of $y = x^n$ do the corresponding tangents have the slopes 1 or n, respectively? $(n \in \mathbb{N})$

138. Prove the statement
 a) All parabolas p_a with equation $y = ax^2 - x + 1$ $(a \neq 0)$ touch each other at a single point B. This point of tangency B is therefore independent of the parameter a.
 b) All cubic parabolas $p_k\colon y = kx^3 + x + 3$ $(k \neq 0)$ touch each other at a single point.

139. At which point K of the curve is the graph of the function defined by the equation $y = \frac{1}{8}x^3$ intersected by its tangent t at the point $x = -2$?

140. The equation for a polynomial function is given. Determine the equation of the tangent at the intersection point of the graph of the function and the y-axis. What do you notice?

a) $y = 4x^2 + 3x - 1$
b) $y = 7x^4 - 8x^3 + 9x^2 - 10x + 11$
c) $y = 2x^5 - 4x^3 + 6x$
d) $y = 101x^{100} - 99$
e) $y = a_n x^n + a_{n-1} x^{n-1} + a_{n-2} x^{n-2} + \ldots + a_2 x^2 + a_1 x + a_0;\ a_n \neq 0,\ n \in \mathbb{N}$

141. The parabola p given by $y = 2 - \frac{1}{2} x^2$ intersects the x-axis at a point R on the right of the origin. At this point R, the tangent t to the parabola and the normal n to the parabola are drawn.

a) At which points T and N do t and n, respectively, intersect the y-axis?
b) At which point S does the normal n intersect the parabola p a second time?

142. The cubic parabola k is given by the equation $y = k(x) = x^3 - x^2 + x - 1$. Determine the equation of the tangent t to the curve at the point $x_0 = 1$ and locate the point of intersection S, where the tangent t intersects the parabola k one more time.

143. At each of the points $U(u \,|\, u^2)$ and $V(v \,|\, v^2)$, which are on the unit parabola, we consider the tangents t_u and t_v, respectively, to the unit parabola. Calculate the abscissa x_S of the intersection point S of t_u and t_v.

144. Let the function f be given by the formula $f(x) = x^2 + 9$. Determine the equations of all the tangents to the graph of f that pass through the origin.

145. Let the function f be given by $y = f(x) = x^2 + 4$ and let P be a point not on the graph G_f. Determine the equation of the tangent to G_f that passes through the point P.

a) $P(0 \,|\, 0)$
b) $P(-2 \,|\, 4)$
c) $P(2 \,|\, 7)$

146. Let the parabola be given by the equation $y = \frac{1}{2} x^2$ and P be a point. Find the equation of the normal to the parabola passing through P.

a) $P(6 \,|\, 0)$
b) $P(-12 \,|\, 6)$
c) $P(0 \,|\, 5)$

147. At what acute angle does the graph of the function f intersect the x-axis?

a) $f(x) = -x^2 + 4$
b) $f(x) = 2x^2 + 6x - 20$
c) $f(x) = 0.2x^3 + 0.8x^2 - x$
d) $f(x) = x^5 - 2x^3 + x$

148. Consider the curve given by the equation $y = \frac{1}{3} x^3$.

a) At which point of the curve does the corresponding tangent form an angle of $45°$ with the x-axis?
b) Justify why the slope of the curve cannot be negative at any point.

149. At what (acute) angle do the two parabolas intersect each other if their equations are as given?

a) $y = f(x) = x^2 - \frac{1}{2};\ y = g(x) = -x^2$
b) $y = f(x) = 0.5x^2 + 3x + 14;\ y = g(x) = 0.5x^2 + 4x + 13$

150. For which parameter $a > 0$ do the parabolas with the given formulas intersect each other orthogonally, i.e. perpendicularly?

a) $f(x) = x^2$
$g(x) = \frac{1}{2} - ax^2$

b) $f(x) = ax^2$
$g(x) = 1 - \frac{1}{3}x^2$

c) $f(x) = ax^2$
$g(x) = -\frac{1}{4}x^2 + a$

151. Let the curve k be given by the equation $y = k(x) = 0.5x^3 - 4x^2 + 9x - 4$ and the descending line g by the equation $y = g(x) = 4 - x$. Which of the following two statements are true?

a) Line g is tangent to the curve k at point $x_1 = 2$.

b) The curve k and the line g intersect orthogonally, i.e. perpendicularly, at point $x_2 = 4$.

152. The parabola p given by the equation $y = p(x) = ax^2$ $(a \neq 0)$ and the line g given by the equation $y = g(x) = mx$ $(m \neq 0)$ both pass through the origin and intersect each other at another point S. Prove that the angle of intersection between the line g and the parabola p at point S only depends on the slope m of the line and not on the parameter a of the parabola. *Note:* There is no need to compute the angle of intersection to prove this.

General Functions

For **153–156**: Determine the equations of both the tangent and the normal at point P of the curve k.

153. a) $k\colon y = \dfrac{2}{x+2}$; $P(-1\,|\,y_P)$ b) $k\colon y = \dfrac{x^2 - 1}{x - 3}$; $P\big(\frac{1}{3}\,|\,y_P\big)$ c) $k\colon y = \dfrac{x}{x^2 + 4}$; $P(-2\,|\,y_P)$

154. a) $k\colon y = \sqrt{3x + 1}$; $P(1\,|\,y_P)$ b) $k\colon y = \frac{1}{x} + \sqrt{x}$; $P(4\,|\,y_P)$ c) $k\colon y = \sqrt{x^2 - 7}$; $P(4\,|\,y_P)$

155. a) $k\colon y = \tan(x)$;
$P\big(\frac{\pi}{4}\,|\,y_P\big)$

b) $k\colon y = \sin(x)$;
$P\big(\frac{\pi}{3}\,|\,y_P\big)$

c) $k\colon y = \sin(x)\cos(x)$;
$P\big(\frac{\pi}{4}\,|\,y_P\big)$

156. a) $k\colon y = 2\mathrm{e}^x + 1$; $P(0\,|\,y_P)$ b) $k\colon y = \ln\big(\frac{x}{4}\big)$; $P(4\,|\,y_P)$ c) $k\colon y = x\ln(x)$; $P(1\,|\,y_P)$

157. Determine the equations of the horizontal tangents to the graph given by $f(x) = 2\sqrt{x} + \frac{1}{\sqrt{x}}$.

158. From an arbitrary point P of the curve given by the equation $y = \frac{1}{x}$ $(x \neq 0)$ we draw perpendicular lines onto the x- and y-axes and denote their perpendicular foots by X and Y, respectively. Show, that the segment $\overline{XY}$ is always parallel to the tangent at point P.

159. Every tangent t to the curve given by the equation $y = \frac{1}{x}$ $(x \neq 0)$ intersects the two coordinate axes at the respective points X and Y. Show that the point of tangency B of the tangent t is always the midpoint of the segment $\overline{XY}$.

160. Let the function f be given by $f(x) = \mathrm{e}^{3x}$. At which point is the tangent to the graph of f parallel to the line $g\colon y = x$? What is the equation of the tangent at this point?

161. Show that the exponential curve given by the equation $y = f(x) = e^{x-2}$ and the parabola given by the equation $y = g(x) = \left(\frac{x}{2}\right)^2$ touch each other at the point $x = 2$. At which point does the tangent common to both curves intersect the x-axis?

162. Consider the tangent t of the curve defined by $y = f(x) = \ln(x)$ at the point $x = e^2$. At which point x_0 does t intersect the x-axis?

163. For which constant q do the two curves defined by the equations $y = s(t) = \sin(t)$ and $y = c(t) = q - \cos(t)$ touch each other in the range $0 \le t \le \frac{\pi}{2}$?

164. Let the function be given by $y = f(x)$ and let P be a point not on the graph G_f. Determine the equation of the tangent to G_f passing through P.

a) $f(x) = \dfrac{4}{x+4};\quad P(0\,|\,0)$

b) $f(x) = \dfrac{x}{2-x};\quad P(-2\,|-5)$

c) $f(x) = e^{\frac{x}{2}};\quad P(0\,|\,0)$

d) $f(x) = \sqrt{2x+3};\quad P(-6\,|\,0)$

165. Determine the points and the angles of intersection of the graph G_f with the coordinate axes.

a) $f(x) = \dfrac{x}{x-1}$

b) $f(x) = \dfrac{1-x^2}{2(x^2+1)}$

c) $f(x) = (x+1)e^x$

d) $f(x) = e^{-x}\cos(x),\ -\pi \le x \le \pi$

166. At which angles do the curves k_1 and k_2 intersect each other?

a) $k_1\colon y = f(x) = \dfrac{2}{x^2+1}$

b) $k_1\colon y = f(x) = \cos(x),\ 0 \le x \le \frac{\pi}{2}$

$k_2\colon y = g(x) = x^2$

$k_2\colon y = g(x) = \sin(x),\ 0 \le x \le \frac{\pi}{2}$

167. Calculate the intersection angles of the curves k_1 and k_2 defined by the given equations.

a) $y = k_1(x) = \dfrac{3-x}{(x-2)^2}$

b) $y = k_1(x) = e^{\frac{-(x+1)}{2}}$

c) $y = k_1(x) = \ln(x+3)$

$y = k_2(x) = x - 3$

$y = k_2(x) = e^{-x^2}$

$y = k_2(x) = \ln(7-x)$

168. For which positive coefficient a do the two graphs defined by the equations $y = f(x) = ax^2$ and $y = g(x) = \frac{1}{x}$ intersect orthogonally, i.e. perpendicularly?

169. The curve k is given by the equation $y = \frac{x^2+1}{x}$ $(x \neq 0)$.

a) Calculate the slope m of the line g that passes through the origin and intersects the curve orthogonally, that is perpendicularly.

b) Show that the line g is an angle bisector of the two asymptotes of the curve k.

170. Show: The graphs of $y = \cos(x)$ and $y = \tan(x)$ intersect perpendicularly.

171. a) At what angles does the graph of the function $y = e^x - e$ intersect the two coordinate axes?

b) Calculate the two respective angles of intersection, σ_1 and σ_2, of the exponential curve $k\colon y = e^x - 1$ with the line $g_1\colon y = 1$, parallel to the x-axis, and the line $g_2\colon x = 1$, parallel to the y-axis.

172. Let the parabola p be given by $p(x) = a - \frac{1}{2}x^2$ and the logarithmic curve ℓ by $\ell(x) = b + \ln(x)$. Show that the curves p and ℓ intersect each other orthogonally, that is perpendicularly, irrespective of the values selected for the two parameters a, $b \in \mathbb{R}$.

173. Let $P(x_0 \,|\, y_0)$ be an arbitrary point of the curve k given by the equation $y = f(x)$.

a) Confirm that the equation for the tangent t to the curve at point P has the following form:
$t\colon y = f(x_0) + f'(x_0) \cdot (x - x_0) = f'(x_0) \cdot x + f(x_0) - x_0 \cdot f'(x_0)$.

b) What is the equation for the normal n to the curve at point P?

174. True or false? At point $(0\,|\,0)$, the graph of $f(x) = |x|$ has

a) a tangent with the equation $y = 0$.

b) infinitely many tangents.

c) no tangents.

d) two tangents with the equations $y = x$ and $y = -x$, respectively.

3.5 Special Points and Properties of Curves

Let f be a function and $P(x_0 \,|\, f(x_0))$ a point on the graph G_f of the function f.

Roots or Zeros

f has a root (or zero) at x_0	$f(x_0) = 0$

Extrema or Stationary Points, Asymptotic Behaviour

	Necessary Condition	Sufficient Condition
f has a (local) maximum at x_0, P is a maximum point of G_f	$f'(x_0) = 0$	$f'(x_0) = 0$ $f''(x_0) < 0$
f has a (local) minimum at x_0, P is a minimum point of G_f	$f'(x_0) = 0$	$f'(x_0) = 0$ $f''(x_0) > 0$
f is monotonically increasing on $]a; b[$	$f'(x_0) \geq 0$ for all $x_0 \in \,]a; b[$	
f is monotonically decreasing on $]a; b[$	$f'(x_0) \leq 0$ for all $x_0 \in \,]a; b[$	
f is strictly increasing on $]a; b[$	Sufficient Condition: $f'(x_0) > 0$ for all $x_0 \in \,]a; b[$	
f is strictly decreasing on $]a; b[$	Sufficient Condition: $f'(x_0) < 0$ for all $x_0 \in \,]a; b[$	

Inflection Points, Curvature

	Necessary Condition	Sufficient Condition
f has an inflection point at x_0, P is an inflection point of G_f	$f''(x_0) = 0$	$f''(x_0) = 0$ $f'''(x_0) \neq 0$
f has a saddle point at x_0, P is a saddle point of G_f	$f'(x_0) = 0$ $f''(x_0) = 0$	$f'(x_0) = 0$ $f''(x_0) = 0$ $f'''(x_0) \neq 0$
G_f is convex (or bending upward) on $]a; b[$	Sufficient Condition: $f''(x_0) > 0$ for all $x_0 \in]a; b[$	
G_f is concave (or bending downward) on $]a; b[$	Sufficient Condition: $f''(x_0) < 0$ for all $x_0 \in]a; b[$	

For the definitions of the terms *pole*, *singularity*, *asymptote* and *asymptotic curve* refer to chapter X.4 «Rational Functions» on page 205.

Conditions for the Change of Sign

Let f be a function and x_0 a point.

- If $f'(x_0) = 0$ holds and f' changes its sign at the point x_0, then f has a (local) extremum at x_0.

Change of sign from $+$ to $-$
$\Rightarrow$ maximum point

Change of sign from $-$ to $+$
$\Rightarrow$ minimum point

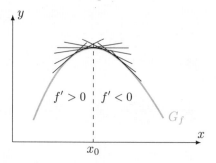

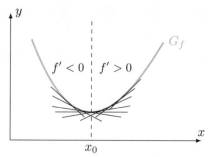

- If $f''(x_0) = 0$ holds and f'' changes its sign at the point x_0, then f has an inflection at x_0.

In this section, we use the word symmetry to exclusively signify a possible symmetry of the graph with respect to the y-axis or the origin.

175. 🗎 The graph G_f of a general function f is given.

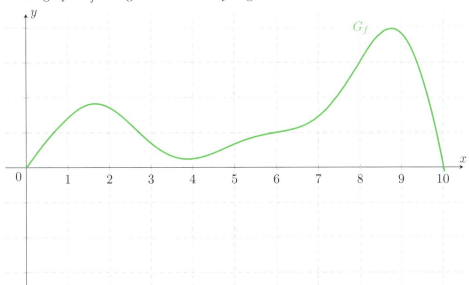

a) Sketch the graphs of the two derivatives f' and f'' into the same coordinate system.

b) How can one characterise the following six properties of the graph G_f in the interval $[0; 10]$ by using the sketches of the two derivatives?

   1) increasing and decreasing   2) convexity and concavity

   3) maximum point   4) minimum point

   5) inflection point   6) saddle point

Polynomial Functions

176. Determine the extrema of the function f given by $f(x) = x^3 + 6x^2 - 15x + 19$.

177. Determine all the minimum points of the function given by the equation $y = \frac{1}{4}x^4 - \frac{1}{8}x^3 - 2x^2 + \frac{3}{2}x + 5$.

178. We consider the graph G_f of a function f on the interval $I = [-3.5; 1.5]$.

a) Locate the extrema (x-coordinates).

b) Locate the global minimum and maximum points in the interval I.

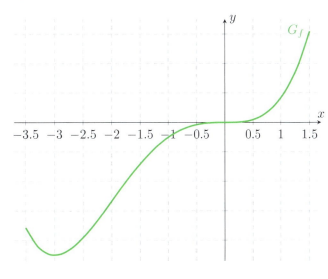

179. Let the function f be given by $f(x) = x^3 + 3x^2 - 4$.

    a) Determine all minimum, maximum and inflection points.

    b) List the intervals on which the graph of f is monotonically increasing or decreasing.

    c) Sketch the graph G_f.

    d) At which points does the function f attain its global maximum and its global minimum, respectively, if the function is restricted to the interval $I = [-2.5; 2.5]$?

180. Let the function p be given by $p(x) = -2x^3 - 6x^2 + 8$. Determine the domain and codomain, the roots, the extrema and the inflection points as well as the symmetry of this function.

181. The function f is given by the equation $y = f(x) = 3x^4 - 8x^3 + 6x^2$.

    a) At every inflection point of f, calculate the equation of its tangent to f.

    b) On which interval is the graph of f convex or concave?

182. For the plotted section of the curve of a function f, indicate if the values of $f(a)$, $f'(a)$ and $f''(a)$ are negative, zero or positive.

a)   b)   c)   d)

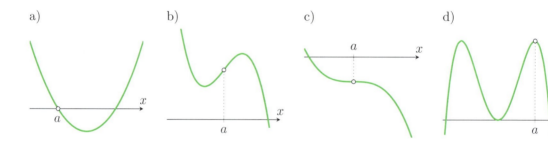

183. For a function g, the graph of its first derivative g' is given. What can you say about the monotonicity of the function g? Where are the extrema?

    a) g' is a polynomial function of degree 3.   b) g' is a polynomial function of degree 4.

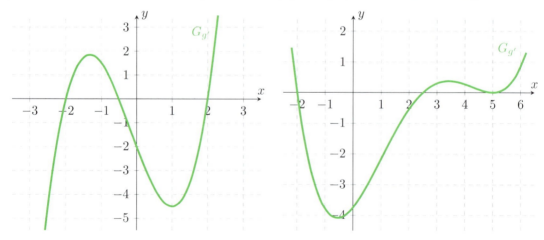

184. For a function f, the graph of its second derivative f'' is given. What can you say about the curvature of f? Where are the inflection points?

a) f'' is a polynomial function of degree 2. b) f'' is a polynomial function of degree 3.

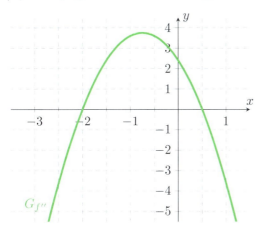
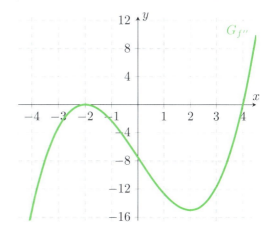

185. For which parameter a does the function f given by $f(x) = ax^4 - 2x^2$ have an inflection point at $x = \frac{2}{3}$?

186. For which values of a does f_a defined by $f_a(x) = x^5 + ax^3 + x$ have a unique inflection point?

187. Why can a function f given by the equation $y = f(x)$ have a point x_0 that is a solution of the equation $f'(x) = 0$ but is not necessarily a local extremum?

188. Are the following statements regarding polynomial functions true or false? Justify.

a) There always is an inflection point between two extrema.

b) There always is an extremum between two inflection points.

c) If the graph is symmetric with respect to the y-axis, $x = 0$ is an extremum.

189. True or false? Explain.

a) If $f''(a) = 0$, the graph of f has an inflection point at $x = a$.

b) If $f'(a) = 0$, the graph of f either has a local maximum or a local minimum at $x = a$.

c) If f has a global minimum at $x = a$, then $f'(a) = 0$.

d) Every saddle point of f is also an inflection point of f.

190. a) Is it possible that a function f has a minimum at some point even though the second derivative f'' is not positive at that point?

b) Firstly, state necessary conditions for the existence of a maximum of a function f at some point and, secondly, sufficient conditions for a maximum.

191. Adi claims that, for determining the inflection points of a function f, it suffices to equate the second derivative f'' to zero. In other words: The condition $f'' = 0$ is sufficient. His sister Ida contradicts him and explains that $f'' = 0$ merely is a necessary condition for the existence of an inflection point but not a sufficient one. Who is right, Adi or Ida?

192. Regarding the function f defined by $y = f(x) = 5.4x^3 - 32.1x^2 + cx + d$, Lena claims that she cannot exactly calculate the inflection point numerically as she does not know the values of the coefficients c and d. Nael counters that knowing those values is not at all necessary and that it is sufficient to know the coefficients of x^3 and x^2, in this case 5.4 and -32.1. Who is right, Lena or Nael?

193. Can the implication be reversed in the mathematical proposition? If not, give a counterexample.

a) If the function f has an extremum at point x_0, then $f'(x_0) = 0$.

b) If the graph of the function f has an inflection point at x_0, then $f''(x_0) = 0$.

194. Prove the statement concerning polynomial functions of degree 3 defined by the equation $y = ax^3 + bx^2 + cx + d$.

a) If the quadratic term bx^2 of a polynomial function of degree 3 is missing, the inflection point of the corresponding graph is on the y-axis.

b) If the linear term cx of a polynomial function of degree 3 is missing and the quadratic term bx^2 is not, then one extremum of the corresponding graph is on the y-axis.

195. a) Prove the proposition: If the function f is a polynomial function of degree 4, the graph of f always has at least one horizontal tangent.

b) Is the converse of the proposition in part a) also true? If not, give a counterexample.

196. The normal to the curve defined by the equation $y = x^3 - 4.2x + 1.5$ at its inflection point intersects the curve at two additional points. Calculate the coordinates of these intersection points and show that the inflection point bisects the segment between them.

197. While examining the function f given by the formula $f(x) = -\frac{1}{3}x^3 - 1.1x^2 + 3.2x + 2$, Luana obtains the following abscissae for the local extrema: $x_1 = -3.2$ and $x_2 = +1.0$. How can she find out which of the two values belongs to the maximum point and which to the minimum point without any additional calculation?

198. Discuss the function with respect to domain and codomain, roots, maximum, minimum, inflection and saddle points, symmetry. Use this information to sketch its graph.

a) $f(x) = \frac{1}{3}x^3 - x$ b) $f(x) = 3x^4 + 4x^3$

c) $f(x) = 2 - \frac{5}{2}x^2 + 4x^4$ d) $f(x) = \frac{1}{10}x^5 - \frac{4}{3}x^3 + 6x$

e) $f(x) = \frac{1}{6}(x + 1)^2(x - 2)$ f) $f(x) = (x - 1)(x + 2)^2$

g) $f(x) = x^3 - 2x^2 + x$ h) $f(x) = \frac{1}{8}x^4 - \frac{3}{4}x^3 + \frac{3}{2}x^2$

General Functions

199. Let the function $f(x) = x^2 + \frac{a}{x}$ with $a \in \mathbb{R}$ be given. For which parameter a does the graph of f have a minimum point at $x = -1$?

200. Let the rational function k be defined by the equation $y = k(x) = \frac{12}{x^2 + 3}$.

a) Discuss the function k (with respect to domain and codomain, roots, extrema and inflection points, symmetry) and sketch the graph of the function k including any asymptotes.

b) Determine the equations of all tangents at inflection points.

201. Let the rational function g be defined by the equation $y = g(x) = \frac{x-2}{x+1}$.

   a) Discuss the function g (with respect to domain and codomain, roots, poles, asymptotes) and sketch the graph of g.

   b) Explain why the function does not have any extrema nor inflection points.

   c) Determine the equation of the tangent t to the curve at the intersection point of the curve with the y-axis and calculate the acute angle of inclination, φ, between the graph of g and the x-axis.

202. Let the rational function g be given by the equation $y = g(x) = \frac{Z(x)}{N(x)}$. This equation contains the polynomial numerator $Z(x) = x^3 + x^2 - 5x + 3$ and the polynomial denominator $N(x) = 2x^2 + 8x$.

   a) Discuss the function g (with respect to roots, poles, extrema and inflection points, asymptotes) and sketch the graph of g.

   b) At which point P does the oblique asymptote intersect the graph of the function?

203. Examine the curve $k : y = \frac{(2x-1)(3-x)}{x^2}$.

   a) Determine the intersection points of k with the x-axis as well as the extrema, the poles and the asymptotes. Sketch the curve.

   b) What are the equations of the tangents to the curve k at its roots?

   c) Calculate the slopes of the tangents to k passing through the origin $(0\,|\,0)$.

204. Let the function f be given by $x \mapsto \frac{x^3}{x^2+12}$.

   a) Determine the domain, the roots, extrema and inflection points together with the asymptotic behaviour and the symmetries of the function f. Sketch the graph G_f and its asymptote(s).

   b) What is the distance from the point $P(\sqrt{2}\,|\,y_P)$ on G_f to the asymptote?

205. For which values of $x > 0$ is the graph of the function f given by $f(x) = \sqrt{x} - \ln(x)$

   a) strictly decreasing?   b) concave?

206. Explain why the function $y = \frac{\cos(x)}{1+\sin(x)}$, $x \neq \frac{3\pi}{2} + 2k\pi$ does not have a single extremum.

For **207–211**: Discuss the function (with respect to domain and codomain, roots, maximum, minimum, inflection and saddle points, poles, symmetry, asymptotic behaviour) and sketch its graph.

207. a) $f(x) = \dfrac{2x}{1+x^2}$ b) $f(x) = \dfrac{4}{1+x^2}$

   c) $g(t) = \dfrac{t}{t+1}$ d) $g(t) = \dfrac{t+1}{t}$

   e) $h(x) = \dfrac{x^3}{x^2-4}$ f) $h(x) = \left(\dfrac{x-1}{x+1}\right)^2$

208. a) $f(x) = \frac{x}{2} + \sqrt{x}$ b) $f(x) = x^2 + \sqrt{x}$

   c) $g(t) = 4t - \frac{1}{\sqrt{t}}$ d) $g(t) = \frac{1}{t} - \frac{1}{\sqrt{t}}$

209. a) $f(t) = \sin(t) + 2\cos(t)$ b) $f(t) = \sin(t) \cdot \cos(t)$

      c) $g(t) = \cos(t) + \cos(2t)$ d) $g(t) = \frac{\cos(t)}{1-\cos(t)}$

210. a) $f(x) = xe^{-x}$ b) $f(x) = e^x(x-1)^2$

      c) $g(t) = te^{-t^2}$ d) $g(t) = \frac{t}{2} + e^t$

211. a) $f(x) = x\ln(x)$ b) $f(x) = \frac{\ln(x)}{x}$ c) $f(x) = \frac{x}{\ln(x)}$

3.6   Determining Equations of Functions

Polynomial Functions

212. The graph of a polynomial function of degree 3 has a saddle point $P(2\,|\,1)$ and intersects the x-axis at point $A(4\,|\,0)$. Determine the equation of the function.

213. The graph of a polynomial function of degree 3 touches the x-axis at $x = 0$. The point $T(3\,|\,9)$ is a saddle point. Determine the equation of the function.

214. The graph of a polynomial function of degree 3 has an inflection point at $W(1\,|\,2)$ and touches the x-axis at $x = 2$. Determine the equation of the function.

215. Determine a polynomial function of degree 3 with a root at $x = -2$. The point $P(0\,|\,y_P)$ is an inflection point and its tangent t has the equation $x - 3y + 6 = 0$.

216. What is the equation of the polynomial function of degree 5, whose graph is symmetric with respect to the origin and has a saddle point at $P(1\,|\,8)$?

217. A polynomial function of degree 3 that is symmetric with respect to the origin has a minimum at the point $M(3\,|-6)$. Determine the equation of the function.

218. The graph of a polynomial function of degree 5 that is symmetric with respect to the origin has an extremum at $P(3\,|\,6)$ and the slope $m = \frac{40}{27}$ at $x = 1$. Determine the equation of the function.

219. A polynomial function of degree 3 that is symmetric with respect to the y-axis passes through the origin and has an inflection point at $W(1\,|\,2.5)$. Determine the equation of the function.

220. The graph of a polynomial function of degree 5 that is symmetric with respect to the origin has a maximum at $M(3\,|\,6)$ and intersects the x-axis at $x = \sqrt{15}$. Determine the equation of the function.

221. A polynomial function of degree 4 that is symmetric with respect to the y-axis touches the x-axis at $x = 4$ and intersects the y-axis at 10.24. Determine the equation of the function.

222. Determine the equation of a cubic polynomial function k that has a root of order 2 and whose graph G_k touches the line g given by the equation $y = 2 - 2x$ at its intersection point with the y-axis. At what angle does G_k intersect the x-axis?

223. The points $A(1\,|\,0)$ and $B(-1\,|\,0)$ are inflection points of the function f defined by $f(x) = x^4 + bx^3 + cx^2 + dx + e$. Determine all the intersection points of this curve with the x-axis as well as its extrema.

224. a) Sketch the graph G_f of the polynomial function of degree 3 with the equation $y = f(x)$ that has the following properties:

    i) $f(-3) = f(0) = 0$ ii) $f(-2) = 4$ iii) $f(-1) = 2$
    iv) $f'(-2) = f'(0) = 0$ v) $f''(-1) = 0$

    b) In part a), significantly more specifications are given for the graph of f than are actually necessary, namely seven. How many specifications would suffice to uniquely determine the equation for f and thus also its graph?

225. Choose a such that the graph of the function f given by $f(x) = \frac{1}{2}(x^4 - ax^2)$ has an inflection point at $x = 1$. Where is the other inflection point? Determine the extrema as well.

226. A polynomial function f has a root of order 2 at the point $x = 1$ and $f''(x) = 6x$ holds. Determine the equation of the polynomial function and sketch its graph.

227. Determine the equation of a function f which has a local maximum at point $(1\,|\,1)$ and a local minimum at point $(-1\,|\,-1)$.

228. The figure shows the graph G_f of a function f. Determine the equation of f, if the transition from the parabolic arc to the segment s at point $(2\,|\,0)$ is kink-free.

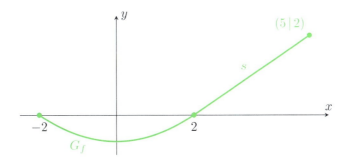

229. Let the straight street segments $u = P(-3\,|\,2)Q(-1\,|\,0)$ and $w = R(1\,|\,0)S(5\,|\,2)$ be given. A curvilinear connecting road v is to be built from Q to R such that the transitions at Q and R are «kink-free». Describe the course of the road v by a cubic polynomial function, which has this property at the two transitions Q and R when connected by straight street segments to P and S, respectively.

230. An aeroplane about to land has the flight altitude H at time $t = 0$ and lands at time $t = T$. The flight altitude $h = h(t)$ while transitioning from the horizontal approach to the horizontal landing, that is in the time interval $0 \le t \le T$, can be modelled by a polynomial function of degree 3 quite well. Show that the flight altitude h during the time interval in question is described by the function:

$$h = h(t) = H \cdot \left(2\left(\frac{t}{T}\right)^3 - 3\left(\frac{t}{T}\right)^2 + 1 \right)$$

231. A road runs in a straight line from $F(-3\,|\,8)$ to $G(0\,|\,2)$, from there curvilinearly to $Z(1\,|\,0)$ and then along the (positive) x-axis.

    a) Develop the equation of a curved arc from G to Z such that the two transitions at G and Z are kink-free.

    b) Improve the curvilinear arc from G to Z such that the two transitions at G and Z can be driven on «jerk-free».

General Functions

232. Determine the two coefficients a and b such that the graph of the function given by $f(x) = \frac{ax^2-1}{bx}$ has slope 16 at the point $P(-1\,|\,-8)$.

233. Let one half of a parabola in quadrant I be given by the equation $y = f(x) = ax^2$ $(a > 0;\ x \geq 0)$ and the exponential curve by $y = g(x) = e^x$ $(x \in \mathbb{R})$.

    a) Determine the coefficient a such that the two curves touch at $x_0 = 2$.

    b) Show that the value found for a can also be determined without giving the point of tangency x_0.

234. Set up an equation of a rational function that has a pole at $x = -1$, a root of order 2 at $x = 3$ and whose graph has slope -11 at point $x = 0$.

235. Let the function f be given by $x \mapsto \frac{9(x^2+ax+b)}{x^2}$.

    a) Calculate the parameters a and b such that f has roots at $x = 4$ and $x = 12$.

    b) Discuss the function f (with respect to poles, extrema, inflection points, asymptotes) and sketch its graph for $x > 0$.

236. What value must the coefficient $a \in \mathbb{R}$ have so that the curve given by the formula $y = f(x) = \frac{ax-1}{x-1}$

    a) has a tangent at $x = 2$ that is parallel to the line $g\colon 3x + y - 5 = 0$?

    b) has a tangent at $x = -1$ that passes through the origin?

237. Determine the equation of the simplest possible function f, which has a pole without change of sign at $x = -1$, a root of order 2 at $x = 2$ and whose graph G_f passes through the origin with a slope of $+45°$.

238. What values must the coefficients a and b have so that the function given by $f(x) = \frac{a}{x^2+b}$ has a maximum with value 4 and an inflection point at $x = 1$?

239. Let the exponential curve and the logarithmic curve be given by the equations $y = f(x) = e^{x-u} + v$ and $y = g(x) = \ln(x)$. Calculate the two parameters u and v such that the two curves touch each other at x_0 and give the equation of the function f.

    a) $x_0 = 1$ b) $x_0 = e$

Which Function Models the Situation?

240. Let the parabola p be given by the equation $y = p(x) = -x^2 + 6x = x(6 - x)$. Let the abscissa of the point $Q(x\,|\,0)$ fulfil the constraint $0 \leq x \leq 6$; point Q thus is on the x-axis somewhere between the two roots $x_1 = 0$ and $x_2 = 6$ of the parabola p. The point $P(x\,|\,y)$ is vertically above Q on the parabola and thus has the same abscissa x as point Q. Now consider the triangle OPQ; here O is the origin of the coordinate system.

a) Determine a formula $F = F(x)$ that calculates the area F of the triangle OPQ as a function of the abscissa x of Q for $0 \leq x \leq 6$. Create a table of values for the integer values $x = 0$, $1, 2, \ldots, 6$.

b) What is the analogous formula $u = u(x)$ for the circumference u of the triangle OPQ?

241. The fuel cost for the propulsion of a cargo ship is approximately proportional to the third power of its velocity v. For a specific type of vessel, experience shows that it amounts to 125 francs per hour, when the ship is travelling with a velocity of $v_0 = 10\,\text{km/h}$. The costs that are independent of the velocity amount to 2000 francs per hour. In the equation $y = B(v)$, the function gives the complete running cost per kilometre travelled as a function of v.

a) Determine the formula for $B(v)$ for the complete running cost.

b) Create a table of values for the complete running cost for $v = 5, 10, 15, 20, 25, 30, 35, 40$ (in km/h) with the accuracy of one franc and sketch the essential progression of the function B including its asymptote and its asymptotic curvilinear approximation.

242. *Subtangent.*   If $P(x\,|\,y)$ is a point on the curve f given by the equation $y = f(x)$, then the x-axis, the tangent t to the curve at the point P and the perpendicular line passing through the point P to the x-axis form a right triangle with one leg of the triangle on the x-axis, the so-called *subtangent*. The length s of this subtangent depends on the equation of the curve and on the abscissa x of the point P. It therefore can be represented as a function of x: $s = s(x)$. Determine the formula $s = s(x)$ that calculates the length of the subtangent for the given equation of the curve.

a) $y = x^2$ b) $y = e^x$ c) $y = \sqrt{x}$ d) $y = \frac{1}{x}$ e) $y = x^n$ f) $y = \sin(x)$

3.7   Optimisation Problems

Simple Functions

243. A rectangle has a circumference of $u = 60$ cm. We attach squares to *both* of the narrow sides and *one* of the long sides facing outwards. For which dimensions of the rectangle does the sum of the areas of the three squares have an extreme value? Is this extremum a minimum or a maximum?

244. A square of side length x is cut off each corner of a rectangular piece of cardboard with side lengths a and b. By folding up the protruding rectangles, the remaing piece can be made into a box with an open top. For which x is the volume of this box largest, when

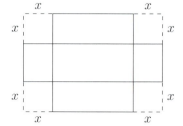

    a) $a = b = 12$ cm?   b) $a = 15$ cm, $b = 24$ cm?

245. A flyer is to be printed showing a piece of text that needs A cm^2 of space. The top and bottom margins should be a cm and the left and right margins b cm wide. For which dimensions of the sheet will the paper consumption be the lowest?

    a) $A = 150$, $a = 3$, $b = 2$ b) general

246. The Ziegler family wants to put up a fence adjacent to a 12 m long barn wall to create a pasture for their goats. For this purpose, they have fencing of length ℓ at their disposal. How should the rectangular pasture be dimensioned so that the Zieglers achieve the largest possible pasture, when naturally no fencing is required along the barn wall?

Solve this optimisation problem for fencings of length

    i) $\ell_1 = 18$ m   ii) $\ell_2 = 30$ m   iii) $\ell_3 = 42$ m

and note that there are two fundamentally different fencing variants.

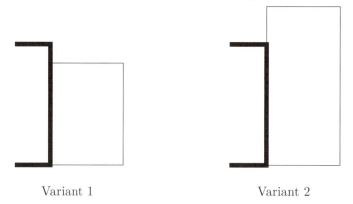

                Variant 1   Variant 2

247. A rectangular box of volume $V = 25$ m^3 with one edge having length $a = 5$ m is to have a minimum surface area. How long are the other two edges?

248. The wire frame model of a cuboid is to be made using a 120 cm long wire, such that one edge is three times as long as another one and the volume is as large as possible. How does one have to choose the lengths of the edges? What is the maximum volume?

249. A gift parcel with a square base must have a volume of $4\,\mathrm{dm}^3$. How should the edge lengths be chosen so that as little twine as possible is needed for tying the parcel? *Note:* The twine is tied crosswise on the square base and top and runs along the other four sides only once.

250. A cylindrical litre container is to be made from as little sheet metal as possible. What dimensions does such a container have

a) including a top?   b) excluding a top?

251. A piece of wire of length ℓ is cut into two parts. One piece of the wire is formed into a square, the other into a circle. How must one cut the wire so that the sum of the areas of both figures is a

a) minimum?   b) maximum?

252. The number $160 = u + v$ is to be written as a sum of two positive summands u and v such that the term $T = u^3 + v^2$ is smallest. Calculate the value of both summands u and v and justify why the solution found yields the minimum value for T. This exercise is to be solved in two ways:

1) The term to be minimised is written as a function of the variable u: $T = T(u)$.

2) Conversely, define the term to be minimised as a function of the variable v: $T = T(v)$.

253. An online retailer can sell a units of a very popular item every month; the net profit is b francs per item sold. Since experience shows that lower prices promote sales, she assumes, for the sake of simplicity, that a price reduction of Fr. 1.–, Fr. 2.–, Fr. 3.–, ... per item would increase her monthly sales proportionally, that is by c units, $2c$ units, $3c$ units, ... With which reduction in price per item can she expect the largest net profit under this simplified assumption?

a) $a = 200$, $b = 10$, $c = 50$ b) $a = 500$, $b = 13$, $c = 50$
c) $a = 1000$, $b = 10$, $c = 100$ d) $a = 1000$, $b = 8$, $c = 100$

254. A discount shop produces lampshades. They use $5.4\,\mathrm{m}$ of wire per lampshade to manufacture a right square prism including a cross-shaped mounting attachment. The square base and the lateral surface of the prism are covered with fabric. Determine the dimensions of such a lampshade if — for purely decorative reasons — as much fabric as possible is to be used.

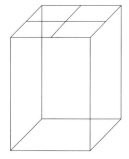

255. The volume of an office building with a flat roof and a square floor plan is to be $12'000\,\mathrm{m}^3$. The heat loss per square metre of roof is three times larger than per square metre of wall. For which dimensions does the building have the smallest heat loss?

256. At which points in the interval $-2 \le x \le 1$ does the tangent to the graph G_f of the function f given by $y = f(x) = 2x^3 + 6x^2 + 12x - 5$ have the greatest slope and at which points the smallest slope?

257. We consider the squares $OPQR$, where the vertex O is the origin. The vertex P is on the line g given by the equation $y = 2.4x + 33.8$. Calculate the smallest possible area A, which such a square $OPQR$ can have.

258. Let the functions f and g be given by the formulas $f(x) = \frac{1}{4}x^2$ and $g(x) = 6$. Their graphs enclose a region. In this region a rectangle is inscribed, whose sides are parallel to the coordinate axes. How long and wide is the rectangle with maximum area? What is the maximum area?

259. The parabola given by the equation $y = \frac{1}{4}x^2 + 4$ and the line parallel to the y-axis at $x_0 = 8$ enclose a curvilinear region in quadrant I. A rectangle with sides parallel to the coordinate axes is inscribed into this region. At which height y_h do we place a horizontal cut in order to maximise the area F of the cut-out rectangle?

260. The (internal) cross-section of a road tunnel has the shape of a rectangle surmounted by a semi-circle; its area is $60\,\mathrm{m}^2$. How wide is this tunnel if the cross-section has the smallest circumference?

261. A cylinder is to be inscribed into a right circular cone with base diameter d and height h. What is the cylinder height relative to the cone height if

a) its volume   b) its lateral surface area

is to be as large as possible?

262. A circular sector with the central angle α is cut out of a circular piece of paper and formed into a paper cone. Which angle α maximises the volume of the conical paper cup?

263. Let the parabola be given by the equation $y = f(x) = 16 - x^2$ and consider the point $A(-1\,|\,0)$. The varying point $B(x\,|\,0)$ is to the right of A on the x-axis (i.e. $-1 < x < 4$) and the point $C(x\,|\,y)$ is above B on the parabola. Finally, $D(-1\,|\,y)$ is the fourth point of the rectangle $ABCD$; which generally is not on the parabola. What is the maximum area F that this rectangle can have?

264. Out of a cylindrical trunk with diameter $d = 30\,\mathrm{cm}$, a beam with a rectangular cross-section is to be sawed, which has the greatest possible load-bearing capacity. Studies have shown that the load-bearing capacity is proportional to the product of the width b and the square of the height h, that is bh^2. How must the width b and the height h of the beam be chosen in order to maximise its load-bearing capacity?

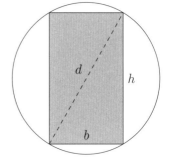

Other Function Types

265. For positive numbers a, the line g given by the equation $y = 2a - (4 + a^3)x$ together with the two coordinate axes encloses a right triangle Δ. The position of g only depends on the parameter a.

a) For which parameter a does the corresponding line g pass through the point $G(2 \,|\, {-8})$?

b) What is the largest area F_Δ of the right triangle Δ as a function of a?

266. A cylinder of radius r and height h is to be placed into a hemispherical airdome with radius $R = 12.3\,\mathrm{m}$. How should r and h be chosen in order to maximise the volume V of the cylinder? This exercise is to be solved in the following two ways:

1) The volume V of the cylinder is considered as a function of its radius r: $V = V(r)$.

2) The volume V of the cylinder is considered as a function of its height h: $V = V(h)$.

267. Let the falling line g be given by the equation $y = g(x) = 16 - 0.5x$ and consider the point $A(-1 \,|\, 0)$. The varying point B is on the x-axis to the right of A and the point C on the falling line g is vertically above B. What is the total length of the shortest path comprised of the two segments $\overline{AC}$ and $\overline{CB}$?

268. A pipeline is to be built from an oil refinery R, which lies on a coastline running straight from west to east, to the distribution centre V in the inland. V is $16\,\mathrm{km}$ east and $12\,\mathrm{km}$ north of R. The pipeline shall first be routed eastwards along the coast and then from a suitable place routed straight in the inland to V. What is the minimum construction cost to be expected if the cost to lay the pipeline along the coast is $15'000$ euros per kilometre and in the inland $25'000$ euros?

269. The factory F is located at a river $120\,\mathrm{m}$ wide. The electric power station E is located on the other side of the river $435\,\mathrm{m}$ upstream. A power cable is to be run from E to F; the first part is routed on land along the bank of the river and then straight through the river. What is the minimum construction cost that must be budgeted in any case if laying the power cable in water costs 273 francs per metre but only 105 francs on land?

270. A comet travels past the earth $E(0 \,|\, 2.6)$ along the orbit given by $k: y = 0.2 \cdot x^2$. Calculate the position(s) where the comet is closest to earth.

271. Which point of the curve given by the equation $y = x^5 + 1$ is closest to the origin?

272. Which point P of the parabola given as indicated has the shortest distance from point Q?

a) $y = 0.25x^2$; $Q(0|3)$ b) $y = 0.5x^2$; $Q(6|0)$

273. A Romanesque church window consists of a rectangle (width $50\,\mathrm{cm}$; height $100\,\mathrm{cm}$) surmounted by a semi-circle. How wide is an inscribed rectangle with

a) maximum circumference?   b) maximum area?

274. Let the three points $U(-6\,|\,0)$, $V(6\,|\,0)$ and $W(0\,|\,w)$ be given and let Y be a fourth point, which lies below W on the y-axis. If Y is connected to each of the three given points by straight line segments, the result is an upside-down Y. At which height h above the x-axis must the «branching point» Y lie in order to minimise the sum ℓ of lengths of the three segments $\overline{YU}$, $\overline{YV}$ and $\overline{YW}$? Answer this question for the two points $W_1(0\,|\,8)$ and $W_2(0\,|\,3)$, that is for $w_1 = 8$ and $w_2 = 3$.

275. An upright cylinder (radius r, height h) is to be inscribed into a hemisphere (radius $R = 12.3\,\text{m}$) such that the total area comprised of the lateral surface and (top) base of the cylinder is an extremum? What is this extreme total area? Is the extremum a minimum or a maximum?

276. An isosceles triangle of circumference $U = 10\,\text{cm}$ is rotated around its altitude. How long is its base line if the resulting cone has a maximum volume?

277. In a square of side length $10\,\text{cm}$, a horizontal segment u is placed in the centre, whose ends, as can be seen in the figure, are connected with the vertices of the square by four segments of equal lengths v. What common angle φ between u and the four segments v minimises the total length of u and all the segments v?

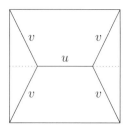

278. Two corridors of widths $2\,\text{m}$ and $6.75\,\text{m}$ join at a right angle. How long can a rod maximally be if it should be able to be transported horizontally around the corner formed by the corridors?

279. A cone is inscribed into a hemisphere of radius R such that the apex S of the cone coincides with the centre of the hemisphere and such that the cone has maximum volume. Calculate the *half-aperture angle* φ of this «voluminous» cone with an accuracy of a tenth of a degree.

280. A rectangle is inscribed into a semicircle (radius $r = 55.9\,\text{cm}$); two corners lie on the diameter of the semicircle and two on its arc. How long and wide is the inscribed rectangle with maximum circumference?

281. A water channel with the largest possible capacity is to be built using four boards of length l and width b each. The two edge boards, one on the left and the other on the right, should be aligned vertically. What are the angles between any two adjacent boards?

282. Imagine that water droplets are hanging from the line g given by the equation $y = \text{e} \cdot x + 5$ in quadrant I and falling vertically parallel to the imagined y-axis onto the logarithmic curve given by the equation $y = \ln(x)$ where they are «absorbed». The falling water droplets travel distances of varying lengths until they are caught by the logarithmic curve. How many length units does the shortest of all possible falling distances measure?

283. Let the vertex P of a rectangle be on the graph of the function $f(x) = \text{e}^{\frac{-1}{3}x+1}$ in quadrant I. Determine the coordinates of the point P such that the rectangle of which two sides lie on the coordinate axes has the largest possible area?

Applications to Economics

Terms

- *Total cost function:* The total cost function K is often modelled as a polynomial function, where $K(x)$ denotes the cost of producing x items. $K(x)$ is made up of the *variable cost* and the *fixed cost*.

  Example: $K(x) = K_v(x) + K_f = 0.1x^3 - 12x^2 + 60x + 98$
  Here $K_v(x) = 0.1x^3 - 12x^2 + 60x$ is the variable cost (e.g. depending on the production volume or the output) and $K_f = 98$ the fixed overhead cost (e.g. rent, wages, insurances).

- *Marginal cost:* The marginal cost describes the cost growth when increasing the quantity produced by a sufficiently small amount. It therefore is the first derivative of the total cost function $(= K'(x))$.

- *Unit cost or average costs:* Total cost divided by the production volume x, i.e. $k(x) = \frac{K(x)}{x}$.

- *Inverse demand function or price function:* The inverse demand function describes the relationship between the price $p(x)$ and the quantity demanded x. Basically, the common assumption is: The higher the price rises, the lower the quantity demanded.

- *Total revenue function:* The total revenue function E denotes the total revenue $E(x)$, which is generated by selling the produced volume x, in a nutshell: total revenue = price per production unit $\cdot$ production volume.

Important note: The models used for the total cost function K in the exercises below are generally very inaccurate, even if they roughly reflect what is being described correctly. This is always related to the fact that many of the possible influencing factors can only be taken into account incompletely in the modelling or are not known at all. Other functions are also subject to the same effect even if the specifically chosen modelling may seem plausible. Or as stated by the British statistician GEORGE BOX (1919–2013): «Essentially, all models are wrong – but some are useful.»

284. A polynomial function K that fulfils the following conditions is called a *total cost function satisfying the law of diminishing returns*.

1) K is strictly increasing for $x \geq 0$.

2) K has an inflection point (K changes from concave to convex) in quadrant I.

3) $K(0) \geq 0$ holds.

Solve the following parts of the exercise.

a) Interpret the above three conditions. What do they mean?

b) Check if the total cost function $K(x) = \frac{1}{5000}x^3 - \frac{1}{50}x^2 + 4x + 200$ satisfies the law of diminishing returns.

c) For which production volume x is the marginal cost a minimum?

d) For which x is the unit cost $k(x) = \frac{K(x)}{x}$ a minimum?

e) Prove the following statement: «The minimum of the unit cost is achieved when it coincides with the marginal cost.» Also show the above statement graphically, by drawing the graphs of the marginal cost function $K'(x)$ and of the unit cost function $k(x)$.

285. The weekly production cost (in francs) for the production of a liquid chemical product (specified in hectolitres) is described by the total cost function (specified in francs) $K(x) = \frac{1}{50}x^3 - 4x^2 + 296x + 4000$, where x denotes the production volume in hectolitres per week. The market price for this liquid product is 200 francs per hectolitre.

a) Use the above specifications to establish the profit function $G(x)$, which gives the weekly profit as a function of the production volume x sold.

b) For which production volume x is the profit per week greatest?

c) Which production volume x yields a loss (negative profit)?

286. A floor panel manufacturer uses the daily total cost function $K(x) = 0.02x^2 + 4x + 720$, where x denotes the number of panels to be produced per day. The manufacturer operates on the market in accordance with the inverse demand function $p(x) = 15 - 0.002x$, where x is the daily quantity of floor panels in demand. How many panels should be produced and sold in order to

a) maximise the total profit per day?   b) maximise the profit per panel per day?

287. Let the total cost function for a monopolist be given by $K(x) = 0.1x^3 - 2.2x^2 + 6x + 44$. The monopolist operates on the market in accordance with the inverse demand function $p(x) = 6 - 0.1x$, where p is the price and x the quantity in demand.

a) What is the price increase if the demand is reduced by 20 units?

b) Determine the amount x of commodity to be produced to minimise the variable cost K_v per produced output unit.

c) What amount x does the monopolist have to produce and sell to maximise the following quantities?

- total revenue
- total profit
- unit profit

3.8   Further Topics

Special Symmetry Properties of Polynomial Functions of Degrees 2, 3 and 4

Oblique Reflection Symmetry

Two figures F_1 and F_2 are *obliquely symmetric* with respect to an axis s in direction of the line $g \nparallel s$ if the connecting segment of corresponding points $P_1 \in F_1$ and $P_2 \in F_2$ (with $P_1 \neq P_2$) is parallel to g and the midpoint of the segment $\overline{P_1P_2}$ is on s. For $g \perp s$ this simply is the ordinary reflection symmetry.

288. The parabola p is given by the equation $y = p(x) = \frac{1}{2}x^2 - 3x + 4$ and the tangent to the parabola at the intersection of p with the y-axis is denoted by t.

a) Let the two lines f and g be parallel to t. Line f passes through the point $F(0 \mid 12)$, line g through the point $G(6 \mid 4)$. Both lines intersect the parabola in two points each, F_1 and F_2 or G_1 and G_2, respectively. Show by calculation that the midpoints of the connecting segments F_1F_2 and G_1G_2 are on the y-axis.

b) Let an arbitrary line b parallel to t intersect the parabola p in the two points B_1 and B_2. Prove in general that the midpoint M_B of the segment $B_1 B_2$ always is on the y-axis, that is the parabola p is obliquely symmetric in direction of the tangent with respect to the y-axis.

289. Let the curve k be given by the equation $y = k(x) = x^3 - 3x^2 - 2x + 1$.

a) The line falling to the right with a $45°$ angle at the inflection point W of k intersects the given curve in two more points S_1 and S_2. Determine the coordinates of these points and show that the inflection point W is the midpoint of the connecting segment $S_1 S_2$.

b) Prove in general that the curve k is symmetric with respect to its inflection point W. *Hint:* Subject the curve k to a translation that moves the inflection point W to the origin $(0\,|\,0)$ and show that the equation corresponding to this translated curve belongs to an odd function.

290. If the unit parabola p given by $y = x^2$ intersects an arbitrary line g, the two intersection points S_1 and S_2 are always the endpoints of a chord of the parabola.

a) Let the four parallel lines $g_1\colon y = x$, $g_2\colon y = x + 2$, $g_3\colon y = x + 6$ and $g_4\colon y = x + 90$ be given. Show that the midpoints M_k ($k = 1, 2, 3, 4$) of the four chords cut out by p are on one line v. What is the equation of this line v?

b) The unit parabola p intersects the general line $g\colon y = mx + q$. Prove that the midpoints of chords of the parabola parallel to g all lie on one vertical line.

c) With respect to which axis and in which direction is the unit parabola p obliquely symmetric?

291. Let the polynomial function f be given by the equation $y = f(x) = \frac{3}{16}x^4 - \frac{3}{2}x^2 + \frac{1}{2}x + 2$.

a) Show that the line d given by the equation $y = d(x) = \frac{1}{2}x - 1$ touches the graph G_f of the given function f twice, namely at the points $x_1 = -2$ and $x_2 = 2$. Moreover, show that the tangent t at the intersection point of G_f with the y-axis is parallel to this so-called «double tangent».

Hint: Sketch the graph G_f including the two tangents d and t.

b) A line g is laid through the two points P and Q on the curve with abscissae $x_P = -u$ and $x_Q = u$. Show that $g \,||\, d$ also holds and this is irrespective of $u \in \mathbb{R}^+$.

c) In which two points S_1 and S_2 does the tangent t from part a) intersect the graph G_f? Show that the point B where the tangent t touches the graph G_f bisects the connecting segment $\overline{S_1 S_2}$ between the intersection points.

d) Calculate the two inflection points W_1 and W_2 and confirm that the connecting segment $v = W_1 W_2$ also is parallel to the double tangent d. Moreover, show that the midpoint M of the segment $\overline{W_1 W_2}$ is on the y-axis.

e) The properties stated and proven in parts a) to d) on the whole express an essential symmetry of the graph G_f. What type of symmetry is it?

f) The connecting segment $v = W_1 W_2$ of the two inflection points intersects G_f in two more points L and R, where L, W_1, W_2 and R should lie on v in this order. Show that $\overline{LW_1} = \overline{W_2 R}$ holds and calculate the distance ratio $\overline{LW_1} : \overline{W_1 W_2}$. What do you notice?

Implicit Differentiation

Not every plane curve in the coordinate system can be described as a graph of a (single) function f. Familiar examples for this situation are the circle or a parabola opening to the right. Nevertheless, the x- and y-coordinates of the points of such curves can generally be described by an implicit equation of the form $F(x, y) = 0$.

Examples:

- Circular line k of radius 5 around the origin: $F(x, y) = x^2 + y^2 - 25 = 0$
- unit parabola p opening to the right with vertex at the origin: $F(x, y) = y^2 - x = 0$
- Semicubical parabola: $F(x, y) = ax^3 - y^2 = 0$, $a > 0$

Even if it is not possible in the examples to explicitly describe the dependence of the y-values on the x-values as a single equation of the form $y = f(x)$, the possible y-values do still depend on the corresponding x-values. That is also why it is possible to determine the derivative $y' = \frac{dy}{dx}$, that is the slope m_t of the tangent at the point $(x \mid y)$ of the curve using the specifically available implicit equation $F(x, y) = 0$ of the curve.

Example: $F(x, y) = y^2 + 4x = 0 \Rightarrow \frac{d}{dx} F(x, y) = \frac{d}{dx}(y^2 + 4x) = 0$

Applying the chain rule on $\frac{d}{dx}(y^2)$ produces the equation $2y \cdot \frac{dy}{dx} + 4 = 0$. Solving for $\frac{dy}{dx}$ directly yields the derivative wanted: $y' = \frac{dy}{dx} = \frac{-2}{y}$. The slope of the tangent t at the point $(x \mid y)$ of the curve therefore has the value $m_t = \frac{-2}{y}$; $y \neq 0$.

Two specific numerical examples: At point $(-1 \mid 2)$ we have $m_t = -1$ and at point $(-9 \mid -6)$ we have $m_t = \frac{1}{3}$.

---

292. Determine the derivative $y' = \frac{dy}{dx}$ by implicit differentation of the given equation $F(x, y) = 0$.

a) $F(x, y) = x^3 - y^2 = 0$ b) $F(x, y) = x^2 - 4xy + y^2 - 4 = 0$

c) $F(x, y) = \sin(x) - 2x + \cos(y) + 3y = 0$ d) $F(x, y) = \sqrt{x} + \sqrt{y} - 1 = 0$

293. The curve implicitly defined and characterised by the equation $F(x, y) = x^3 + y^3 - 3axy$ is called the «Folium of Descartes». Determine the derivative $y' = \frac{dy}{dx}$ for the parameter value $a = 1$.

294. Let the unit parabola p opening to the right with vertex at the origin be given: $F(x, y) = y^2 - x = 0$. Calculate the slope of the tangents to the curve at the two points $P(4 \mid -2)$ and $Q(1 \mid 1)$.

295. The general line g is given by the the equation $F(x, y) = ax + by + c = 0$. Determine the slope $m = y' = \frac{dy}{dx}$ of the line g without solving the equation $F(x, y) = 0$ for y.

296. The circle k with centre at the origin $O(0\,|\,0)$ passes through the point $K(8\,|\,1)$.

a) Show numerically that the two points $P_1(8\,|\,-1)$ and $P_2(-7\,|\,4)$ are on the circle k and calculate the slopes $m_{1,2}$ of the tangents $t_{1,2}$ to the circle at the points P_1 and P_2.

b) Determine the slope m_t of the tangent t at the general point $P(x\,|\,y)$ of the circle and verify that the tangent t is always normal to the corresponding radius of contact, $r = \overline{OP}$.

297. In this exercise, we examine the curve k consisting of all points $P(x\,|\,y)$ whose coordinates satisfy the following equation: $F(x,y) = x^2 - 2xy + y^2 - 2x - 2y + 1 = 0$.

a) Show numerically that the ten points $A(16\,|\,25)$, $B(9\,|\,16)$, $C(4\,|\,9)$, $D(1\,|\,4)$, $E(0\,|\,1)$, $F(1\,|\,0)$, $G(4\,|\,1)$, $H(9\,|\,4)$, $I(16\,|\,9)$ and $J(25\,|\,16)$ all lie on k, and draw them into a coordinate system. What type of curve might k be?

b) Determine $\frac{\mathrm{d}}{\mathrm{d}x}F(x,y)$ and use it to deduce a formula for the derivative $y' = \frac{\mathrm{d}y}{\mathrm{d}x}$. What are the slopes m of k at each of the points A, C, D, F, G and I given above?

c) Derive the equations for the lines t_A and t_G that are tangent to the curve at the points A and G, respectively.

d) What is the equation for the normal n_H to the curve at point H?

e) What is the equation for the normal n_D to the curve at point D?

f) Substantiate why the curve k is symmetric with respect to the angle bisector ω of quadrant I.

g) Determine the intersection point S of the curve with this angle bisector ω and calculate the slope m_S of the curve at point S.

h) Is there a point K on the curve where the curve k has slope 1?

298. The parabola p opening to the left and symmetric with respect to the x-axis is given by the equation $F(x,y) = y^2 + 2x - 4 = 0$.

a) Show that the given parabola is symmetric with respect to the x-axis and calculate the coordinates of its vertex S.

b) Deduce the formula for the slope of the tangent $y' = \frac{\mathrm{d}y}{\mathrm{d}x}$ using the derivative of $\frac{\mathrm{d}}{\mathrm{d}x}F(x,y)$.

c) The parabola is intersected twice by the y-axis under equally sized acute angles σ. Calculate this intersection angle σ, accurate to one decimal place.

d) Let m_n denote the slope of the normal n to the curve at point $P(x_P\,|\,y_P)$ of the parabola. Show that the slope of the normal n to the curve always equals the ordinate of the point $P(x_P\,|\,y_P)$ of the parabola and we thus always have: $m_n = y_P$. At which point Q of the curve does the normal incline with an angle of $-45°$? And at which point R of the curve does the normal incline with an angle of $+60°$?

Approximation Method of Newton

Newton's method, named after ISAAC NEWTON (1643–1727), is a procedure for solving equations of the form $f(x) = 0$ by iteratively calculating approximations of the root x^*. For this, one chooses an initial guess x_0 near the root x^* and considers the tangent t of the graph G_f at point x_0. One then considers the tangent t instead of the graph G_f and the intersection of t with the x-axis at x_1 is the next improved approximation of the root x^*. This procedure is first repeated with x_1, then with x_2, x_3 and so on.

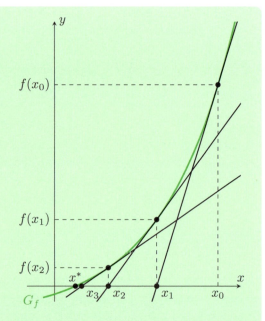

Recursion formulas:

- initial guess: x_0

- $x_{k+1} = x_k - \dfrac{f(x_k)}{f'(x_k)};\ \ k \in \mathbb{N}_0$

The idea of Newton's method is to linearise the function f at the initial guess x_0, i.e. to replace the graph of f with the tangent t at point x_0. The calculated approximations $x_1, x_2, x_3, \ldots$ generally converge very rapidly to the root sought. The initial guess x_0 should be chosen as close to the root as possible. The method needs to be repeated for each further root of f.

Example: Calculation of all roots of the function f with $f(x) = x^3 + x - 1$: A figure shows that f has precisely one root, located between 0.5 and 1. Using the initial guess $x_0 = 1$ and $f'(x) = 3x^2 + 1$ yields $x_1 = x_0 - \frac{f(x_0)}{f'(x_0)} = 1 - \frac{f(1)}{f'(1)} = 1 - \frac{1}{4} = \frac{3}{4} = 0.75$. Repeating this step provides $x_2 = 0.686046\ldots$, $x_3 = 0.682339\ldots$, and so on. The calculator returns the value $0.682327\ldots$

299. Calculate the root of the function f approximately using the given initial guess x_0.

    a) $f\colon x \mapsto x^3 + x^2 - 1$; initial guess $x_0 = 1$; two iterations

    b) $f\colon x \mapsto x^4 - x^3 - 1$; initial guesses $x_0 = 1$ and $x_0 = -1$; two iterations each

    c) $f\colon x \mapsto x^5 - 100$; choose an initial guess and the number of iterations yourself

300. Approximately solve the equation $x^3 = 4x^2 + 2$ with Newton's method. Choose an initial guess x_0 and perform two iterations.

301. Find approximations of the roots of $f(x) = x^3 - 3x + 6$ with Newton's method. Describe the behaviour of the approximations with initial guess x_0.

    a) $x_0 = -2$ b) $x_0 = -1$ c) $x_0 = -0.5$

302. Calculate an approximation of $\ln(2)$ accurate to four decimal places with Newton's method. *Hint:* Find a function f that has a root at $\ln(2)$.

303. What happens with Newton's method for determining a root of the function f if the initial guess x_0 is chosen as a point with $f'(x_0) = 0$? ($f(x_0) \neq 0$)

304. *When the method fails.* Consider the function f given by $f(x) = \frac{e^x - 1}{e^x + 1}$ whose only root is at $x = 0$.

a) Perform four iterations of Newton's method each, first for the initial guess $x_0 = 2$, and then for $x_0 = 2.5$.

b) Explain the behaviour of the approximations observed in part a) using the graph G_f.

305. True or false? Explain.

a) For $f(x) = ax + b$, Newton's method yields the exact solution in one step.

b) For $f(x) = x^2 - a$, $a > 0$, the approximation x_{k+1} of Newton's method is the arithmetic mean of the approximation x_k and its reciprocal $\frac{1}{x_k}$.

c) For $f(x) = x^3 - 9x$, all successive approximations of Newton's method with initial guess $x_0 = \frac{3}{2}$ equal the root -3 of f.

Linear Approximation

The tangent t to the graph G_f of a function f at point $(x_0 \,|\, f(x_0))$ ideally describes the shape of the curve near x_0. Therefore, the tangent t can be used as a replacement function for f at x_0. The images of t and f agree with each other better and better when evaluated closer and closer to x_0.

The tangent t is the *linear approximation* of f at point x_0. If f is *linearised* at point x_0, then the term $f(x)$ is replaced by $f(x_0) + f'(x_0) \cdot (x - x_0)$ of the tangent t, i.e. $f(x) \approx f(x_0) + f'(x_0) \cdot (x - x_0)$ for $x \approx x_0$.

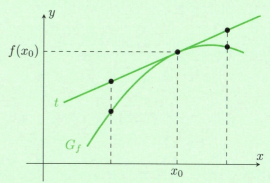

Examples:

- *Linearising* the function f with $f(x) = \ln(x)$ at the point $x_0 = 1$:

  $f(1) = \ln(1) = 0$, $f'(x) = \frac{1}{x}$, $f'(1) = 1$ $\Rightarrow$ $\ln(x) \approx f(1) + f'(1)(x-1) = x - 1$ for $x \approx 1$

- *Calculating an approximation* of $\ln(1.01)$:

  By means of linearisation, we have $\ln(1.01) \approx 1.01 - 1 = 0.01$.

306. Linearise the function f at the point x_0 and use this to calculate an approximation of the image $f(x_1)$.

a) $f: x \mapsto e^x$; $x_0 = 0$; $x_1 = 0.01$

b) $f: x \mapsto (1+x)^3$; $x_0 = 0$; $x_1 = 0.1$

c) $f: x \mapsto \sqrt{x}$; $x_0 = 4$; $x_1 = 4.05$

d) $f: x \mapsto \sin(x)$; $x_0 = \pi$; $x_1 = \frac{5\pi}{6}$

307. Calculate approximately by means of a suitable linearisation.

a) $\sqrt{9.5}$

b) 99.9^3

c) $\sqrt{3.98}$

d) $\sqrt[3]{1001}$

308. The value $e^{0.5}$ is approximated by means of the tangent to the graph of $f(x) = e^x$ at point $(0\,|\,1)$. Which of the three following approximations is the correct one?

i) 0.5   ii) $1 + e^{0.5}$ iii) $1 + 0.5$

309. Linearise $f(x) = \sin(x)$, $g(x) = \tan(x)$ and $h(x) = \ln(x+1)$ each at the point $x_0 = 0$. What do you notice? Which linearisation yields the most precise approximation at the point 0.1?

310. Determine a linearisation of $f(x)$ at point $x_0 = 0$ by means of the linearisation $(1+x)^r \approx 1 + rx$ for $x \approx 0$.

a) $f(x) = (1-x)^6$ b) $f(x) = \frac{2}{1+x}$ c) $f(x) = \frac{1}{\sqrt{1+x}}$ d) $f(x) = (4+3x)^{\frac{1}{3}}$

311. Determine the limit by means of a suitable linearisation.

Example: $\lim\limits_{x \to 0} \frac{\cos(x)-1}{x}$ $\Rightarrow$ Linearisation of $\cos(x)$ at point $x_0 = 0$: $\cos(x) \approx 1$ for $x \approx 0$

$\Rightarrow \lim\limits_{x \to 0} \frac{\cos(x)-1}{x} = \lim\limits_{x \to 0} \frac{1-1}{x} = \lim\limits_{x \to 0} \frac{0}{x} = \lim\limits_{x \to 0} 0 = 0$

a) $\lim\limits_{x \to 1} \frac{\ln(x)}{1-x}$ b) $\lim\limits_{x \to 0} \frac{\sin(x)}{x}$ c) $\lim\limits_{x \to 0} \frac{e^x - e^{-x}}{x}$ d) $\lim\limits_{x \to 0} \frac{(1+3x)^{30}-1}{x}$

312. When the temperature of an object of length ℓ_0 is raised by the temperature difference ΔT, it expands to the length ℓ_1 given by the formula $\ell_1 = \ell_0 \cdot \sqrt[3]{1 + \gamma \cdot \Delta T}$. Here, γ denotes the thermal coefficient of volume expansion which depends on the material. Determine an approximation formula in the case of small temperature changes by means of linearisation.

313. Approximation of square roots

a) Calculate an approximation of $\sqrt{50}$ by linearising $f(x) = \sqrt{x}$ at point $x_0 = 49$.

b) How can Basil use the approximation of $\sqrt{50}$ from part a) to derive a good approximation of $\sqrt{2}$?

c) Bettina has already linearised $f(x) = \sqrt{x}$ at the point $x_0 = 4$ and obtained $t\colon y = \frac{x}{4} + 1$. How can she use this to produce a good approximation for $\sqrt{50}$?

d) Which of the two approximations for $\sqrt{50}$ from parts a) and c) is generally more precise?

314. Approximation of logarithms: Calculate an approximation of $\ln(2)$ without using a calculator by linearising $f(x) = \ln(x)$ at a suitable point.

Taylor Series and Taylor Polynomials

315. The following functions are given: $f(x) = 3x^4 + x^3 - 5x^2 - x + 2$, $f_1(x) = -x + 2$, $f_2(x) = -5x^2 - x + 2$ and $f_3(x) = x^3 - 5x^2 - x + 2$.

a) Calculate $f(0)$, $f_1(0)$, $f_2(0)$, $f_3(0)$. What do you notice?

b) Calculate $f'(0)$, $f_1'(0)$, $f_2'(0)$, $f_3'(0)$. What do you notice?

c) Calculate $f''(0)$, $f_1''(0)$, $f_2''(0)$, $f_3''(0)$. What do you notice?

d) Calculate $f'''(0)$, $f_3'''(0)$. What do you notice?

e) Calculate $f^{(4)}(0)$ and compare it with the coefficient of x^4.

f) Write a formula of the function $f(x)$ only using the derivatives of f at the point 0. Use parts a) to e).

If two functions f and g not only have the same image at a point x_0 but also coinciding first and second derivatives, then we know that their graphs G_f and G_g coincide at the point x_0, have the same slope and also the same curvature or an inflection point there. All in all, the two graphs G_f and G_g have a very similiar shape near x_0. The following can therefore be assumed as a generalisation: The shape of the graphs G_f and G_g near x_0 match better still if we additionally demand that:

$f^{(3)}(x_0) = g^{(3)}(x_0)$, $f^{(4)}(x_0) = g^{(4)}(x_0)$, $f^{(5)}(x_0) = g^{(5)}(x_0)$, $\ldots$

This idea shall be illustrated by means of the function f with $f(x) = \sin(x)$. As the sine function is an odd function, we choose g as an odd polynomial function, for example of degree 7, so $g(x) = ax^7 + bx^5 + cx^3 + dx$, and choose $x_0 = 0$. Now, the coefficients a, b, c and d must be determined such that the following equations hold:

$\sin(0) = g(0)$, $\sin'(0) = g'(0)$, $\sin''(0) = g''(0)$, $\sin'''(0) = g'''(0)$, $\ldots$, $\sin^{(7)}(0) = g^{(7)}(0)$.

For the left-hand sides of these eight equations we obtain:

$\sin(0) = 0$, $\sin'(0) = \cos(0) = 1$, $\sin''(0) = -\sin(0) = 0$, $\sin'''(0) = -\cos(0) = -1$,
$\sin^{(4)}(0) = \sin(0) = 0$, $\sin^{(5)}(0) = \cos(0) = 1$, $\sin^{(6)}(0) = -\sin(0) = 0$ and
$\sin^{(7)}(0) = -\cos(0) = -1$.

To determine the right-hand sides of the eight equations, we first calculate the derivatives of the polynomial function g:

$g'(x) = 7ax^6 + 5bx^4 + 3cx^2 + d$, $g''(x) = 42ax^5 + 20bx^3 + 6cx$, $g'''(x) = 210ax^4 + 60bx^2 + 6c$,
$g^{(4)}(x) = 840ax^3 + 120bx$, $g^{(5)}(x) = 2520ax^2 + 120b$, $g^{(6)}(x) = 5040ax$, $g^{(7)}(x) = 5040a$.

In accordance with the derivatives of the sine function, we have:

$g(0) = g''(0) = g^{(4)}(0) = g^{(6)}(0) = 0$.

For the other derivatives, we obtain the following conditions step by step:

$g'(0) = d = 1$, $g'''(0) = 6c = -1 \Rightarrow c = -\frac{1}{6} = \frac{-1}{3!}$, $g^{(5)}(0) = 120b = 1 \Rightarrow b = \frac{1}{120} = \frac{1}{5!}$
and $g^{(7)}(0) = 5040a = -1 \Rightarrow a = \frac{-1}{5040} = \frac{-1}{7!}$.

This finally yields the following equation for the polynomial function g:

$g(x) = \frac{-x^7}{7!} + \frac{x^5}{5!} - \frac{x^3}{3!} + x$,

which generally is written in reverse order:

$g(x) = x - \frac{x^3}{3!} + \frac{x^5}{5!} - \frac{x^7}{7!}$.

In the immediate neighbourhood of $x_0 = 0$ we thus have: $\sin(x) \approx x - \frac{x^3}{3!} + \frac{x^5}{5!} - \frac{x^7}{7!}$.

316. Sketch the graphs of the function $y = \sin(x)$ and the four following functions into the same coordinate system:

- $y = x$
- $y = x - \frac{x^3}{3!}$
- $y = x - \frac{x^3}{3!} + \frac{x^5}{5!}$
- $y = x - \frac{x^3}{3!} + \frac{x^5}{5!} - \frac{x^7}{7!}$

For **317–319**: The function g always denotes the one deduced in the grey box overleaf with the equation $g(x) = x - \frac{x^3}{3!} + \frac{x^5}{5!} - \frac{x^7}{7!}$.

317. Since $\sin'(x) = \cos(x)$, the derivative g' should yield a function which can be used to approximate the cosine function. Show that $g'(x) = 1 - \frac{x^2}{2!} + \frac{x^4}{4!} - \frac{x^6}{6!}$ and sketch the graphs of the following functions into the same coordinate system:

- $y = 1$
- $y = 1 - \frac{x^2}{2!}$
- $y = 1 - \frac{x^2}{2!} + \frac{x^4}{4!}$
- $y = 1 - \frac{x^2}{2!} + \frac{x^4}{4!} - \frac{x^6}{6!}$
- $y = \cos(x)$

318. Determine the second derivative g'' of the polynomial function g and figure out for which function this second derivative g'' may be an approximation.

319. Which polynomial functions h or k would be even more suitable than g for approximating the sine function in the neighbourhood of $x_0 = 0$?

320. As is well known, all derivatives of the exponential function e^x coincide. We thus have: $e^x = (e^x)' = (e^x)'' = (e^x)''' = \ldots$ For which polynomial function p with $p(x) = a + bx + cx^2 + dx^3 + ex^4 + fx^5 + gx^6$ do the following conditions hold at $x_0 = 0$: $p(0) = e^0$, $p'(0) = e^0$, $p''(0) = e^0$, $p'''(0) = e^0$, $\ldots$, $p^{(6)}(0) = e^0$?

Note: By means of higher mathematics, one can show that the following applies:

- $\sin(x) = x - \frac{x^3}{3!} + \frac{x^5}{5!} - \frac{x^7}{7!} + \frac{x^9}{9!} - \frac{x^{11}}{11!} \pm \ldots = \sum\limits_{k=0}^{\infty} (-1)^k \, \frac{x^{2k+1}}{(2k+1)!}$

- $e^x = 1 + x + \frac{x^2}{2!} + \frac{x^3}{3!} + \frac{x^4}{4!} + \frac{x^5}{5!} + \frac{x^6}{6!} + \ldots = \sum\limits_{k=0}^{\infty} \frac{x^k}{k!}$

Such series are commonly called *Taylor series* (BROOK TAYLOR, 1685–1731).

Taylor Polynomials

If the function $f \colon [a; b] \to \mathbb{R}$ is continuous and sufficiently often differentiable, the associated *Taylor polynomials* of degrees 1, 2, 3 and n at point x_0 are defined by:

- $T_1(x) = f(x_0) + \frac{f'(x_0)}{1!}(x - x_0)$

- $T_2(x) = f(x_0) + \frac{f'(x_0)}{1!}(x - x_0) + \frac{f''(x_0)}{2!}(x - x_0)^2$

- $T_3(x) = f(x_0) + \frac{f'(x_0)}{1!}(x - x_0) + \frac{f''(x_0)}{2!}(x - x_0)^2 + \frac{f'''(x_0)}{3!}(x - x_0)^3$

- $T_n(x) = f(x_0) + \frac{f'(x_0)}{1!}(x - x_0) + \ldots + \frac{f^{(n)}(x_0)}{n!}(x - x_0)^n = \sum\limits_{k=0}^{n} \frac{f^{(k)}(x)}{k!}(x - x_0)^k$

$y = T_1(x)$ is the equation of the tangent t to the graph G_f at point x_0. Analogously, $y = T_2(x)$ is the equation of the parabola p, which touches the graph G_f at point x_0 and also has the same second derivative there as the function f; i.e. $T_2''(x_0) = f''(x_0)$ holds. For the nth Taylor polynomial at point x_0, we have: $T_n(x_0) = f(x_0)$, $T_n'(x_0) = f'(x_0)$, $T_n''(x_0) = f''(x_0)$, $T_n^{(n)}(x_0) = f^{(3)}(x_0), \ldots, T_n^{(n)}(x_0) = f^{(n)}(x_0)$. Apart from the two images, the first to the nth order derivatives also coincide at the point x_0.

321. Let the function f be given by the equation $y = f(x) = \frac{1}{4}x^4 - 2x^2 + 4$.

    a) Determine the Taylor polynomial of degree 2 at point $x_0 = 0$ associated with f.

    b) What is the Taylor polynomial of degree 2 at point $x_0 = 2$ associated with f?

322. a) What is the Taylor polynomial of degree 3 at point $x_0 = 0$ associated with $f(x) = \sin(x)$?

    b) Determine the Taylor polynomial of degree 4 at point $x_0 = 0$ associated with $f(x) = \cos(x)$ and compare the result for $T_4(x)$ with the formulas of the function from exercise 317.

323. Using the Taylor polynomials of degrees 1 to 4 at point $x_0 = 25$ associated with the root function $f(x) = \sqrt{x}$, calculate the approximations $T_1(27)$, $T_2(27)$, $T_3(27)$ and $T_4(27)$. Compare these four numerical approximations with the exact value $5.1961524227\ldots$ of $\sqrt{27}$.

324. The shape of the function f given by $f(x) = \frac{1}{x}$ is to be approximated by the associated Taylor polynomials of degrees 1 to 4 at point $x_0 = 1$. Evaluate the polynomials $T_1(x)$, $T_2(x)$, $T_3(x)$ and $T_4(x)$ for $x = \frac{5}{4}$. Then show that $T_n\left(\frac{5}{4}\right)$ converges to $f\left(\frac{5}{4}\right) = \frac{4}{5} = 0.8$ when n converges to ∞.

3.9 Miscellaneous Exercises

For Chapter 3.1: Introduction

325. a) Let the function f be given. Calculate the average rate of change of $f \colon y = x^2$ in the intervals $[0; 1]$, $[1; 2]$, $[0; 2]$, $[-1; 0]$, $[a; b]$ and $[a; a + h]$.

    b) Show that the function f is strictly increasing on the interval I if and only if $\frac{f(x_2) - f(x_1)}{x_2 - x_1} > 0$ holds for all $x_1, x_2 \in I$ with $x_1 \neq x_2$.

326. The square $OPQR$ with side length a lies in quadrant I of the xy-coordinate system such that the two vertices $P(a \mid 0)$ and $R(0 \mid a)$ are diagonally opposite each other. The side length a of the square however is not constant but increases continually, i.e. is a function of time t: $a = a(t)$, t in seconds. Make a sketch.

    a) Determine the instantaneous rate of change in the area of the square $OPQR$ at time t_0, if $a(t_0) = 12.3\,\text{cm}$ and $a = a(t)$ increases uniformly by $0.1\,\text{cm/s}$. How can this rate of change be illustrated approximately in the sketch?

    b) For which side length is the instantaneous rate of change in the area of the square $+6.54\,\text{cm}^2/\text{s}$?

327. For a given function f, the symmetric difference quotient is defined by $\frac{f(x+h) - f(x-h)}{2h}$, $h \neq 0$.

    a) Calculate the symmetric difference quotient for the following three functions: $f(x) = x^2$, $g(x) = x^2 + x - 1$ and $k(x) = ax^2 + bx + c$. What do you notice?

    b) Write a version of the symmetric difference quotient that has an h instead of a $2h$ as the denominator.

    c) Show: $f'(x) = \lim\limits_{h \to 0} \frac{f(x+h) - f(x-h)}{2h}$.

328. Differentiate the function given by its equation by determining the limit of the difference quotient with respect to its argument and use it to evaluate the indicated terms.

a) $f(x) = (1-x)^2 - 3$;

$f'(4),\ f'(-3),\ f'(1-\sqrt{2})$

b) $q(t) = \dfrac{1-t}{2t-3}$;

$q'(2),\ q'(-1),\ q'\left(\frac{3}{2} - \sqrt{2}\right)$

c) $w(\ell) = \sqrt{2\ell - 3}$;

$w'(2),\ w'\left(\frac{11}{7}\right),\ w'(4)$

329. A square with side length s_1 is enlarged to a square with side length s_2. The change in area ΔF is approximated by the differential $\mathrm{d}F$. How can $\mathrm{d}F$ be calculated in this case?

- $\mathrm{d}F = 2s_1(s_2 - s_1)$
- $\mathrm{d}F = 2s_2(s_2 - s_1)$
- $\mathrm{d}F = s_2^2 - s_1^2$
- $\mathrm{d}F = (s_2 - s_1)^2$

330. True or false?

a) $f'(x) = \frac{\mathrm{d}}{\mathrm{d}x} f(x) = \frac{\mathrm{d}f}{\mathrm{d}x}(x) = \frac{\mathrm{d}f(x)}{\mathrm{d}x}$

b) $f'(x) = \lim\limits_{\Delta x \to 0} \frac{f(x+\Delta x) - f(x)}{\Delta x} = \lim\limits_{h \to 0} \frac{f(x) - f(x-h)}{h}$

c) $f'(x) = \lim\limits_{\Delta x \to 0} \frac{\Delta f}{\Delta x}(x)$

d) $f'(x) = \lim\limits_{h \to 0} \frac{f(x+h) - f(x-h)}{h}$

331. Reason with purely geometric arguments, why $g(x) = f(x - c)$ necessarily implies that also $g'(x) = f'(x - c)$ holds. *Hint:* How are the two graphs G_f and G_g related?

332. The starting point for this exercise are the graphs G_f and G_g of the two functions f and g (with coincident domains $D_f = D_g = D$) as well as the corresponding graphs $G_{f'}$ and $G_{g'}$ of their derivatives. Is the statement true or false?

a) The graphs G_f and $G_{f'}$ are never congruent, so by no means can they coincide.

b) It is possible that the graph $G_{f'}$ can emerge by a pure translation (parallel shift) of the graph G_f.

c) If the formulas for f and g fulfil the condition $f(x) > g(x)$ for all $x \in D$, then the graph $G_{f'}$ always is above the graph $G_{g'}$.

For Chapter 3.3: Differentiation Rules

333. Differentiate with respect to x.

a) $ax^2 + bx + c$

b) $x^n + nx + \frac{1}{n}$

c) $\mathrm{e}^x + \mathrm{e}x - \frac{1}{x}$

d) $t^5 x + t^3 x^2 - tx^3$

e) $au^3 + bu^2 + cu + d$

f) $(x-1)^2 + (m+3)^2$

334. The area F of a circle is a function of the radius r and is given by the well-known formula: $F = F(r) = \pi r^2$.

a) Differentiate F with respect to r. What do you notice?

b) It may happen that the radius varies with time t, that is: $r = r(t)$. In this case, the area F of the circle also varies with time t. Determine $\frac{\mathrm{d}F}{\mathrm{d}t}$ in general and for $r = r(t) = 3 + 2.1t$ $(t \geq 0)$.

335. The radius r of a sphere uniformly decreases with time t (in seconds); the constant rate of change is $-1\,\mathrm{cm/s}$. Calculate the instantaneous rate of change for the volume $V = V(t)$ of the sphere in m^3/s in the case that the radius of the sphere is $1.02\,\mathrm{m}$.

336. The volume of an upright cuboid of height h with a square base of side s can be calculated by the formula $V = s^2 h$. If the dimensions $s = s(T)$ and $h = h(T)$ are affected by the ambient temperature T, then this also holds for the volume of the cuboid. Therefore, $V = V(T)$ is also a function of the ambient temperature T. Determine $\frac{\mathrm{d}V}{\mathrm{d}T}$.

337. A balloon filled with gas continuously and constantly loses $1.2\,\mathrm{m}^3$ of its content per minute, without changing its spherical shape; its volume V is a function of time: $V = V(t)$. The decrease of V over time also leads to the continuous decrease of the radius $r = r(t)$ of the sphere over time.

   a) Calculate the instantaneous rates of change of the radius r of the balloon in centimetres per minute, if the content of the balloon is $V_1 = 123.6\,\mathrm{m}^3$ and $V_2 = 43.7\,\mathrm{m}^3$, respectively.

   b) For which volume of the balloon is the instantaneous rate of change of the radius of the balloon $-1.2\,\mathrm{cm/min}$?

338. Prove the power rule $(x^n)' = n \cdot x^{n-1}$ for natural exponents $n \in \mathbb{N}$ by mathematical induction.

339. Justify why the derivatives of a quadratic function at its roots are always opposite numbers.

340. True or false?

   a) If f is periodic, then so is f'.

   b) If f is even, then so is f'.

   c) If f is odd, then f' is even.

341. Let the function f be given with $f(1) = 1$, $f'(1) = 3$. Calculate $\frac{\mathrm{d}}{\mathrm{d}x}\left(\frac{f(x)}{x^2}\right)$ at the point $x = 1$.

342. Calculate the 99th derivative of $f(x) = \cos(3x) + x^{100}$.

343. a) Show: For two functions f and g we have: $f(x) \cdot g(x) = \left(\frac{f(x)+g(x)}{2}\right)^2 - \left(\frac{f(x)-g(x)}{2}\right)^2$.

   b) Differentiate the right-hand side of part a) with respect to x. The result is the product rule.

344. A lamp standing on the floor illuminates a wall at a horizontal distance of $20\,\mathrm{m}$. Tim, $1.5\,\mathrm{m}$ tall, walks from the lamp to the wall with a constant horizontal velocity of $2\,\mathrm{m/s}$ and sees his shadow on the wall becoming smaller. At what velocity does his shadow on the wall decrease when he is still $10\,\mathrm{m}$ away from the wall?

345. Anna thinks that the relationship $(\sin(x))' = \cos(x)$ only holds if the angle x is given in radians. Zoe however claims that $(\sin(x))' = \cos(x)$ always holds regardless of whether x is given in radians, degrees or any other kind of angular units. Who is right?

346. Use $|x| = \sqrt{x^2}$ as well as the chain rule to show that $\frac{\mathrm{d}}{\mathrm{d}x}|x| = \frac{x}{|x|} = \frac{|x|}{x}$.

For Chapter 3.4: Tangent, Normal and Intersecting Angle

347. The curve defined by the equation $y = x^m$ passes throught the point $P(1\,|\,1)$ for any exponent $m \in N$.

a) Locate the two points X and Y, where the tangent t to the curve at point P intersects the two coordinate axes.

b) At which points X and Y does the normal n to the curve at point P intersect the two coordinate axes?

c) The normal n and the tangent t at the point P each enclose with the coordinate axes a right triangle, Δ_n and Δ_t, respectively. What is the difference d in area of these two triangles Δ_n and Δ_t as a function of the exponent m?

348. Let G_f be the graph of the function f defined by the equation $y = f(x) = x^3 - 2x + 1$.

a) Show by evaluating the limit of the difference quotient that the slope m_t of the tangent t to the graph G_f at the point x_0 can be determined by the formula $m_t = 3x_0^2 - 2$.

b) What is the slope m_t of the tangent at point $P(-1\,|\,2)$ of G_f?

c) Calculate the coordinates of all the points $Q(x_Q\,|\,y_Q) \in G_f$, at which the tangents to the graph G_f form an angle of $+45°$ with the x-axis.

349. A mini robot equipped with a laser moves on the floor of a room with the corners $O(0\,|\,0)$, $P(6\,|\,0)$, $Q(6\,|\,8)$ and $R(0\,|\,8)$. It is programmed such that $t \geq 0$ seconds after the start it is located at the point $S\big(2t\,|\,6t - 2t^2\big)$.

a) Show that the mini robot travels on a parabolic arc from O to P in the first three seconds after the start. Determine the simplest possible equation of the form $y = f(x)$ for this parabolic arc.

b) At which point S^* is the robot if the laser, whose beam always runs in the direction of the movement and thus tangentially to the robot's path, is directed exactly to the corner Q? How many seconds after the start is the robot at this point S^*?

350. Let the function f be given by the equation $y = f(x) = 2\sqrt{x} + \frac{1}{x}$; $x > 0$. B. Haupt claims that the graph G_f of this function has horizontal tangents at least two points. B. Streit vehemently denies this statement; as she claims she can prove that the graph G_f does not have any horizontal tangents at all. Who is right – if anyone?

351. By which amount (or by how many units) must the cosine curve be moved upwards so that it touches the sine curve?

352. Show: All curves p given by $p_a\colon y = \sqrt{a - 2x}$ intersect all exponential curves e given by $e_b\colon y = e^{x+b}$ at a right angle, and this holds independently of the parameters a and b.

353. *Deducing the equation of the tangent from the equation of the secant.* Let the function f and the points a and b, $a \neq b$, be given.

a) Determine the equation of the secant, which intersect the graph G_f at the points a and b.

b) Determine the equation of the tangent at point a by evaluating the limit of the equation of the secant for $b \to a$.

354. Show that the family of parabolas given by the equation $f_k(x) = -(x - k)^2 + k$, $k \in \mathbb{Z}$, have a common tangent.

355. a) At which point in the interval $[0; 2]$ is the differential quotient of $f(x) = x^2 + 1$ equal to the difference quotient of f on the interval $[0; 2]$?

b) At which point in the interval $[0; 3]$ is the slope m_t of the tangent to the graph of $f(x) = x^2 - 3x + 1$ equal to the slope m_s of the secant through the points $(0 \mid f(0))$ and $(3 \mid f(3))$ of the graph of f?

c) At which point in the interval $[0; s]$ is the instantaneous rate of change of $f(x) = ax^2 + bx + c$ equal to the average rate of change of f on the interval $[0; s]$?

For Chapters 3.5, 3.6 and 3.7: Extrema and More

356. Examine the function f in the neighbourhood of its singularities and sketch its graph for the given interval I.

a) $f(x) = \dfrac{x^3}{|x|}$; $I = [-2; 2]$
b) $f(x) = \left| \dfrac{x}{x - 2} \right|$; $I = [-1; 5]$

357. a) The family of curves f_b is given by $f_b(x) = x^2 + bx + 1$. On which curve do all the local minimum points of f_b lie, if b runs through all of $\mathbb{R}$?

b) The family of curves f_b is given by $f_b(x) = ax^2 + bx + c$. On which curve do all the local extreme points of f_b lie, if a and c are fixed and b runs through all of $\mathbb{R}$?

c) The family of curves f_a is given by $f_a(x) = ax^2 + x + 1$. On which curve do all the local extreme points of f_a lie, if a runs through all of $\mathbb{R}$?

d) The family of curves f_a is given by $f_a(x) = ax^2 + bx + c$. On which curve do all the local extreme points of f_a lie, if b and c are fixed and a runs through all of $\mathbb{R}$?

358. On which curve do all the local extreme points of the family of curves defined by $f_a(x) = x - ax^3$ lie, if $a \in \mathbb{R}^+$?

359. Let the function $s(x) = |x - 3| + |x + 1|$ be given. Determine all values of x for which $s(x)$ is smallest. Use a geometric argument.

360. The reaction r of the body to a dose of a medication can often be modelled by an equation of the form $r = r(m) = m^2 \left(c - \frac{1}{3}m \right)$, where c is a positive constant and m describes the amount of the administered medication which is absorbed in the blood. The derivative $\frac{\mathrm{d}}{\mathrm{d}m} r(m) = \frac{\mathrm{d}r}{\mathrm{d}m}$ is called the sensitivity of the body to the medication. At what quantity of administered medication does the body react most sensitively?

361. Let the function A be defined by $A(t) = 3 \cdot e^{\frac{-1}{100}t^2 + \frac{1}{10} \cdot t}$. The function A describes the population of a bacterial species at time t in a lethal environment for this species.

a) In which time interval do the bacteria multiply at the beginning?

b) At which time is the mortality rate of the bacteria greatest?

c) Show that $A'(t) = f(t) \cdot A(t)$ and determine $f(t)$.

362. Let the segment $A(-6\,|\,2)B(-2\,|\,2)$ as well as the segment $C(2\,|-2)D(6\,|-2)$ be given.

    a) Determine the simplest possible smooth (kink-free) transition curve $k\colon y = f(x)$ between the two segments.

    b) Determine the simplest possible jerk-free (without jumps in the curvature) transition curve $k\colon y = f(x)$ between the two segments.

363. Let the line $g\colon y = \frac{3}{4}x + 5$ and the point $P(13\,|-4)$ be given. Determine the distance of the point P from the line g in two different ways:

    i) Determine the perpendicular foot F of the normal to g through P and calculate the distance $\overline{PF}$.

    ii) Chose an arbitrary point $G(x\,|\,y)$ on g and determine x such that the distance from P to $G \in g$ is as small as possible.

364. Let the unit parabola p be given by the equation $y = x^2$. At the point $P\big(u\,|\,u^2\big)$ of the parabola construct the normal n to the parabola, where $u \in \mathbb{R}$ may be chosen arbitrarily.

    a) At which point X does this normal n to the curve intersect the x-axis? $(u \neq 0)$

    b) Characterise the part of the y-axis that is intersected by none of the normals of the curve for any $u \neq 0$.

    c) Calculate the coordinates of the point S at which the normal to the curve intersects the parabola a second time. $(u \neq 0)$

    d) The parabola p and the normal n at P to p enclose a parabolic chord $\overline{PS}$. What is its minimum length?

365. Let the two points $A(1\,|-2)$ and $B(3\,|\,4)$ be given. Determine the position of the point C on the x-axis for which the difference in distances $d = \overline{CB} - \overline{CA}$ is as large as possible.

366. Two lines of positive slopes m and $2m$ both pass through an arbitrary point in the coordinate system. Emma has drawn many different pairs of such lines and claims that the (acute) angle φ between the two lines measures always less than $20°$. Hugo claims that the maximum angle $\varphi_{\max}$ definitely is at least $20°$ or might even be $24°$, that is an integer fraction of $360°$. He says that this is typical in exercises of this kind and is much more likely. Who is right?

367. Is it true that the sum of any positive number z and its reciprocal is never less than 2?

368. The Bernese mathematician JAKOB STEINER (1796–1863) asked himself the following question: «Which positive number, when radicated with itself, gives the largest root?»

    Note: This is about the largest value of $\sqrt[x]{x}$ for $x \in \mathbb{R}^+$.

369. *Method of least squares.* An unknown quantity x_0 is measured n times under equal conditions. The measurements are x_1, x_2, x_3, ..., x_n. According to CARL FRIEDRICH GAUSS (1777–1855), the value x which minimises the sum s of the squares of the deviations, that is $s = (x_1 - x)^2 + (x_2 - x)^2 + \ldots + (x_n - x)^2$ is as small as possible, is the best estimate of x_0. Show that, in this sense, the arithmetic mean $\bar{x} = \frac{1}{n}\sum_{k=1}^{n} x_k$ is the best estimate.

370. The following exercises are from the chapter «De maximis et minimis» of the book *Institutiones calculi differentialis* by LEONHARD EULER (1707–1783) published in 1755.

   a) Invenire casus, quibus haec functio ipsius x; $x^4 - 8x^3 + 22x^2 - 24x + 12$ fit maximum vel minimum.

   b) Proposita sit haec functio $y = x^5 - 5x^4 + 5x^3 + 1$; quae quibus casibus fiat maximum minimumve, quaeritur.

   c) Invenire casus, quibus formula $\dfrac{xx - x + 1}{xx + x - 1}$ fit maximum vel minimum.

371. On a horizontal surface, a weight G is moved uniformly due to the tension F. The tension acts at an angle of inclination α above the horizontal and is affected by the friction coefficient μ. The tension F is smallest, when $\tan(\alpha) = \mu$ holds. Explain this physical fact.

372. In quadrant I draw the radius $r = \overline{OA}$ from the origin O to a point A on the unit circle with equation $x^2 + y^2 = 1$. From point A construct the normal $\overline{AB}$ onto the x-axis and from point B on the x-axis draw the normal $\overline{BC}$ onto $\overline{OA}$. What is the angle of inclination φ of the radius $r = \overline{OA}$ that produces the triangle ABC with the largest possible area?

373. Let the parabola, which opens downward, be given by the equation $y = 1 - ax^2$, where $a > 0$. Rectangles are inscribed into the segment of the parabola that is located above the x-axis.

   a) Calculate the abscissa of the right upper vertex of the rectangle with maximum area for the parameter $a = \frac{1}{3}$.

   b) Carry out an analogous calculation for the rectangle with maximum circumference for $a = 2$.

   c) For which value of a are these two rectangles identical?

3.10   Review Exercises

For Chapters 3.1 and 3.2: Introduction and Graphing Derivatives

374. Let the function f be given by the equation $y = \frac{1}{2}x^3 - 3x^2 - 3x + 4$.
   a) Calculate the average rate of change of the function f on the interval $[-2; 0]$.
   b) What is the instantaneous rate of change of the function f at the point $x = 2$?

375. On the star Sirius, the falling distance s of an object in free fall is given by the following formula: $s = 0.8 \cdot t^2$ (distance fallen in metres, time in seconds).

   a) What is the average velocity of an object in free fall on Sirius in the time intervals between 4 and 5 s, between 4.5 and 5 s, between 4.9 and 5 s as well as between 4.99 and 5 s?

   b) What is the instantaneous velocity this object will approximately have after 5 s according to the results of part a)?

376. Nora wants iced tea. For this, she boils rosehip tea and lets it cool. The function $T(t)$ describes the temperature of the tea at time t, measured from when it finished boiling. Express the following quantities formally:

   a) decrease in temperature during the time interval $[t_1; t_2]$

   b) average decrease in temperature per time unit in the time interval $[t_1; t_2]$

   c) instantaneous rate of decrease in temperature at time t_2

377. The water volume left in a swimming pool, the drain of which has been opened, can be approximately described by the function V given by the formula $V(t) = 200 \cdot (50 - t)^2$ (t in minutes, V in litres).

   a) Calculate $\frac{V(5)-V(0)}{5-0}$. What does this numerical value mean?

   b) Calculate $V'(t)$. What does this derivative mean? What do $V'(t) < 0$ or $V'(t) > 0$ indicate?

   c) Calculate $V'(5)$. What does this value tell us?

378. Determine the derivative f' of the function f at point a by means of the difference quotient.

   a) $f(x) = \frac{1}{4}x^2$, $a = 6$ b) $f(x) = \frac{-1}{x}$, $a = 3$

379. The function f is given by its formula. Determine its derivative f' by evaluating the limit of the difference quotient.

   a) $f(x) = x^2 - 2x$ b) $f(x) = \dfrac{x}{1-x}$

380. a) Determine the slope function f' of $y = f(x) = \frac{1-x}{x-2}$ by concretely calculating the limit of its difference quotient.

   b) At which points do the tangents to the graph of f have an angle of inclination of $\pm 45°$?

381. Max and Moritz, each for himself, had to differentiate partly the same and partly totally different functions for practise. When comparing the solutions for exercise 5 they noticed that they both obtained the same derivative as the result: $g'(x) = f'(x)$. Moritz and Max where somewhat irritated when they noticed that they weren't differentiating the same function. Is it possible at all that two different functions $g \neq f$ have identical derivatives $g' = f'$?

382. Sketch the graph of the derivative f' for the plotted function f.

   a)
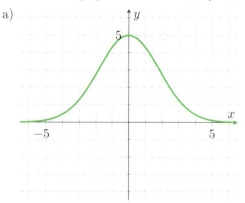

   b)
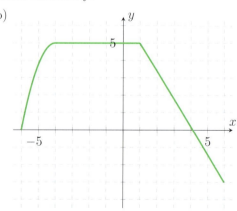

c)

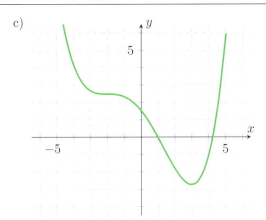

d)

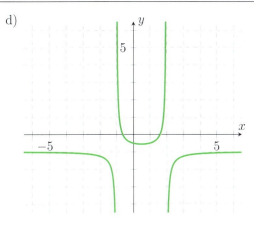

For Chapter 3.3: Differentiation Rules

383. True or false? $\frac{d}{dz}\left(z^7\right)$ is equal to:

a) z^6 b) $7z^6$ c) 0 d) z^7

384. Determine $f'(x)$.

a) $f(x) = 4x^5$ b) $f(x) = \frac{1}{3}x^6 - 2x$ c) $f(x) = \frac{-2}{x^3}$

d) $f(x) = \frac{1}{x} - \frac{1}{x^2}$ e) $f(x) = 4\sqrt{x}$ f) $f(x) = 8 + \sqrt[3]{x}$

385. Rewrite the formula of the function in a way such that you could determine the derivative y' using merely the sum, factor and power rules. The derivative itself need not be determined.

a) $y = \frac{x^2}{4} + 13$ b) $y = \frac{-1}{x}$ c) $y = \frac{\sqrt{x}}{7}$

d) $y = 3x \cdot (2x - 1)$ e) $y = (x - 2)^2$ f) $y = \frac{5x^{-4} + 3x}{21}$

386. Determine the derivative f'.

a) $f(x) = x^2(x^2 - 3x + 1)$ b) $y = \frac{x}{x + 2}$ c) $f(x) = \sqrt{8x^3} + 2$

387. Show step by step that it does not matter when differentiating the function f defined by $f(x) = (3x - 1)^2$ if (1) the formula for the function is first expanded and then differentiated, (2) the formula for the function is written as a product so that the product rule can be applied, or (3) the chain rule is directly applied to differentiate $f(x)$.

388. Determine $f'(x)$.

a) $f(x) = \sin(x) + \cos(x)$ b) $f(x) = x - 2 \cdot \sin(x)$ c) $f(x) = 3x^2 - 10 \cdot \cos(x)$

389. Determine the derivative y'.

a) $y = x \cdot \cos(x)$ b) $y = \sin(2x + 1) - 3$ c) $y = x^2 \cdot \sin(x) + 2x \cdot \cos(x)$

390. Determine $\frac{d}{dt}s(t)$.

a) $s(t) = \dfrac{\sin(t)}{1 - \cos(t)}$ b) $s(t) = \tan(t) - 2t$ c) $s(t) = \dfrac{\sin(t) - 1}{\cos(t)}$

391. Determine the fourth and fifth derivatives $y^{(4)}$ and $y^{(5)}$ of the given function $y = y(x)$.

a) $y = 6 \cdot \cos(x)$ b) $y = -5 \cdot \sin(x)$

392. Differentiate the function f with respect to its argument for the given formula of f.

a) $f(x) = e^{3x}$ b) $f(x) = x - 2 \cdot e^x$ c) $f(x) = 2^x$

d) $f(t) = 3 \cdot \ln(21t)$ e) $f(t) = 2 \cdot \sqrt{t} \cdot \ln(t)$ f) $f(t) = \ln\left(\frac{t-1}{t}\right)$

393. Determine $\frac{dy}{dx}$.

a) $y = 2x \cdot 10^x$ b) $y = \frac{3^x}{x}$ c) $y = x^2 \cdot 2^x$

394. Using the two functions f and g, three new functions are formed, namely: $d(x) = f(x) - g(x)$, $p(x) = f(x) \cdot g(x)$ and $q(x) = \frac{f(x)}{g(x)}$.

Evaluate the three derivatives $d'(x_0)$, $p'(x_0)$ and $q'(x_0)$ by means of the following additional information regarding the given functions f and g:

- $f(x_0) = 4$ - $g(x_0) = -1$ - $f'(x_0) = -2$ - $g'(x_0) = 3$

395. Determine the derivative of $y = (1 + (2 + x^3)^4)^5$.

396. True or false? Let the functions f and g be given and $h = f \circ g$. Then, $h'(2)$ equals:

a) $(f' \circ g')(2)$ b) $f'(2) \cdot g'(2)$ c) $f'(g(2)) \cdot g'(2)$ d) $f'(g'(2)) \cdot g'(2)$

397. True or false? Let $f(x)$ and $c \in \mathbb{R}$ be given. Then, we have:

a) $\frac{d}{dx}(c \cdot f(x)) = c \cdot \left(\frac{d}{dx}f(x)\right) + f(x) \cdot \left(\frac{d}{dx}c\right)$ b) $\frac{d}{dx}(c \cdot f(x)) = c \cdot \left(\frac{d}{dx}f(x)\right)$

c) $\frac{d}{dx}(c \cdot f(x)) = \left(\frac{d}{dx}c\right) \cdot \left(\frac{d}{dx}f(x)\right)$

398. Let the composite functions F_1 and F_2 be given as follows: $F_1(x) = f(u(x)) = \frac{u(x)}{3} + 21$, with $u(x) = 3x - 63$, and $F_2(x) = g(v(x)) = 2 - \frac{1}{v(x)}$, with $v(x) = \frac{1}{2-x}$.

a) Determine both derivatives $\frac{d}{dx}F_1(x)$ and $\frac{d}{dx}F_2(x)$ by applying the chain rule.
b) Show that $F_1(x)$ and $F_2(x)$ describe the same function.

For Chapter 3.4: Tangent, Normal and Intersecting Angle

399. Let the parabola be given by the equation $y = \frac{-1}{2}x^2 + 4x$. Determine the equations of the tangent t and the normal n to the parabola at the point x_0.

a) $x_0 = 0$ b) $x_0 = 6$ c) $x_0 = 4$

400. Let the function f be given by the equation $f(x) = \frac{x^2+1}{x+1}$. Determine the equations of the tangent t and the normal n to the graph of f at point x_0.

a) $x_0 = 0$ b) $x_0 = 2$ c) $x_0 = -3$

401. At which points of the curve given by the equation $y = \frac{1}{3} \cdot x^3 - x^2 + x - 4$ do the corresponding tangents to the curve have the prescribed slope m?

    a) $m = 0$ b) $m = 4$ c) $m = -1$

402. At which point P of the parabola given by the equation $y = 2x^2 - 8x$ is the tangent to the parabola parallel to the x-axis?

403. The curve w is defined by its equation $y = \sqrt{x}$ $(x \geq 0)$. What are the equations of the tangent t and the normal n to the curve at the point $x_0 = 4$?

404. For which value of the parameter $a \in \mathbb{R}$ does the graph of the function f defined by the equation $f(x) = \frac{1}{2}x^3 + ax^2 + 32x - 48$ touch the x-axis at the point $x_0 = 4$?

405. The tangent t to the graph of the function f defined by $f(x) = x^2 - x + 16$ passes through the point P. Determine the equation of the tangent t.

    a) $P(0\,|\,0)$ b) $P(-2\,|\,6)$

406. At what acute angle does the graph G_f intersect the x-axis?

    a) $f(x) = 1.2\sqrt{x} - 3.4$ b) $f(x) = e^{2x} - 1$

407. At what acute angle do the graphs of the functions f and g intersect each other?

    a) $f(x) = \frac{1}{2}x^2$ and $g(x) = \frac{1}{2}x^2 - 2x + 6$ b) $f(x) = \frac{2}{x^3}$ and $g(x) = 2x^3$

408. At what acute angle do the hyperbolic branch given by the equation $y = h(x) = \frac{1}{x}$ $(x > 0)$ and the parabolic arc defined by the equation $y = p(x) = \sqrt{x}$ $(x \geq 0)$ intersect each other?

For Chapter 3.5: Special Points and Properties of Curves

409. Determine all minimum points of the function given by the formula $f(x) = x^4 - 4x^3 + 4x^2$.

410. Let the function f be given by $y = x + \frac{1}{x+1}$. Determine all maximum and minimum points of the graph of the function.

411. Let the function f be given by $y = f(x) = \frac{1}{6}x^4 + \frac{2}{3}x^3$.

    a) Calculate all maximum, minimum and saddle points.

    b) At which point in the interval $[-4; 1]$ is the absolute maximum point and at which the absolute minimum point of the function f?

412. Discuss the functions (with respect to domain and codomain, roots, maximum and minimum points, inflection and saddle points, symmetry) and sketch their graphs.

    a) $f(x) = x^3 - 3x$ b) $f(x) = x^4 - 2x^2$ c) $f(x) = 3x^2 - x^3$ d) $f(x) = 2x^3 - x^4$

413. For which of the plotted curves do $f(a) > 0$, $f'(a) = 0$, $f''(a) = 0$ and $f'''(a) > 0$ hold?

i)   ii)   iii)   iv)

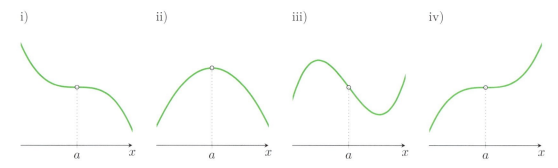

414. True or false? Let the function f be given on the interval $[a; b]$. If $f'(a) = 0$, then

    a) f must have a global minimum at the point $x = a$.

    b) f may have a global maximum or a global minimum at the point $x = a$.

    c) f cannot have a global maximum at the point $x = a$.

For Chapter 3.6: Determining Equations of Functions

415. The parabola defined by the equation $y = ax^2 + bx + c$ passes through the point $(1 \mid 2)$ and touches the line $y = x$ at the origin. Calculate the parameters a, b and c.

416. A polynomial function of degree 3 and symmetric with respect to the origin has an extremum at the point $P(2 \mid -9)$. Determine the formula of the function.

417. Which polynomial function of degree 3 has a graph that is symmetric with respect to the origin and has a local minimum at the point $(-2 \mid -4)$?

418. The curve k of a polynomial function of degree 3 passes horizontally through the origin; the tangent t to the curve at the point $K(3 \mid 9)$ on k also passes through the origin. Determine the equation $y = k(x)$ of this curve k. k descends into quadrant IV at another point on the positive x-axis. Calculate the angle of intersection σ between k and the x-axis.

419. For which integer values of n is the curve defined by the equation $y = x^n + a \cdot x$ symmetric with respect to the origin?

420. A polynomial function of degree 4 is symmetric with respect to the y-axis, passes through the point $P(-1 \mid 9)$ and has slope $m = 4$ at the point $(2 \mid 3)$. What is the equation of the function?

For Chapter 3.7: Optimisation Problems

421. At which point does the curve defined by the equation $y = x^3 + 3x^2 + 2x + 4$ have the smallest slope?

422. Cutting away four squares of equal size at each corner of a rectangle of length $l = 16\,\text{cm}$ and width $b = 10\,\text{cm}$ and folding up the protruding rectangles yields a box with an open top. What side length of the cut-off squares maximises the capacity of the box?

423. The Haas family would like to provide a rectangular outdoor enclosure for their four pairs of bunnies, in the area of a recessed corner of the house so that no fence has to be installed on two sides. With the fencing of length $\ell = 40\,\text{m}$, they also want to subdivide the enclosure into four compartments of equal size, one for each pair of bunnies. What is the maximum number of square metres available to each pair of bunnies?

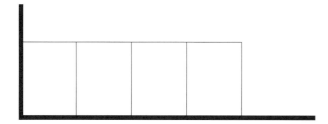

424. The three points $F(-20\,|\,0)$, $G(70\,|\,0)$ and $Z(0\,|\,40)$ represent villages that are to be connected to one another by a water pipeline. From a point X on the straight line connecting F and G, a branch runs directly to the point Z. Where should the branching point X be placed in order the minimise the total construction cost, if the construction costs along the segments $\overline{FX}$, $\overline{XG}$ and $\overline{XZ}$ are expected to be 600 francs, 900 francs and 500 francs, respectively, per length unit?

425. A wall mirror framed with metal has the shape of a rectangle surmounted by a semi-circle. Which dimensions does the mirror with largest possible area have, if the perimeter of the mirror is 300 cm?

426. A region in quadrant I is enclosed by the coordinate axes, the parabola $p\colon y = x^2 + 3$ and the line $g\colon x = 5$. A rectangle with sides parallel to the axes is inscribed in this region. What is the largest area of this rectangle?

427. How high is a cylinder of maxium lateral surface that is inscribed into a sphere (radius R)?

4   Integral Calculus

4.1   Integration as the Inverse of Differentiation

1. *Can accidental differentation be reversed without difficulty?*   Many mathematical operations
have an inverse. For instance, subtraction is the inverse of addition, root extraction the inverse
of exponentiation and many functions have an inverse function. Can differentiation be inverted,
that is can the original function f be calculated from f'?

a) Which function f has the derivative f'? Verify by differentiation.

   i) $f'(x) = 2x$ ii) $f'(x) = 2x - 1$ iii) $f'(x) = 4x^3$ iv) $f'(x) = x^n$

   v) $f'(x) = \cos(x)$ vi) $f'(x) = e^x$ vii) $f'(x) = \frac{1}{x^2}$ viii) $f'(x) = x^{-m}$

b) Are the results from part a) unique? Justify.

c) Which of the following functions f have the derivative f' with $f'(x) = 2x$?

   i) $f(x) = x^2 + 1001$ ii) $f(x) = x^2 - 1$ iii) $f(x) = x^2 + x$ iv) $f(x) = -x^2$

d) The function f given by $f(x) = x^2$ has the derivative or slope function f' with $f'(x) = 2x$.
Which other functions have the same derivative?

e) Describe all functions f whose derivative f' is as given.

   i) $f'(x) = \sin(x)$ ii) $f'(x) = 3x^2$ iii) $f'(x) = e^x + 2x$

Integration and the Antiderivative

If, for a given function f, we search for a function F which fulfils $F' = f$, we call this process
antidifferentiation and call F an *antiderivative* of f.

2. Determine an antiderivative F of the given function f. Pay careful attention to the function
argument.

a) $f(x) = 4x$ b) $f(x) = x^3$ c) $f(x) = -2.34 x^5$ d) $f(x) = x^n$

e) $f(t) = \frac{-3}{4} t^8$ f) $f(t) = 1001$ g) $f(t) = 0$ h) $f(t) = g t$

3. Determine an antiderivative F of the given function f.

a) $f(x) = 4x + 321$ b) $f(x) = 6x^2 - 4$ c) $f(x) = 3 - \frac{1}{x^2}$ d) $f(x) = x^3 + x^{-3}$

4. Determine an antiderivative F of the given function f by use of appropriate algebraic manipulations.

a) $f(t) = \dfrac{1}{2\sqrt{t}}$ b) $f(t) = \dfrac{-3}{\sqrt{t}}$ c) $f(t) = 7\sqrt[4]{t^3}$ d) $f(t) = \dfrac{1}{\sqrt[3]{t^2}}$

5. Determine an antiderivative F of the function f.

a) $f(x) = \sin(x) + \cos(x)$ b) $f(x) = 3\sin(x) - 2\cos(x)$ c) $f(x) = \dfrac{1}{\cos^2(x)}$

d) $f(x) = 2e^x$ e) $f(x) = 2^x$ f) $f(x) = a^x, \ a > 0, a \neq 1$

6. Determine the equation of the antiderivative F of the function f, whose graph G_F passes through
the origin.

a) $f(x) = 3x^2 + 4x - 1$ b) $f(x) = \dfrac{3x^4 - x^2}{x^2}$ c) $f(x) = \dfrac{1}{(x-2)^2}$

7. The first derivative f' and an additional constraint of a function f are known. What is the equation of f?

a) $f'(x) = 3x^2 - 14$
   $f(5) = 67$

b) $f'(x) = 1 - \frac{1}{x^2}$
   $f(3) = 4$

c) $f'(x) = 5 - x$
   $f(-2) = -f(2)$

d) $f'(x) = 1 + \sin(x)$
   $f(0) = 0$

e) $f'(x) = -3\cos(x)$
   $f\left(\frac{\pi}{2}\right) = 1$

f) $f'(x) = \pi e^x$
   $f(1) = 1$

8. Determine the antiderivative F for the given function f, whose graph passes through the point P.

a) $f(x) = \dfrac{1}{\sqrt{x}}; \ P(9\,|\,2)$

b) $f(x) = \frac{1}{2}x^2 - 2x; \ P\left(1\,|\,\frac{7}{3}\right)$

c) $f(x) = \dfrac{2x^3 + 5x^2}{x^2}; \ P\left(\frac{5}{2}\,\middle|\,\frac{21}{4}\right)$

9. Determine the equation $y = f(x)$ of the curve in the xy-plane that passes through the point $P(9\,|\,4)$ and whose slope at the point x is given by $3\sqrt{x}$.

For **10–13**: Graphical integration

10. The graph G_f of a function f is given. Which of the graphs a, b or c could be the graph of an antiderivative F of f?

a)

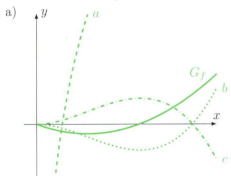

b)

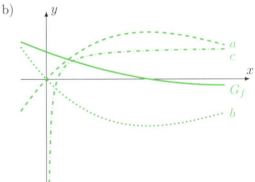

11. The graph G_f of a function f is given. Sketch the graph of the antiderivative F of f that fulfils $F(0) = 0$.

For exercise 11:

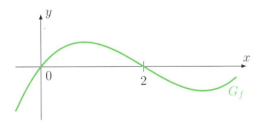

For exercise 12:

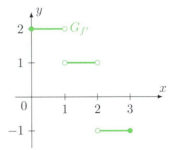

12. The graph $G_{f'}$ of the derivative f' is given, see above. Sketch the graph of the function f which fulfils the following two conditions: f is continuous and $f(0) = -1$.

13. Sketch the graph of an antiderivative F of the function f, whose graph G_f is displayed in a coordinate system. Draw the solution that passes through the origin.

a) 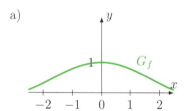 b) 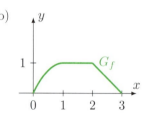 c)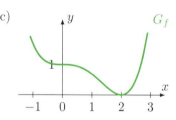

The Indefinite Integral

If F is an antiderivative of f, then the set of all antiderivatives is called the *indefinite integral* of f:

$$\int f(x)\,\mathrm{d}x = F(x) + C, \quad C \in \mathbb{R}.$$

$\int$ is called the *integral symbol*, $f(x)$ the *integrand*, $\mathrm{d}x$ the *differential* and C the *constant of integration*. For the origin of the integral symbol see the box on page 129.

Some Indefinite Integrals

$\int x^r\,\mathrm{d}x = \frac{1}{r+1}x^{r+1} + C,\, r \in \mathbb{R}\backslash\{-1\}$

$\int \frac{1}{x}\,\mathrm{d}x = \ln|x| + C$

$\int \sin(x)\,\mathrm{d}x = -\cos(x) + C$

$\int \cos(x)\,\mathrm{d}x = \sin(x) + C$

$\int \tan(x)\,\mathrm{d}x = -\ln|\cos(x)| + C$

$\int \frac{1}{2\sqrt{x}}\,\mathrm{d}x = \sqrt{x} + C$

$\int \sqrt{x}\,\mathrm{d}x = \frac{2}{3}\sqrt{x^3} + C$

$\int \mathrm{e}^{kx}\,\mathrm{d}x = \frac{1}{k} \cdot \mathrm{e}^{kx} + C$

$\int a^x\,\mathrm{d}x = \frac{1}{\ln(a)} \cdot a^x + C,\, a > 0,\, a \neq 1$

$\int \ln(x)\,\mathrm{d}x = x\ln(x) - x + C$

Sum and Factor Rules

Sum rule: $\displaystyle\int (f(x) \pm g(x))\,\mathrm{d}x = \int f(x)\,\mathrm{d}x \pm \int g(x)\,\mathrm{d}x$

Factor rule: $\displaystyle\int (c \cdot f(x))\,\mathrm{d}x = c \cdot \int f(x)\,\mathrm{d}x$

For **14–16**: Calculate the indefinite integral.

14. Polynomial functions

a) $\displaystyle\int 4x\,\mathrm{d}x$

b) $\displaystyle\int 1\,\mathrm{d}x$

c) $\displaystyle\int 0\,\mathrm{d}x$

d) $\displaystyle\int 6x^3\,\mathrm{d}x$

e) $\displaystyle\int (3x^2 - 1)\,\mathrm{d}x$

f) $\displaystyle\int (4x^3 + x^2 - 7)\,\mathrm{d}x$

g) $\displaystyle\int (5x^4 + \frac{1}{2}x^3 - \frac{7}{3}x^2)\,\mathrm{d}x$

h) $\displaystyle\int (3x + 4)(x^2 - 1)\,\mathrm{d}x$

i) $\displaystyle\int (3x^2 - 1)^2\,\mathrm{d}x$

15. Rational functions

a) $\int \dfrac{1}{x^2}\, dx$
b) $\int \dfrac{2}{x^3}\, dx$
c) $\int \left(\dfrac{6}{x^4} - \dfrac{1}{x^2} \right) dx$

d) $\int \left(\dfrac{6}{x} - \dfrac{5}{x^2} + \dfrac{4}{x^3} \right) dx$
e) $\int \dfrac{x^3 - x + 5}{x}\, dx$
f) $\int \dfrac{10}{x+2}\, dx$

16. Roots, trigonometric functions and exponential functions

a) $\int \sqrt{x}\, dx$
b) $\int \cos(x)\, dx$
c) $\int 5^x\, dx$

d) $\int \sqrt{x} \cdot (x+5)\, dx$
e) $\int (e^x + \sin(x))\, dx$
f) $\int \frac{1}{x} \cdot \sqrt{x}\, dx$

g) $\int (x + \sqrt[4]{x})\, dx$
h) $\int (2 \cdot \sin(x) - 3 \cdot \cos(x))\, dx$
i) $\int \pi \cdot \ln(x)\, dx$

j) $\int (1 + \tan^2(x))\, dx$
k) $\int (t + \cos(t))\, dt$
l) $\int x^{1-2t}\, dx$

17. Justify.

a) $\int \sin^2(x)\, dx = \frac{1}{2}(x - \sin(x) \cdot \cos(x)) + C$
b) $\int \cos^2(x)\, dx = \frac{1}{2}(x + \sin(x) \cdot \cos(x)) + C$

c) $\int \tan^2(x)\, dx = \tan(x) - x + C$
d) $\int x e^x\, dx = (x-1)e^x + C$

18. Prove the integration formula $\int \ln(x)\, dx = x \ln(x) - x + C$ given in the green box above.

19. a) Are both F with $F(x) = \frac{1}{2}x^2 + 1$ and G with $G(x) = 4 - \sin^2 x + \frac{1}{2}x^2 - \cos^2 x$ antiderivatives of the same function f with $f(x) = x$?

b) Are both F with $F(x) = \sqrt{x+1}$ and G with $G(x) = \frac{x}{1+\sqrt{x+1}}$ antiderivatives of the same function f?

20. Verify the given integration formula.

a) $\int \dfrac{2x^2 + 1}{\sqrt{x^2 + 1}}\, dx = x \cdot \sqrt{x^2 + 1} + C$
b) $\int e^{\sqrt{x}}\, dx = 2e^{\sqrt{x}}(\sqrt{x} - 1)$

21. a) Basil has noticed while differentiating $f(x) = x e^x$ that $f'(x) = e^x + x e^x$ again contains the term $x e^x$. How can he use this discovery to calculate $\int \ln(x)\, dx$?

b) Bettina has noticed while differentiating $f(x) = x \ln(x)$ that $f'(x) = \ln(x) + 1$ again contains the term $\ln(x)$. How can she use this observation to calculate $\int \ln(x)\, dx$?

Integration by Simple Substitution

If F is an antiderivative of f, we have for $a \neq 0$ and $b \in \mathbb{R}$:

$$\int f(ax + b)\, dx = \frac{1}{a} \cdot F(ax + b) + C.$$

Here, $F(ax + b)$ is the value of the antiderivative F at the point $(ax + b)$.

For **22–25**: Determine the indefinite integral of the function f.

22. a) $f(x) = (2x + 3)^5$
     b) $f(x) = 7(3x - 1)^6$
     c) $f(x) = (12 - x)^3$
    d) $f(x) = \left(3 - \frac{1}{2}x\right)^3$
     e) $f(x) = 12\,(3x - 9)^3$
     f) $f(x) = \left(3 - x\sqrt{2}\right)^2$

23. a) $f(x) = (4x + 3)^{-2}$
     b) $f(x) = 2(1 - x)^{-3}$
     c) $f(x) = \left(4 - \frac{1}{2}x\right)^{-4}$
    d) $f(x) = \dfrac{4}{(2x - 1)^3}$
     e) $f(x) = \dfrac{2}{\left(2 - \frac{1}{3}x\right)^5}$
     f) $f(x) = \dfrac{1}{\sqrt{2x + 1}}$

24. a) $f(t) = \sin(2t)$
     b) $f(t) = \cos(\pi t)$
     c) $f(t) = 5 \cdot \cos(10t)$

25. a) $f(x) = 4e^{-4x}$
     b) $f(x) = e^x - e^{-x}$
     c) $f(x) = 9 \cdot 2^{-3x}$
    d) $f(x) = \dfrac{1}{2x - 1}$
     e) $f(x) = \dfrac{2}{3 - 2x}$
     f) $f(x) = 12 \cdot \ln(3x)$

26. Prove the formula in the green box on the previous page.

Distance as the Antiderivative of Velocity

27. *Distance travelled as an antiderivative.* From the chapter on differential calculus, we know that the following equations hold (in physics, the derivative with respect to the time variable t is often denoted by a superimposed dot instead of a prime mark):

$$v(t) = \frac{\mathrm{d}}{\mathrm{d}t}s(t) = s'(t) = \dot{s}(t), \qquad a(t) = \frac{\mathrm{d}}{\mathrm{d}t}v(t) = v'(t) = \dot{v}(t) = \ddot{s}(t).$$

If we know the distance travelled $s = s(t)$ as a function of time t, we can obtain the velocity function $v = v(t)$ by differentiating once and the acceleration function $a = a(t)$ by differentiating twice.

In many cases, however, we are in the reverse situation: In experiments, it is often easy to measure the force $F = F(t)$, e.g. with a spring scale. Because of Newton's second law, $F = m \cdot a$ and this can be used to determine the acceleration $a = a(t)$. How can we now conversely use the acceleration $a = a(t)$ to determine the velocity $v = v(t)$ or the position $s = s(t)$?

a) What are the formulas for $v(t)$ and $s(t)$ of the general velocity and position functions if the acceleration has the constant value a?

b) An object at rest is at its initial position on a horizontal, frictionless track. An acceleration of $a = 5\,\mathrm{m/s^2}$ is imparted on it by means of a force. What is its velocity after 1 second? Where is it after 1 second? When does it reach the target destination 20 metres away?

c) *Free fall:* A freely falling object is dropped from a height of 20 metres. The gravitational force $F = mg$ acts on it, where $g = 9.81\,\mathrm{m/s^2}$ is the gravitational acceleration. What is its velocity after 1 second? Where is it after 1 second? When does it reach the ground?

28. The graph G_v of the velocity function v of a car is pictured here. Sketch the graph of the position function s with $s(0) = 0$.

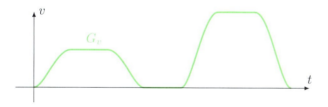

Distance as an Area in the v, t-diagram

Uniform motion: In physics, one learns that, for uniform motion, the distance s travelled is calculated as follows:

$$s = v \cdot t \; (v \cong \text{konst.}).$$

This can be interpreted as the area of the rectangle of width t and height v, see figure on the right.

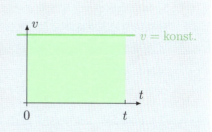

Generalisation $(v = v(t))$: During a very short time interval $[t_1; t_2]$, one may consider the velocity as being almost constant. If we replace the velocity in this interval by the mean value $v_m = \frac{v_1 + v_2}{2}$ for example, the distance travelled

$$\Delta s = v_m \cdot \Delta t$$

in this time span of $\Delta t = t_2 - t_1$ can be interpreted as the area of the rectangular strip, see figure on the right.

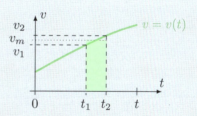

For very short time intervals $[t_1; t_2]$, this area is approximately the same size as the strip between the velocity curve $v = v(t)$ and the t-axis, see figure on the right.

The distance travelled during the whole time interval $[0; t]$ can therefore be interpreted as the area between the velocity curve and the time axis, see figure on the right.

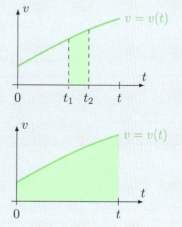

29. a) *Uniformly accelerated motion* ($a \cong$ konst.). Determine the formula for the distance s travelled at time t geometrically using the figure on the left for a uniformly accelerated motion. Compare with the formula known from physics.

For a)

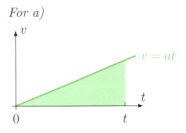

For b)

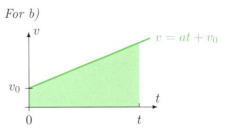

b) If the uniformly accelerated motion starts with an initial velocity $v(0) = v_0$, the graph of the velocity function looks like the figure on the right.

Again, determine the formula for the distance s travelled at time t geometrically.

4.2   Calculation of Areas by Means of Lower and Upper Sums

30. *Calculation of the distance travelled.* The following diagram shows a velocity-time graph. The horizontal axis depicts the time and the vertical axis the instantaneous velocity.

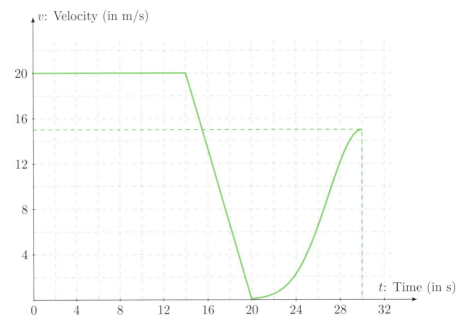

a) Determine the distance travelled in the first 14 seconds, i.e. during the time interval $[0\,\text{s}; 14\,\text{s}]$.

b) What distance is travelled during the interval $[14\,\text{s}; 20\,\text{s}]$?

c) Interpret the results geometrically, that is, consider how the results of parts a) and b) can be visualised in the diagram.

d) Determine an approximation of the distance travelled during the interval $[20\,\text{s}; 30\,\text{s}]$.

31. 📄 A wheelbarrow is pushed uphill over a distance of 20 metres. The force exerted on it as a function of distance is given by the following graph.

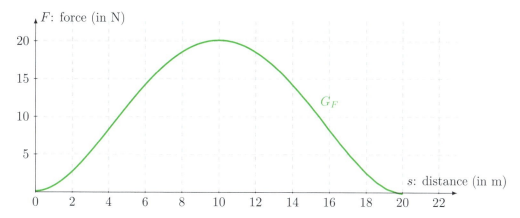

We wish to determine the work done for this case. From physics we know that:

$$\text{work} = \text{force} \times \text{distance}$$
$$\Delta W = F \quad \cdot \quad \Delta s$$

However, this formula only holds if the force is constant and applied in the direction of motion. In the depicted situation, the force is, however, obviously not constant.

a) How can you read off the total work done from the graph?

b) Try to determine an approximation for the total work done as follows: For every displacement of 2 metres, the force applied there is replaced by a constant average force. Geometrically, this can be achieved, by replacing the every curve segment by a horizontal line segment, such that the area of the rectangle formed by this is the best possible match for the area of the strip beneath the curve. Draw these horizontal segments in the graph as precisely as possible and calculate an approximation of the total work W done using this figure.

Calculation of Areas by Means of Lower and Upper Sums

Let f be a continuous function on the interval $[a;b]$. We partition the interval $[a;b]$ into n subintervals $[x_{k-1};x_k]$ of equal length $\Delta x = \frac{b-a}{n}$ by the points $x_k = a + k\Delta x$ for $k = 0,\dots,n$ (especially, $x_0 = a$, $x_n = b$). Using these subintervals, we can define the following two areas.

1) Lower sum

We define the *lower sum* U_n as the area of the step function inscribed below the graph of the function, whose steps coincide with, or lie below, the graph of the function:

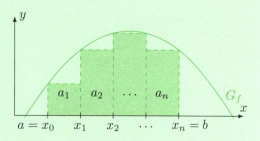

$$U_n = \sum_{k=1}^{n} a_k,$$

$a_k = \underline{f_k} \cdot \Delta x$, where $\underline{f_k}$ is the minimum value of $f(x)$ for $x \in [x_{k-1};x_k]$.

2) Upper sum

We define the *upper sum* O_n as the area of the step function circumscribed around the graph of the function, whose steps coincide with, or lie above, the graph of the function:

$$O_n = \sum_{k=1}^{n} A_k,$$

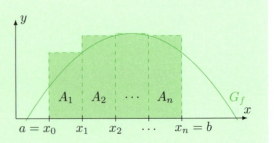

$A_k = \overline{f_k} \cdot \Delta x$, where $\overline{f_k}$ is the maximum value of $f(x)$ for $x \in [x_{k-1}; x_k]$.

The following sum formulas may be useful in calculating lower and upper sums:

$$\sum_{k=1}^{n} k = \frac{n(n+1)}{2}; \quad \sum_{k=1}^{n} k^2 = \frac{n(n+1)(2n+1)}{6}; \quad \sum_{k=1}^{n} k^3 = \frac{n^2(n+1)^2}{4};$$

$$\sum_{k=1}^{n} k^4 = \frac{n(n+1)(6n^3+9n^2+n-1)}{30}; \quad \sum_{k=1}^{n} k^5 = \frac{n^2(n+1)^2(2n^2+2n-1)}{12}$$

32. 📄 *The method of upper and lower sums.* The graph G_f of f given by $f(x) = x^2$ is depicted in the interval $[0; 2]$. The following steps show how the exact area $F(0,2)$ betweeen G_f and the x-axis in the interval $[0; 2]$ may be calculated.

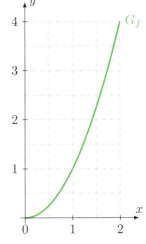

a) Draw the fourth upper sum O_4 of the function f into the coordinate system, while giving the corresponding partition of the interval $[0; 2]$:

$x_0 = a = \dots$ $x_1 = \dots$ $x_2 = \dots$

$x_3 = \dots$ $x_4 = b = \dots$ and determine Δx.

b) Give O_4 as an expanded sum and calculate its value.

c) Rewrite O_4 using the summation symbol $\sum$.

d) Calculate O_{50} using a calculator.

e) Write O_n using the summation symbol $\sum$ and calculate its value as a function of n (by hand or with a calculator).

f) How may the area $F(0, 2)$ be calculated exactly?

33. Calculate the fourth upper and lower sums O_4 and U_4 of the function f defined by $f(x) = -x^3 + 5$ on the interval $[-2; 0]$ by hand.

34. Sketch the graph of the function f with equation $f(x) = \frac{1}{2}x^2 + 1$ on the interval $[0; 2]$.

a) Calculate the lower sum U_6.

b) Calculate the upper sum O_6.

c) Deduce a statement about the area of the region enclosed by the graph of f, the vertical lines $x = 0$ and $x = 2$ and the x-axis.

35. We start from the graph G_f of the function f defined by $f(x) = \frac{1}{100}x^3$ on the interval $[0; 10]$.

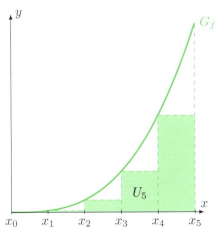 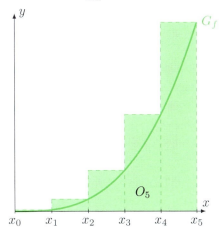

a) Calculation of lower sums

   i)   Partition the interval $[0; 10]$ into 5 parts and calculate the area of the shaded steps U_5 by hand, as shown in the figure on the left.

   ii) Partition the interval $[0; 10]$ into 10 parts and calculate the lower sum U_{10} with a computer.

   iii) Partition the interval $[0; 10]$ into n parts and calculate the lower sum U_n using a sum formula from page 126.

b) Calculation of upper sums

   i)   Partition the interval $[0; 10]$ into 5 parts and calculate the upper sum O_5 by hand, as shown in the figure on the right.

   ii) Partition the interval $[0; 10]$ into 10 parts and calculate the upper sum O_{10} with a computer.

   iii) Partition the interval $[0; 10]$ into n parts and calculate the upper sum O_n using a sum formula from page 126.

c) Joint consideration of the lower and upper sums

   i)   Explain why $U_n \geq U_{n-1}$, $O_n \leq O_{n-1}$ and $U_n \leq A \leq O_n$ always hold, where A is the area of the region enclosed by the curve, the vertical lines $x = 0$ and $x = 10$ and the x-axis.

   ii) As is known, taking the limit yields $\lim\limits_{n\to\infty} U_n \leq A \leq \lim\limits_{n\to\infty} O_n$. Calculate both limits. What do you notice? What does this imply for A?

Remark: This so-called *method of exhaustion*, where the area sought is exhausted from within or enclosed from outside, was used by the Greek scholar ARCHIMEDES (287 BC – 212 BC) to calculate the areas of a circle and a parabolic segment.

36. The graph of the function f defined by the equation $f(x) = \frac{1}{2}x^2 + 1$ is given on the interval $[0; b]$.

   a) Determine the upper sum O_n of f for n equal subintervals.

   b) Calculate the area $A = \lim_{n\to\infty} O_n$.

37. Draw the curve given by the equation $y = f(x)$ and consider the area A enclosed by this curve, the x-axis and the lines $x = a$ and $x = b$ parallel to the y-axis. Cut the area into n strips of width $\frac{b-a}{n}$. Replace these strips by inscribed and circumscribed rectangles ($n = 3, 5, 10, 20, 100$) and calculate the lower and upper bounds for the area A.

   a) $f(x) = \frac{1}{x}$; $a = 1, b = 2$ b) $f(x) = e^{-x}$; $a = 0, b = 1$

   c) $f(x) = \sin(x)$; $a = 0, b = \frac{\pi}{2}$ d) $f(x) = \ln(x)$; $a = 2, b = 3$

38. To calculate the area of a circle of radius r, the circular disc is partitioned into concentric rings of width $d = \frac{r}{n}$. The areas of the rings can be approximated by the formula $F_i = 2\pi r_i d$. If the r_i are the radii of the inner circles of the rings, the sum of the approximated areas of the rings is a lower sum for the circle. If the r_i are the radii of the outer circles of the rings, the summation yields an upper sum.

   a) Calculate the upper and lower sums for arbitrary $n \in \mathbb{N}$.

   b) Show that the upper and lower sums converge to the same limit A as n tends to infinity and from this derive the formula for the area A of the circle.

39. The figure shows the top view of a kidney-shaped swimming pool, whose width is measured every 2 metres. The numerical data are given in metres.

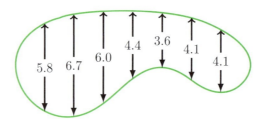

Estimate the area A of the pool using lower and upper sums.

Instructions: For the lower sum, use rectangles with a base of 2 metres and the smaller of the two heights in each strip, for the upper sum, rectangles with the larger of the two heights.

The Definite Integral

It can be proven that the sequence of upper sums (O_n) and the sequence of lower sums (U_n) converge to the same limit for every continuous function f on the interval $[a; b]$. That is
$$\lim_{n \to \infty} U_n = \lim_{n \to \infty} O_n.$$

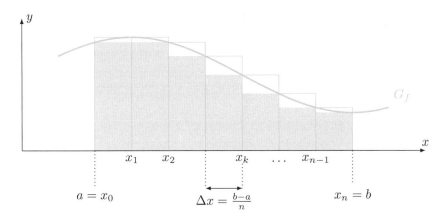

The lower sum consists of the sum of the shaded rectangles.
The upper sum consists of the sum of the outlined rectangles.

If f is a continuous function, the *definite integral* of f over the interval $[a; b]$ is defined as

$$\int_a^b f(x)\, dx = \lim_{n \to \infty} U_n = \lim_{n \to \infty} O_n = \lim_{n \to \infty} \sum_{k=1}^n f(\overline{x}_k) \cdot \Delta x, \quad \text{with } \overline{x}_k \in [x_{k-1}; x_k].$$

Remark: The final term indicates that, instead of the minimum or maximum of f in $[x_{k-1}; x_k]$, one may also use the value of the function at any arbitrary point $\overline{x}_k \in [x_{k-1}; x_k]$.

The integral symbol $\int$ is derived from the letter « $\int$ » (long s) as an abbreviation for the Latin word ſumma.

a) If $f(x) > 0$ on $[a; b]$,   b) If $f(x) < 0$ on $[a; b]$,   c) If f has a single root in $]a; b[$,

    then $\int\limits_a^b f(x)\, dx = A > 0$.   then $\int\limits_a^b f(x)\, dx = -A < 0$.   then $\int\limits_a^b f(x)\, dx = -A_1 + A_2$.

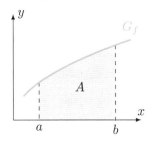 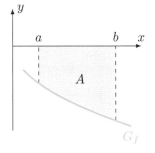 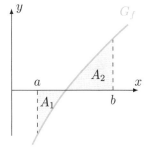

Note: A, A_1 and A_2 denote areas, i.e., A, A_1 and A_2 are positive.

Integration Rules

Exchanging the limits of integration: $\displaystyle\int_b^a f(x)\,\mathrm{d}x = -\int_a^b f(x)\,\mathrm{d}x$

Partitioning the limits of integration: $\displaystyle\int_a^b f(x)\,\mathrm{d}x = \int_a^c f(x)\,\mathrm{d}x + \int_c^b f(x)\,\mathrm{d}x$

It does not matter whether $c \in [a; b]$ or $c \notin [a; b]$ holds.

---

40. Calculate by means of upper and lower sums.

a) $\displaystyle\int_0^2 3x^2\,\mathrm{d}x$
b) $\displaystyle\int_2^4 (x^2 + 3)\,\mathrm{d}x$

41. *Calculation of area.* The graph G_f of the function f given by the equation $f(x) = -x^2 + 2x + 3$ and the x-axis enclose a finite region A.

a) Determine the roots of f and sketch the graph G_f in the range between the roots.

b) In which interval of $\mathbb{R}$ is f monotonically decreasing?

c) Determine an approximation of the area of A by partitioning A into 10 strips and calculating the lower sum U_{10}.

d) Determine a formula for U_n.

e) Calculate the exact value of the area of the region.

f) Write the area as a definite integral.

42. For the functions f and g, we have: $\displaystyle\int_3^8 f(x)\,\mathrm{d}x = 6$ and $\displaystyle\int_3^8 g(x)\,\mathrm{d}x = 11$. Calculate the three integrals.

i) $\displaystyle\int_3^8 (f(x) + g(x))\,\mathrm{d}x$
ii) $\displaystyle\int_3^8 (3f(x) - 2g(x))\,\mathrm{d}x$
iii) $\displaystyle\int_8^3 (g(x) - f(x))\,\mathrm{d}x$

43. For the function f, we have: $\displaystyle\int_{-1}^1 f(x)\,\mathrm{d}x = -2$ and $\displaystyle\int_1^4 f(x)\,\mathrm{d}x = 5$. Calculate the four integrals.

i) $\displaystyle\int_4^1 f(x)\,\mathrm{d}x$
ii) $\displaystyle\int_{-1}^4 f(x)\,\mathrm{d}x$
iii) $\displaystyle\int_1^1 f(x)\,\mathrm{d}x$
iv) $\displaystyle\int_{-1}^1 4f(x)\,\mathrm{d}x$

44. For the function g and for $a < b < c$, we have: $\displaystyle\int_a^b g(x)\,\mathrm{d}x = p$ and $\displaystyle\int_b^c g(x)\,\mathrm{d}x = q$. Calculate the four integrals.

i) $\displaystyle\int_b^a g(x)\,\mathrm{d}x$
ii) $\displaystyle\int_a^c g(x)\,\mathrm{d}x$
iii) $\displaystyle\int_b^b g(x)\,\mathrm{d}x$
iv) $\displaystyle\int_b^c -2g(x)\,\mathrm{d}x$

45. The graph G_f of the function f is given. Determine

a) $\displaystyle\int_1^7 f(x)\,\mathrm{d}x$.
b) $\displaystyle\int_0^1 f(x)\,\mathrm{d}x$.
c) $\displaystyle\int_0^7 f(x)\,\mathrm{d}x$.

d) the area of the region enclosed by the graph G_f and the x-axis on the interval $[0; 7]$.

For exercise 45:   *For exercise 46:*

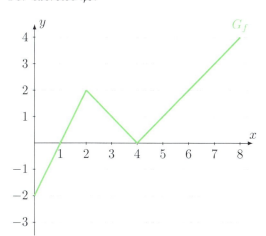   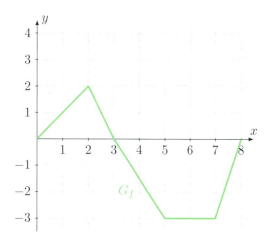

46. The graph G_f of the function f is given. Determine the integral.

a) $\displaystyle\int_0^3 f(x)\,\mathrm{d}x$ b) $\displaystyle\int_3^8 f(x)\,\mathrm{d}x$ c) $\displaystyle\int_0^8 f(x)\,\mathrm{d}x$

Integral Function

Let the function f be continuous. Then, the *integral function* F_a with fixed lower limit a and variable upper limit x is defined by

$$F_a(x) = \int_a^x f(t)\,\mathrm{d}t.$$

---

47. Let the function f be given by $f(t) = t - 1$.

a) Determine the upper sum O_n of the function f over the interval $[0; x]$.

b) Use this result to determine the equation of $F_0(x)$ for the integral function of f.

c) Solve parts a) and b) for the function f with $f(t) = t^2$.

48. Determine the equations of $F_0(x)$ $(x \geq 0)$ for the integral functions of the following functions f.

Hint: All of these functions are linear. For this reason, the integral function can be determined by the application of geometrical considerations to the graph of the function.

a) $f(t) = 1$ b) $f(t) = 2$ c) $f(t) = c;\ c \in \mathbb{R}_0^+$

d) $f(t) = t$ e) $f(t) = 2t$ f) $f(t) = mt;\ m \in \mathbb{R}_0^+$

g) $f(t) = t + 1$ h) $f(t) = 2t + 1$ i) $f(t) = mt + c;\ m, c \in \mathbb{R}_0^+$

49. Determine the equations of $F_0(x)$ for the integral function of f for $x \geq 0$ by applying the upper sum O_n. Use the sum formulas from page 126.

a) $f(t) = t^2$; see exercise 47 c)

b) $f(t) = t^3$ c) $f(t) = t^4$ d) $f(t) = t^5$ e) $f(t) = \frac{1}{2}t^2 + 1$

50. Compare each f with its F_0 in both of the previous exercises 48 and 49. What do you notice?

4.3   Fundamental Theorem of Calculus

Fundamental Theorem of Calculus

Let f be a continuous function on the interval $[a; b]$. Then the following assertions hold:

1) The function F_a defined by $F_a(x) = \int_a^x f(t)\,dt$, $a \le x \le b$ is differentiable on $[a; b]$ and fulfils $F_a'(x) = f(x)$.

2) If F is an antiderivative of f, that is a function with $F' = f$, we have

$$\int_a^b f(x)\,dx = F(b) - F(a) = [F(x)]_a^b.$$

51. Calculate the definite integral using the fundamental theorem and compare the results. Explain your observations by means of sketches.

a) Compare $\int_0^2 x^2\,dx$ with   i) $\int_0^2 (x^2+1)\,dx$;   ii) $\int_0^2 (x+1)^2\,dx$;   iii) $\int_{-1}^1 (x+1)^2\,dx$.

b) Compare $\int_0^4 \sqrt{x}\,dx$ with   i) $\int_0^4 \sqrt{x+1}\,dx$;   ii) $\int_{-1}^3 \sqrt{x+1}\,dx$;   iii) $\int_1^5 \sqrt{x-1}\,dx$.

52. Calculate the definite integral using the fundamental theorem and determine the cases where the value of the integral coincides with the area A of the region enclosed by the function and the x-axis.

a) $\int_{-2}^3 (-(x-3)(x+2))\,dx$ b) $\int_{-3}^3 (3-x)^2\,dx$ c) $\int_0^3 (3-x)^2\,dx$

d) $\int_{-2}^2 x(x-2)(x+2)\,dx$ e) $\int_{-1}^2 (6+3x-3x^2)\,dx$ f) $\int_{-1}^2 (6x+3x^2-3x^3)\,dx$

For **53–59**: Calculate the definite integral using the fundamental theorem. *Remark:* For the most common antiderivatives, see page 120 or consult a mathematical formulary.

53. Polynomial functions

a) $\int_0^2 4x^3\,dx$ b) $\int_2^4 \frac{1}{5}x^3\,dx$ c) $\int_{-1}^2 (x^2+3)\,dx$

d) $\int_3^5 \left(\frac{1}{2}x - 2\right)\,dx$ e) $\int_0^3 2x(1-x)\,dx$ f) $\int_{-3}^3 (x-1)^2\,dx$

54. Rational functions

a) $\int_2^4 \frac{1}{x^2}\,dx$ b) $\int_{-1}^{-0.5} \frac{3}{x^5}\,dx$ c) $\int_1^2 \left(2x - \frac{2}{x}\right)\,dx$

d) $\int_1^3 \left(4x - \frac{3}{x} + \frac{2}{x^2}\right)\,dx$ e) $\int_3^5 \frac{x-2}{x}\,dx$ f) $\int_2^3 \frac{x^2+4x+3}{x}\,dx$

55. Root functions

a) $\displaystyle\int_1^4 3\sqrt{x}\,\mathrm{d}x$

b) $\displaystyle\int_1^4 \frac{1}{\sqrt{x}}\,\mathrm{d}x$

c) $\displaystyle\int_0^8 \sqrt[3]{x}\,\mathrm{d}x$

d) $\displaystyle\int_0^1 x\sqrt{x}\,\mathrm{d}x$

e) $\displaystyle\int_1^4 \left(2x + 6\sqrt{x}\right)\,\mathrm{d}x$

f) $\displaystyle\int_1^4 \sqrt{\frac{3}{x^3}}\,\mathrm{d}x$

56. Trigonometric functions

a) $\displaystyle\int_0^{\frac{\pi}{2}} \sin(t)\,\mathrm{d}t$

b) $\displaystyle\int_{\pi}^{2\pi} \cos(t)\,\mathrm{d}t$

c) $\displaystyle\int_0^{\frac{\pi}{3}} -2\sin(t)\,\mathrm{d}t$

d) $\displaystyle\int_0^{\pi} \left(4\sin(t) - 3\cos(t)\right)\,\mathrm{d}t$

e) $\displaystyle\int_0^{\frac{\pi}{4}} \frac{1}{\cos^2(t)}\,\mathrm{d}t$

f) $\displaystyle\int_{\frac{\pi}{4}}^{\frac{\pi}{3}} \tan(t)\,\mathrm{d}t$

57. Exponential functions

a) $\displaystyle\int_{-1}^1 \mathrm{e}^{x+1}\,\mathrm{d}x$

b) $\displaystyle\int_{-1}^0 \left(2x - \mathrm{e}^x\right)\,\mathrm{d}x$

c) $\displaystyle\int_0^1 \pi \cdot \mathrm{e}^x\,\mathrm{d}x$

d) $\displaystyle\int_0^2 \left(\mathrm{e}^x - x^{\mathrm{e}}\right)\,\mathrm{d}x$

e) $\displaystyle\int_1^2 \left(\mathrm{e}^x + \frac{1}{x}\right)\,\mathrm{d}x$

f) $\displaystyle\int_0^1 10^x\,\mathrm{d}x$

58. Integration by simple substitution

a) $\displaystyle\int_0^5 (2x - 3)^4\,\mathrm{d}x$

b) $\displaystyle\int_{-1}^6 (5 - 2x)^3\,\mathrm{d}x$

c) $\displaystyle\int_2^3 \frac{4}{(1-x)^2}\,\mathrm{d}x$

d) $\displaystyle\int_0^{\frac{\pi}{2}} \sin(3t)\,\mathrm{d}t$

e) $\displaystyle\int_{\frac{-1}{4}}^{\frac{3}{4}} \cos(\pi t)\,\mathrm{d}t$

f) $\displaystyle\int_0^{\pi} \left(\frac{1}{\pi} + \cos(-2t)\right)\,\mathrm{d}t$

g) $\displaystyle\int_0^5 \frac{2}{3}\mathrm{e}^{4x}\,\mathrm{d}x$

h) $\displaystyle\int_{-6}^6 \mathrm{e}^{6-x}\,\mathrm{d}x$

i) $\displaystyle\int_{-\pi}^{\pi} \left(\mathrm{e}^x - \mathrm{e}^{-x}\right)\,\mathrm{d}x$

j) $\displaystyle\int_{12}^{24} \frac{\mathrm{d}x}{\sqrt{2x+1}}$

k) $\displaystyle\int_2^5 \frac{9}{3x - 5}\,\mathrm{d}x$

l) $\displaystyle\int_{-3}^0 \ln(2) \cdot 2^{\frac{-1}{3}x}\,\mathrm{d}x$

59. Absolute value function

a) $\displaystyle\int_{-1}^5 |x|\,\mathrm{d}x$

b) $\displaystyle\int_0^2 |x^2 - 1|\,\mathrm{d}x$

60. Calculate.

a) $\displaystyle\int_{-1}^2 (t^2 x^2 - x)\,\mathrm{d}x$

b) $\displaystyle\int_{-1}^2 (t^2 x^2 - x)\,\mathrm{d}t$

61. For which value of the parameter k is the integral equation true?

a) $\displaystyle\int_0^k x^2\,\mathrm{d}x = 9$

b) $\displaystyle\int_2^4 (6x^2 + 3x + k)\,\mathrm{d}x = 140$

62. For the function f, we have: $\int_1^9 f(t)\,\mathrm{d}t = -1$ and $\int_7^9 f(t)\,\mathrm{d}t = 5$. All calculations should be done mentally.

a) $\displaystyle\int_1^9 -2 \cdot f(t)\,\mathrm{d}t$

b) $\displaystyle\int_1^7 f(t)\,\mathrm{d}t$

c) $\displaystyle\int_7^9 f(t)\,\mathrm{d}t - \int_9^7 f(t)\,\mathrm{d}t$

63. Determine the extremum points of F without integrating.

a) $F(x) = \displaystyle\int_1^x t(t-2)\,\mathrm{d}t$

b) $F(x) = \displaystyle\int_\pi^x (t-3) \cdot \sin(3t)\,\mathrm{d}t$

64. Is the integral equation solvable? If so, determine the equation of the function f.

a) $\displaystyle\int_{-1}^x f(t)\,\mathrm{d}t = x^3 - 2x^2 + x + 4$

b) $\displaystyle\int_0^x f(t)\,\mathrm{d}t = x^3 - 2x^2 + x - 3$

c) $\displaystyle\int_2^x f(t)\,\mathrm{d}t = x^3 - 2x^2 + x + c$

65. Let $f(x) = \sin(2x)$ be given.

a) Determine an antiderivative F of f that only attains positive function values.

b) Determine an antiderivative F of f, whose graph is tangent to the x-axis at infinitely many points.

66. The figure shows the velocity profile of two cars A and B, that are standing next to each other at the start and start from rest.

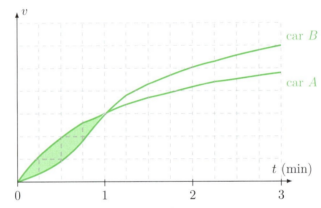

a) Which car is leading after 1 minute? Justify.

b) How can we interpret the shaded area?

c) Which car is leading after 3 minutes? Justify.

d) At what time is car A overtaken by car B? Estimate.

67. True or false? Justify or disprove by giving a counterexample.

If f and g are continuous on $[a; b]$, we have:

a) $\displaystyle\int_a^b (f(x) + g(x))\,\mathrm{d}x = \int_a^b f(x)\,\mathrm{d}x + \int_a^b g(x)\,\mathrm{d}x.$

b) $\displaystyle\int_a^b x \cdot f(x)\,\mathrm{d}x = x \cdot \int_a^b f(x)\,\mathrm{d}x.$

c) $\displaystyle\int_a^b f(x) \cdot g(x)\,\mathrm{d}x = \int_a^b f(x)\,\mathrm{d}x \cdot \int_a^b g(x)\,\mathrm{d}x.$

68. Let $\int_0^2 f(x)\,\mathrm{d}x = 4$, $\int_0^1 f(x)\,\mathrm{d}x = 1$, $\int_1^2 g(x)\,\mathrm{d}x = -1$, $\int_0^1 g(x)\,\mathrm{d}x = 3$ be given. Calculate.

a) $\displaystyle\int_0^1 (2f(x) + g(x))\,\mathrm{d}x$

b) $\displaystyle\int_0^2 g(x)\,\mathrm{d}x$

c) $\displaystyle\int_1^2 (-5f(x))\,\mathrm{d}x$

4.4   Different Interpretations of the Integral

Area

Area Between a Curve and the x-Axis

If f is a continuous function on the interval $[a; b]$, then the area of the region between its graph G_f and the x-axis in the interval $[a; b]$ is given by the integral

$$A = \left| \int_a^{x_S} f(x)\,dx \right| + \left| \int_{x_S}^b f(x)\,dx \right|,$$

where x_S is the only root of f in $]a; b[$.

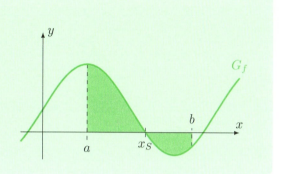

69. Verify that the graph G_f of the function f is not below the x-axis at any point for $a \leq x \leq b$. Then, calculate the area A of the region enclosed by the graph G_f, the x-axis and the two lines $x = a$ and $x = b$.

a) $f(x) = x^2 + 2$; $a = -3$, $b = 3$
b) $f(x) = 31 - x^2$; $a = -1$, $b = 5$
c) $f(x) = x^3 - 6x^2 - 4x + 51$; $a = -2$, $b = 0$
d) $f(x) = 1 - \frac{3}{x^2}$; $a = 2$, $b = 6$
e) $f(x) = 12 - \frac{3}{x}$; $a = 4$, $b = 5$
f) $f(x) = \sqrt{2x - 4}$; $a = 3$, $b = 6$

70. Show that the function f, given by its equation, has precisely one root in the interior of the interval $[a; b]$. What is the area A of the region G enclosed by the graph of f, the x-axis and the two lines $x = a$ and $x = b$?

a) $f(x) = x^2 - 6x + 5$; $a = 2$, $b = 8$
b) $f(x) = \frac{1}{2}x^3 - 2x^2 - \frac{7}{2}x + 5$; $a = -1$, $b = 3$
c) $f(x) = 2e^x - 1$; $a = -2$, $b = 1$
d) $f(x) = \cos(3x)$; $a = 0$, $b = \frac{\pi}{2}$

71. Determine the area of the region enclosed by the graph G_f and the x-axis.

a) $f(x) = -2x^2 + 18$
b) $f(x) = x(x^2 - 1)$
c) $f(x) = (x - 1)^2 - 1$
d) $f(x) = x^4 - 4x^2$
e) $f(x) = x^4 - 10x^2 + 9$
f) $f(x) = x + x^{-1} - 4$

72. In quadrant I the parabola p: $y = ax - x^2$ and the x-axis enclose a region with area $A = 36$. Determine a and sketch the parabola.

73. Determine k, such that the regions enclosed by the graph of f in the interval $[0; k]$ above and below the x-axis have equal areas. Make a sketch.

a) $f(x) = -0.5x + 2$
b) $f(x) = (x - 1)^2 - 4$
c) $f(x) = x^3 - 1$
d) $f(x) = x^3 + \frac{3}{2}x - 2$

74. Determine the total area of all finite regions that the curve with equation $y = f(x)$ encloses with the x-axis.

a) $f(x) = x^2 - 5x + 4$
b) $f(x) = 4x^3 - 12x^2 + 8x$
c) $f(x) = x^3 - 2x^2 - x + 2$

d) $f(x) = x^2 - 2$
e) $f(x) = \begin{cases} 4x^3 + 4, & x \leq 1 \\ 10 - 2x, & x > 1 \end{cases}$

75. Determine a such that the line $g\colon x = a$, the graph G_f of the function f with $f(x) = \sqrt{x}$ and the x-axis enclose a region of area $A = 18$.

76. a) For $t > 0$, we are given the function $f_t(x) = \frac{t}{x^2}$. The graph of f_t encloses a region of area $A(t)$ with the x-axis in the interval $[1; 2]$. Determine $A(t)$ as a function of t and calculate t such that this area is 8.

b) The function $h_t(x) = x^2 - t^2$ is given for $t > 0$. The graph of h_t encloses a region of area $A(t)$ with the x-axis. Determine $A(t)$ as a function of t and calculate t such that this is 36.

77. Determine the value of the parameter a such that in quadrant I, the curve k given by the equation $y = -2x^3 + ax$ and the x-axis enclose a region of area 9.

78. A parabola with the equation $y = ax^2 + bx + c$ passes through the origin $O(0\,|\,0)$ and the point $P(3\,|\,0)$. The segment $\overline{OP}$ is a side of a square lying in quadrant I. How must the coefficients a, b and c be chosen, such that the arc of the parabola over $\overline{OP}$ divides the square in half?

79. The graph of a polynomial function of degree 3 has slope 5 at the point $P(1\,|\,3)$ and is tangent to the x-axis at the point $(0\,|\,0)$.

a) Determine the formula for the polynomial function and sketch its graph.

b) Calculate the area of the region contained by the graph and the x-axis.

80. The graph of a polynomial function of degree 4 is symmetric with respect to the y-axis, is tangent to the x-axis at the origin and has a root at $x = 3$. The graph and the x-axis enclose a region with area 16.2 in quadrant I. Determine the formula of the polynomial function and sketch its graph.

Area Between Two Curves

If f and g are two continuous functions on the interval $[a; b]$, then the area of the region enclosed by their graphs in the interval $[a; b]$ is given by the integral

$$A = \int_a^{x_S} (f(x) - g(x))\,\mathrm{d}x + \int_{x_S}^b (g(x) - f(x))\,\mathrm{d}x.$$

Here, x_S is the only intersection of f and g in $]a; b[$.

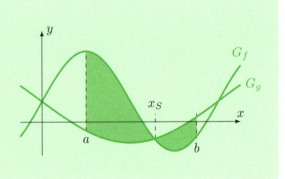

81. Calculate the area A of the finite region enclosed by the curves k_1 and k_2.

a) $k_1\colon y = x^2$
   $k_2\colon y = 8 - x^2$

b) $k_1\colon y = \frac{1}{2}x^2$
   $k_2\colon y = 16 - \frac{1}{2}x^2$

c) $k_1\colon y = x^3 - 5x^2 + 6x$
   $k_2\colon y = x^3 - 7x^2 + 12x$

d) $k_1\colon y = \frac{1}{4}x^3$
   $k_2\colon y = \sqrt{2x}$

82. Two functions are given by their equations. Their graphs enclose a finite region. What is the area A of this region?

a) $f(x) = \frac{1}{2}x^2$ and $g(x) = 4 - x$ b) $f(x) = -x + 10$ and $g(x) = -\frac{1}{2}x^2 + 4x + 2$

83. Calculate the area A of the region between the graphs of the functions f and g in the interval I. Sketch the curves including the enclosed region.

a) $f(x) = 0.1x^2 + 2$, $g(x) = x - 1$, $I = [-1; 2]$ b) $f(x) = (-x)^{-2}$, $g(x) = x^2$, $I = [-4; -2]$

84. Calculate the area A of the region enclosed by the graphs of the functions f and g. Sketch the curves including the enclosed region.

a) $f(x) = x^2$, $g(x) = -x^2 + 4x$ b) $f(x) = (x - 2)^2$, $g(x) = \sqrt{x - 2}$

85. The functions with the equations $f_a(x) = \frac{1}{a}x^2 - a$ and $h_a(x) = \frac{1}{3}x^2 - \frac{1}{3}a^2$ are given for $a > 0$, see the figure for $a = 2$.

a) Determine the roots of f_a in terms of a.

b) Show that the graphs G_{f_a} and G_{h_a} are symmetrical with respect to the y-axis.

c) Show that the graphs of f_a and h_a intersect at points on the x-axis.

d) The graphs G_{f_a} and G_{h_a} enclose a finite region. For which $a \in]0; 3[$ is its area $A(a)$ an extremum? What type of extremum is it?

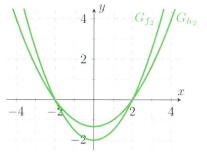

86. What is the area A of the shaded region?

a) $f_1: y = \frac{1}{x}$, $f_2: y = 1$ b) $f_1: y = \frac{1}{x}$, $f_2: y = \frac{3}{x}$ c) $f_1: y = e^x$, $f_2: y = \ln(x)$

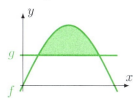

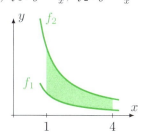

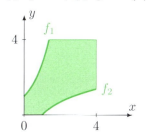

87. Calculate the area A of the shaded region.

a) $f: y = \sin(x)$ b) $f: y = \cos(x)$ c) $f: y = \sin(x)$
   $g: y = \frac{1}{2}$ g: line PQ $g: y = 1 - \sin(x)$

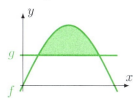

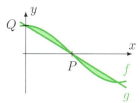

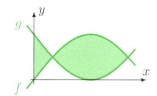

88. Calculate the area A of the shaded region.

a) $f: y = \cos(x)$
   $g: y = \sin(x)$
   $h: x = \frac{\pi}{2}$

b) $f: y = \sin(\pi x)$
   $g: y = x^2 - x$

c) $f: y = \sin^2\left(\frac{\pi}{2}x\right)$
   $g: y = \cos^2\left(\frac{\pi}{2}x\right)$

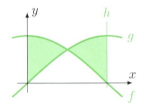   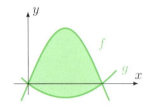   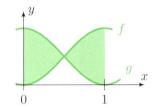

89. First determine the values of the required parameters and then calculate the area of the shaded region.

a) $k_1: y = \cos(x)$
   $k_2: y = x^2 + a$

b) $k_1: y = e^{|x|}$
   $k_2: y = ax^2$

c) $k_1: y = \pm x^2$
   $k_2: x = \pm y^2$

d) $k_1: y = e^x$
   $k_2: y = ax + b$

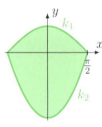   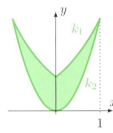   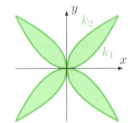   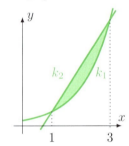

90. Consider the parabola $p: y = x^2$. Determine the equation of the tangent t_a to the graph of p at $x = a$. Likewise, let t_b be the tangent to the graph of p at $x = b$. Calculate the area A of the region enclosed by the graph of p and the two tangents t_a and t_b for $a < b$.

91. Calculate the area A of the region enclosed by the y-axis, the graph of $y = \ln(x)$ and the two horizontal lines $y = 0$ and $y = 2$.

92. The tangent to the curve $k: y = x^3 - 4x^2 + 6x$ at $P(1 \,|\, y_P)$ intersects the curve at another point. What is the area A of the region between the curve and the tangent?

93. Calculate the area of the region enclosed by the curve $f: y = \sin(x)$ and its tangents at its two roots in the interval $[0; \pi]$.

94. What is the area of the region enclosed by the curve $k: y = x^3 - 2x$ and its normal at its inflection point?

95. What is the area of the region enclosed by the x-axis, the curve $f: y = \sqrt{x}$ and its normal at the point $P(1 \,|\, y_P)$ of the curve?

96. A section with area $\frac{5}{3}$ is cut off of the region that is enclosed by the curves $f: y = \frac{1}{x}$ and $g: y = \sqrt{x}$ as well as the x-axis by a vertical cut. Where is this vertical cut?

97. a) Determine $a > 0$, such that the region between the graphs of the functions f and g with $f(x) = ax^2$ and $g(x) = x$ has the area 24.

b) Determine the slope m of a line g through the origin such that the region enclosed by g and the curve $f \colon y = \sqrt{x}$ has the area 4.5.

98. a) The parabola f and the secant s through two points P and Q of the parabola f enclose a shaded region; the so-called *parabolic segment*. The difference of the x-coordinates of the points P and Q is called the *width* of the parabolic segment. Prove that all segments of width 2 of the parabola $f(x) = -x^2 + 3$ have the same area.

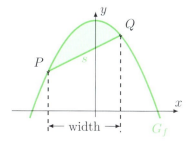

b) Do the same question for an arbitrary width b.

99. A curve k inside the unit square $Q = \{(x \mid y) \mid 0 \le x \le 1,\ 0 \le y \le 1\}$ is given by its equation. Determine the value of the parameter p, q or r, respectively, such that the curve divides the area of the square Q in half.

a) $k \colon y = p \cdot e^x,\ p \in\ ]0; 1[$ b) $k \colon y = \ln(\frac{x}{q}),\ q \in\ ]0; 1[$ c) $k \colon y = r \cdot e^{rx},\ r \in\ ]0; 1[$

d) What do you notice when comparing parts a) and b)? Explain and comment.

100. The parabola defined by the equation $y = x^2$ and the x-axis enclose a region on the interval $[0; 2]$. Which parallel line to the y-axis divides its area in half?

101. The region G enclosed by the parabola $p \colon y = -4x(x-3)$ and the x-axis is subdivided into two regions of equal area by a line G, which passes through the origin. Determine the slope of g.

102. In what ratio does the graph of $f \colon y = x^3$ divide the finite region below the graph of $g \colon y = 8x - x^3$ in quadrant I?

103. Let the curve $k \colon y = 3x^2 - x^3$ and the horizontal line $g \colon y = 5$ be given. If one draws a vertical line through the right root of k, this line forms a rectangle together with the line g and the coordinate axes. What percentage of the area of this rectangle lies below the curve k?

104. The parabola $p_1 \colon y = -x^2 + 6x$ forms a finite region with the x-axis, a parabolic segment. A second parabola $p_2 \colon y = 0.5x^2 + 4.5$ cuts off a small region of this segment. Does the area of the cut off region make up more or less than 5% of the area of the whole segment?

105. For which a is the area between the two parabolas $p_1 \colon y = ax^2$ and $p_2 \colon y = 1 - \frac{x^2}{a}$ largest?

106. Let A_D denote the area of the lightly shaded triangle and A_S the area of the two darkly shaded segments for every positive number a. Show that the ratio of the two areas, $v = A_D : A_S$, is equal for all values of a.

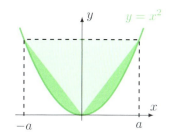

Volume

107. To determine the volume of a solid we cut it into slices (layers) using parallel planes and calculate
the volume of the corresponding stepped solid. As a first example, we consider a right square
pyramid with

- a: length of the sides of the base
- h: height of the pyramid
- x: distance of slice to apex of pyramid
- Δx: thickness of slice
- $Q(x)$: cross-sectional area of slice

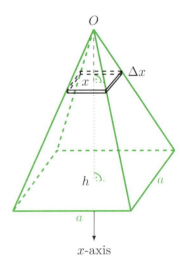

Express the following quantities in terms of a, h, x and Δx.

a) length s of the side of the cross-section

b) area $Q(x) = s^2$ of the cross-section

c) volume $\Delta V = Q(x) \cdot \Delta x$ of a rectangular cuboidal slice

d) volume V_T of the stepped solid. Place a suitable cross-section of the pyramid into the coordinate system for this.

e) volume V_P of the pyramid as the limit $\lim\limits_{\Delta x \to 0} V_T$

Solid of Revolution: Disc Method

curve segment to be rotated	rotation around the x-axis	rotation around the y-axis

If a curve segment is rotated around one of the two coordinate axes, the volume of the solid of
revolution is calculated as follows:

For rotation around the x-axis: $V = \displaystyle\int_a^b \pi \left(f(x) \right)^2 \, \mathrm{d}x = \pi \int_a^b y^2 \, \mathrm{d}x.$

For rotation around the y-axis: $V = \displaystyle\int_c^d \pi \left(f^{-1}(y) \right)^2 \, \mathrm{d}y = \pi \int_c^d x^2 \, \mathrm{d}y$, if f is strictly monotone.

Toroidal Solids: Washer Method

If two graphs G_f and G_g on the interval $[a;b]$ are rotated around the x-axis, this creates a toroidal solid, whose volume is calculated as follows:

$$V = \pi \int_a^b \left((\text{outer function})^2 - (\text{inner function})^2 \right) dx.$$

If e.g. $f(x) \geq g(x)$ holds in the interval $[a;b]$, we have

$$V = \pi \int_a^b \left((f(x))^2 - (g(x))^2 \right) dx.$$

Analogous formulas hold for the rotation around the y-axis.

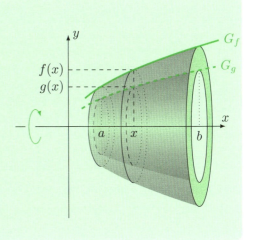

108. The region between the x-axis and the curve defined by the equation $y = f(x)$, $a \leq x \leq b$ is rotated around the x-axis. What is the volume of the solid of rotation?

   a) $f(x) = x + 1$, $a = 1$, $b = 3$ b) $f(x) = \frac{1}{4}(x^2 - 4)$, $a = -2$, $b = 2$

109. Sketch the graph G_f of the function f. The graph and the x-axis enclose a finite region, which is rotated around the x-axis. Calculate the volume of the solid of rotation.

   a) $f(x) = -x^2 + 4$ b) $f(x) = 3x - \frac{1}{2}x^2$ c) $f(x) = x^2(x + 2)$
   d) $f(x) = (x^2 - 1)^2$ e) $f(x) = x\sqrt{4 - x}$ f) $f(x) = \sqrt{1 - x^2}$

110. The region between the two curves f and g is rotated around the x-axis. Calculate the volume of the solid of rotation.

   a) $f: y = 3x^2 - x^3$, $g: y = x^2$ b) $f: y = 3x^2 - x^3$, $g: y = 2x$

111. Calculate the volume of a wooden golf tee (golf tee = peg, which is poked into the ground to hold the golf ball for the first shot). It has the approximate measurements given in the figure and has rotational symmetry. The following functions give the outer and inner contour lines (measurements in cm):

$$f(x) = \begin{cases} \frac{1}{2}x & \text{for } 0 \leq x \leq \frac{1}{2} \\ \frac{1}{4} & \text{for } \frac{1}{2} \leq x \leq \frac{7}{2} \\ \frac{1}{4}\left(1 + (x - \frac{7}{2})^2\right) & \text{for } \frac{7}{2} \leq x \leq \frac{9}{2} \\ \frac{1}{2} & \text{for } \frac{9}{2} \leq x \leq 5 \end{cases} \quad \text{and} \quad g(x) = \begin{cases} 0 & \text{for } 0 \leq x \leq \frac{9}{2} \\ x - \frac{9}{2} & \text{for } \frac{9}{2} \leq x \leq 5 \end{cases}$$

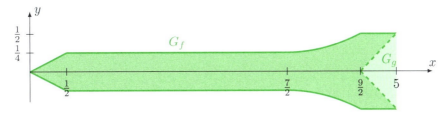

112. The graph of $p\colon y = \frac{1}{4}x^3$, its tangent at $P(2\,|\,y_P)$ and the x-axis enclose a region in quadrant I, which is rotated around the x-axis. What is the volume of this solid of rotation?

113. The region enclosed by the parabola $p\colon y = \frac{1}{6}x^2$ and its normal at $P(3\,|\,y_P)$ is rotated around the x-axis. What is the volume of this solid of rotation?

114. The region below the curve defined by the equation $f(x) = \cos(2x)$, $0 \le x \le \frac{\pi}{4}$ is rotated around the x-axis. What is the volume of the solid of rotation?

115. The region between the graphs of the two functions f and g over the interval $[a;b]$ is rotated around the x-axis. Calculate the volume of the solid of rotation.

a) $f(x) = \frac{1}{\cos(x)}$, $g(x) = 1$, $[-1;1]$

b) $f(x) = \cos(x) + \sin(x)$, $g(x) = \cos(x) - \sin(x)$, $[0;\frac{\pi}{4}]$

116. Determine the volume of the solid which is formed by rotating the depicted region around the x-axis.

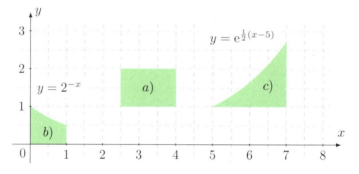

117. The region between the y-axis and the curve defined by the equation $y = f(x)$, $c \le y \le d$ is rotated around the y-axis. What is the volume of the solid of rotation?

a) $f(x) = \frac{1}{2}x^2 + 1$; $c = \frac{3}{2}$, $d = \frac{11}{2}$

b) $f(x) = h(1 - \frac{x}{r})$, $h, r > 0$; $c = 0$, $d = h$

118. The two curves f and g enclose a region in quadrant I which is rotated around the y-axis. Calculate the volume of the solid of rotation.

a) $f\colon y = x^2 + 3$, $g\colon y = -2x^2 + 6$

b) $f\colon y = x$, $g\colon y = x^3$

119. The region between the two curves is rotated around the y-axis. Calculate the volume of the solid of rotation.

a) $f\colon y = x^2$, $g\colon y = \sqrt{x}$

b) $f\colon y = 3\sqrt{4 - x}$, $g\colon y = 3\sqrt{x - 2}$

120. If the curve given by $f\colon y = 2 + x^2 - 2x^4$ is rotated around the y-axis, a hollow depression is formed. Calculate the volume of the depression.

121. Derive the following established volume formulas by means of integration.

   a) Cylinder: $V = \pi R^2 H$, hollow cylinder: $V = \pi(R^2 - r^2)H$
     with R = outer radius, r = inner radius and H = height

   b) Cone: $V = \frac{1}{3}\pi R^2 H$, conical frustum: $V = \frac{1}{3}\pi(R^2 + Rr + r^2)h$
     with R = base radius, r = top radius, H = total height and
     h = height of frustum

   c) Sphere: $V = \frac{4}{3}\pi R^3$, spherical cap: $V = \frac{1}{3}\pi h^2(3R - h) = \frac{1}{6}\pi h(3r^2 + h^2)$
     with R = radius of sphere, r = radius of the base of the cap and h = height of the cap

122. A (lying) container is formed by rotation of the graph of the function f with $f(x) = \sqrt{x}$ around the x-axis. This container is placed in an upright position and filled with a liquid. Up to what height does the liquid in the container reach if its volume is 30?

123. The parabola $p\colon y = 4 - x^2$ and the x-axis enclose a finite region of area A, a so-called parabolic segment.

   a) What is the ratio of the areas A_1 and A_2 of the two subregions that result by dividing the area A by the line $g\colon y = 3x$?

   b) These two subregions of area A_1 and A_2 are rotated around the x-axis. Calculate the difference in volume of the two solids of rotation.

   c) At what height h must a parallel q to the x-axis be drawn, such that this parallel divides the area of the parabolic segment in half?

Mean of a Function

124. The length of day from sunrise to sunset at latitudes like that of Zürich can be approximated roughly by the sine curve given by the equation $T(x) = 4 \cdot \sin\left(\frac{2\pi}{365}(x - 80)\right) + 12$, where x is the number of the day in the year and T is the length of day in hours.

   a) Sketch the graph of the function.

   b) What is the average length of day T_M over the course of a year?

   c) Generally the *mean* of a positive, continuous function $y = f(x)$, $x \in [a; b]$, is understood to be the value y_M such that the area of the rectangle of height y_M over the interval $[a; b]$ is equal to the area of the region between the graph f and the x-axis in the interval $[a; b]$.

     Write this definition as an equation.

   d) Calculate the mean length of day using the equation from part c) and compare the result with the result from b).

Mean of a Function

The *mean* $\overline{f}$ of a function f on an interval $[a; b]$ is defined as

$$\overline{f} = \frac{1}{b - a} \int_a^b f(x)\,\mathrm{d}x.$$

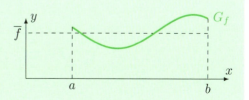

125. Calculate the mean $\overline{f}$ of the function f on the interval $[a;b]$. At which points does the mean occur as the value of the function?

a) $f(x) = x^2$; $a = 0$, $b = 4$

b) $f(t) = 2t + 1$; $a = 1$, $b = 5$

c) $f(u) = \sqrt{2u}$; $a = 0$, $b = 8$

d) $f(x) = e^{\frac{x}{2}+1}$; $a = 0$, $b = 10$

126. Calculate the mean of the function f.

a) $f(x) = 2x + 5$ on the interval $[-1;1]$

b) $f(x) = \sqrt{3x}$ on the interval $[0;3]$

127. a) Calculate the mean of $f(x) = (x-3)^2$ on the interval $[2;5]$.

b) Determine the value $c \in \mathbb{R}$ with $\overline{f} = f(c)$ using the result from part a).

c) Find a non-constant function g with $\overline{g} = -1$, $\overline{g} = 0$, $\overline{g} = c \in \mathbb{R}$ on some interval.

128. Find a function f on $[a;b]$, whose mean $\overline{f} = \frac{1}{b-a}\int_a^b f(x)\,dx$ on $[a;b]$ equals the average slope $\frac{f(b)-f(a)}{b-a}$ of f on $[a;b]$.

129. Is the mean of f' on $[a;b]$ equal to the average rate of change of f on $[a;b]$? Justify.

Stock and Storage Cost

In economics, the mean of a function is used to calculate the following quantities.

The *stock* $I(t)$ is the number of units of a product that the company has in stock at time t (in days). The *average daily stock* $\overline{I}$ for the time period $[0;T]$ is then

$$\overline{I} = \frac{1}{T} \int_0^T I(t)\,dt.$$

If k denotes the cost in francs to store a unit of a product for one day in the warehouse, the *average daily storage cost* for the time period $[0;T]$ is given by $k \cdot \overline{I}$.

---

130. A wholesaler receives a delivery of 1200 boxes of chocolate bars each month (30 days). He supplies the retailers with a constant rate of 40 boxes/day, such that the stock function is $I(t) = 1200 - 40t$, $0 \le t \le 30$. The storage for one box costs 3 Rappen per day.

a) What is the average daily stock for a period of one month?

b) What is the average daily storage cost during this period?

131. A wholesaler receives a delivery of 450 barrels of plastic granules each month (30 days). Experience shows that the number of barrels in storage decreases more slowly at first, but then more rapidly towards the end of each month. Therefore, the stock function I as a function of number of days t since the delivery can be modelled quite well by $I(t) = 450 - \frac{1}{2}t^2$, $0 \le t \le 30$. Storage for one barrel costs 2 Rappen per day.

a) What is the average daily stock for a period of one month?

b) What is the average daily storage cost during this period?

4.5   Improper Integrals

132. *Two typical questions.* The graph of the function $f(x) = \frac{1}{x}$, $x \in [1; \infty[$ extends to infinity. In the following, we examine the area A of the region G between the graph of f and the x-axis on the one hand and the volume V of the solid generated by rotating G around the x-axis on the other hand.

a) Determine $A(t) = \int_1^t f(x)\, \mathrm{d}x$ and $V(t) = \pi \int_1^t (f(x))^2\, \mathrm{d}x$ for $t \in [1; \infty[$.

b) Determine $\lim\limits_{t \to \infty} A(t)$ and $\lim\limits_{t \to \infty} V(t)$, if the limits exist.

c) Does the region G have a finite or infinite area A? Does the solid of rotation of G have a finite or infinite volume V? Explain.

Improper Integrals with Limits $\pm\infty$

Let f be a continuous function.

a) We define $\displaystyle\int_a^\infty f(x)\, \mathrm{d}x = \lim_{t \to \infty} \int_a^t f(x)\, \mathrm{d}x$, provided that the limit exists.

b) We define $\displaystyle\int_{-\infty}^b f(x)\, \mathrm{d}x = \lim_{t \to -\infty} \int_t^b f(x)\, \mathrm{d}x$, provided that the limit exists.

c) We define $\displaystyle\int_{-\infty}^\infty f(x)\, \mathrm{d}x = \int_{-\infty}^a f(x)\, \mathrm{d}x + \int_a^\infty f(x)\, \mathrm{d}x$, provided that both integrals exist for any arbitrary $a \in \mathbb{R}$.

For **133–135**: Calculate the improper integral if it exists.

133. a) $\displaystyle\int_1^\infty \frac{1}{x^4}\, \mathrm{d}x$ b) $\displaystyle\int_{-\infty}^{-1} \frac{1}{x^3}\, \mathrm{d}x$ c) $\displaystyle\int_1^\infty \frac{2x+4}{x^3}\, \mathrm{d}x$ d) $\displaystyle\int_2^\infty \frac{x^3-5}{x^6}\, \mathrm{d}x$

134. a) $\displaystyle\int_1^\infty \frac{2}{x^2}\, \mathrm{d}x$ b) $\displaystyle\int_2^\infty \frac{1}{x-1}\, \mathrm{d}x$ c) $\displaystyle\int_{-\infty}^{-2} \frac{1}{(x+1)^3}\, \mathrm{d}x$ d) $\displaystyle\int_3^\infty \frac{5+t}{t^3}\, \mathrm{d}t$

135. a) $\displaystyle\int_0^\infty \mathrm{e}^{-x}\, \mathrm{d}x$ b) $\displaystyle\int_{-\infty}^0 \mathrm{e}^{-x}\, \mathrm{d}x$ c) $\displaystyle\int_{-\infty}^1 \mathrm{e}^{2t+1}\, \mathrm{d}t$ d) $\displaystyle\int_{-\infty}^\infty \mathrm{e}^{-|t|}\, \mathrm{d}t$

136. Examine if the shaded region extending to infinity has a finite area A. If so, determine the area.

a) $f(x) = -4x^{-3}$ b) $f(x) = \frac{1}{(x+1)^2}$ c) $f(x) = \mathrm{e}^{\frac{-1}{2}x}$

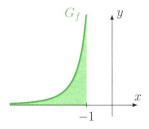
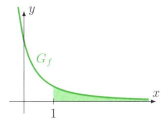
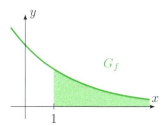

137. The equations of the two functions f and g, whose graphs G_f and G_g intersect in quadrant I, are given. Sketch the basic shape of the curves and calculate the area A of the region above the x-axis and below both graphs G_f and G_g.

a) $f(x) = \mathrm{e}^{x-2}$ and $g(x) = \mathrm{e}^{2-x}$

b) $f(x) = \mathrm{e}^{x-1}$ and $g(x) = \mathrm{e}^{3-x}$

138. For $a, b > 0$, we define the two rational functions f and g by $f(x) = \frac{a}{(x-b)^2}$, $x \leq 0$ and $g(x) = \frac{a}{(x+b)^2}$, $x \geq 0$.

Make a rough sketch of the graphs of f and g for $a = 36$ and $b = 3$ and then show in general that the following equation relating the two improper integrals is always fulfilled: $\int\limits_{-\infty}^{0} f(x)\,\mathrm{d}x = \int\limits_{0}^{\infty} g(x)\,\mathrm{d}x$.

139. Consider the function f in the indicated interval. Its graph G_f and its asymptote g for $x \to \infty$ or $x \to -\infty$, respectively, enclose an infinite region. Is its area finite? If so, calculate the area and make a sketch.

a) $f(x) = \frac{1}{2}x + \frac{2}{x^2}$, $[2; \infty[$

b) $f(x) = \frac{-1}{3}x + \mathrm{e}^x$, $]-\infty; 1]$

140. Calculate the area of the depicted infinite region if $k_1 : y = \sin(ax)$ and $k_2 : y = \frac{1}{x^2}$.

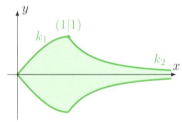

141. Prove that the two curves $v : y = \mathrm{e}^x$ $(x \in \mathbb{R})$ and $w : y = \sqrt{2\mathrm{e}x}$ $(x \geq 0)$ are tangent at the point $x_0 = \frac{1}{2}$. Determine the area of the region extending to infinity and is enclosed by the two curves and the negative x-axis.

142. The curve $f : y = \frac{1}{x^2}$ and the line $g : x = 1$ are given.

a) What is the area of the region in quadrant I, enclosed by the x-axis, the curve f and the line g?

b) What are the equations of the two lines parallel to the y-axis that trisect this region into three parts of equal area?

143. a) What percentage of $\int\limits_{1}^{\infty} \mathrm{e}^{-x}\,\mathrm{d}x$ is $\int\limits_{1}^{a} \mathrm{e}^{-x}\,\mathrm{d}x$? Calculate the numerical value of the percentage for the specific values $a = 2, 5, 10, 20, 50, 100$.

b) Repeat part a) for the integrand x^{-2} instead of e^{-x}.

144. Calculate the area of the region enclosed by the curve given by the equation $g(x) = \mathrm{e}^{x-2}$, the tangent t to the curve at point $x = 2$ and the x-axis.

145. Let the half of the parabola in quadrant I with equation $y = f(x) = ax^2$ $(x \geq 0)$ and the exponential curve with equation $y = g(x) = e^x$ be given.

   i) Determine the coefficient a such that the two curves are tangent at point $x_0 = 2$.

   ii) Prove that this value of a can also be determined without specifying the value of x_0.

   iii) Calculate the area of the (improper) region A that is enclosed by the negative x-axis and the two curves up to their point of tangency.

   iv) This region A is rotated around the x-axis. Determine the (improper) volume of this solid of rotation that extends to infinity.

146. A boat sails with a velocity of $v_1(t) = 2\,\text{m/s}$ for 4 seconds. Then, it enters a bed of reeds, where it moves along unpowered. For the sake of simplicity, we assume that its velocity is modelled by $v_2(t) = \frac{32}{t^2}$. What distance does the boat travel if we wait indefinitely?

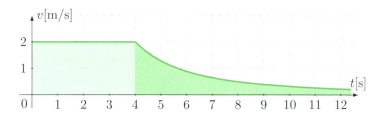

147. A freight car coasts to a standstill on a horizontal, straight track. Its initial velocity is 100 metres per minute and its velocity at time t (in minutes) is given by $v(t) = 100 \cdot e^{-2t}$ (in metres per minute). What is the maximum distance the wagon travels?

148. Let $f(x)$ denote the number of people in Switzerland with an income of x francs. Interpret the expression $\int_0^\infty f(x)\,\mathrm{d}x$ and give an expression for the average income per person in Switzerland.

4.6   Wide-ranging Applications

Geometry

149. The graph k of a polynomial function of degree 3 is tangent to the line t defined by the equation $y = 2 - x$ at $T(0\,|\,2)$ and intersects it at $S(2\,|\,0)$ orthogonally, that is at a right angle.

   a) Determine the equation of k as well as its roots.

   b) The line t and the tangent s to the curve at point S together with k each enclose a finite region A_t and A_s, respectively. Calculate the ratio of the areas $A_t : A_s$.

150. The curve $k\colon y = 9.36 + 0.1x^2 - 0.01x^4$ and the line $g\colon y = 8.4$ enclose a finite region. If this finite region is rotated around the x-axis, an annular solid is formed, a lovely ring! How many cm^3 of precious metal are necessary for its fabrication and what is the maximum outer diameter of this ring if the unit of the coordinate system is $1\,\text{mm}$? (Accuracy: to two significant digits)

151. A cylindrical hole ($r = 5\,\text{cm}$) is drilled through the centre of a wooden ball ($R = 13\,\text{cm}$) of homogeneous density. By what percentage does this decrease its original weight?

Note: Place the xy-plane such that the origin lies in the centre of the sphere and the x-axis coincides with the axis of the cylinder. Draw the cross-section and calculate the volume using an integral.

152. The polynomial function p: $y = x^3 - x^2 + x - 1$ is given.

i) Determine the equation of the tangent t_1 to the curve at the intersection point S_1 of the graph of p with the x-axis and calculate the point S_2, where the tangent t_1 intersects the graph of p.

ii) What is the equation of the tangent t_2 to the curve at the intersection point S_2? At which point S_3 does the tangent t_2 intersect the graph of p?

iii) Determine the equation of the tangent t_3 to the curve at the intersection point S_3 and calculate the coordinates of the intersection point S_4 where t_3 intersects the graph of the given polynomial function.

iv) The tangents t_1, t_2 and t_3 each enclose a finite region A_1, A_2 and A_3, respectively, with the graph of p. What are the two ratios of area $A_1 : A_2$ and $A_2 : A_3$?

153. The exponential curve f: $y = e^{x-n}$ and the power curve g: $y = (\frac{x}{n})^n$, with $n \in \mathbb{N}$, are given.

a) Show that f and g are tangent at the point $x_0 = n$.

b) The exponential curve and the mutual tangent t enclose a finite area in quadrant I. Calculate the area A_n of this region and determine the limit $\lim\limits_{n \to \infty} A_n$.

154. The graph G_f of the function f: $y = \frac{3}{16}x^4 - \frac{3}{2}x^2 + \frac{1}{2}x + 2$ and the tangent t: $y = \frac{1}{2}x + 2$ (see exercise 291 in chapter 3.8 on page 97) enclose two finite regions A_1 and A_2. What is the ratio $A_1 : A_2$ of these areas?

155. A hemispherical glass bowl on a garden table is filled to the brim with water. By what angle must this bowl be tilted so that exactly half of its volume flows out?

156. The cross-section of a watering trough has the shape of a parabola p: $y = \frac{20}{9}x^2$ with abscissa between -0.6 and 0.6 (units in metres).

a) Determine the capacity of the trough if its length is given as 2 metres.

b) How much water must be filled into the trough so that the water in the trough is 0.5 metres deep?

c) How deep is the water in the trough if 500 litres are filled into the empty trough?

Physics

157. The runway of an airport extends from A to Z. It is 2 kilometres long. A landing aeroplane touches down at the point L and immediately starts to brake. Let $s(t)$ be the distance (in metres) between the aeroplane and point A t seconds after touchdown. Determine the formula for $s(t)$ for an aeroplane that touches down with a velocity of $120\,\text{m/s}$ and then brakes with a constant deceleration of $-6\,\text{m/s}^2$ if it is not to coast beyond the runway.

158. A bee flies from a tulip to a rose in a straight line. Its velocity $v = v(t)$ in m/s can be described by a polynomial of degree 3. The bee starts from rest (i.e. $v(0) = 0$) and lands after 3 seconds (i.e. $v(3) = 0$). It attains its maximum velocity $v_{max} = 8\,m/s$ after 2 seconds.

a) Which function v models the velocity of the bee for $0 \le t \le 3$ (time in seconds) according to the specifications given above?

b) How far apart are the two flowers?

159. For calculating the required safe trailing distance while driving on the motorway, one must consider the least favourable conditions: optimal braking acceleration of $a_1 = -8\,m/s^2$ for the front car, worst braking acceleration of $a_2 = -4\,m/s^2$ for the rear car. What is the safe trailing distance d between the two cars if both are travelling at $v = 120\,km/h$, in order to prevent a collision if

a) both drivers brake simultaneously?

b) the rear driver starts to brake after a reaction time of one second?

c) Compare the results with the safe trailing distance that the rule of thumb known as the «Halbe-Tacho-Regel» (literal translation: half speedometer rule) recommends.

Remark: The «Halbe-Tacho-Regel» is the following trailing distance recommendation: Half of the velocity (in km/h) $\widehat{=}$ minimum trailing distance (in metres).

160. A skydiver jumps from a plane at an altitude of 2500 metres. His descent velocity can be modelled approximately by $v(t) = 50(1 - e^{-0.175t})$, when his parachute is closed (time t in seconds after jumping, $v(t)$ in metres per second).

Remark: The velocity v of the skydiver initially increases rapidly due to the Earth's gravity. However, the increasing velocity also increases air resistance, which is the reason why the velocity only increases slowly towards the end. A differential equation shows that it can be modelled with the exponential function given above.

a) Calculate the average descent velocity during the first 10 seconds.

b) Determine a formula which gives the altitude of the skydiver at time t.

c) A second skydiver also jumps from an altitude of 2500 metres 5 seconds later. His descent velocity can be modelled approximately by $w(t) = 45(1 - e^{-0.35t})$, when his parachute is closed (time t in seconds after his jump, $w(t)$ in metres per second). How can we interpret the area between the two graphs of v and w in the interval $[0; 10]$?

Work as a Definite Integral

If a force acts at a point $x_k \in [a; b]$, overcoming this force along a small displacement Δx requires the small amount $\Delta W \approx F(x_k) \cdot \Delta x$ of work. Here, $F(x_k)$ is the component of force in direction of Δx. The total work required in the interval $[a; b]$ is thus given by the formula:

$$W = \lim_{n \to \infty} \sum_{k=1}^{n} \Delta W = \lim_{n \to \infty} \sum_{k=1}^{n} F(x_k)\Delta x = \int_{a}^{b} F(x)\,dx$$

161. A point mass of mass m and of positive electrical charge Q_1 sits in the field of a negative electrical point charge Q_2. What is the work required to move the point mass from distance r to distance R? What is the acceleration at distance r?

162. The elongation of a spring requires a force F proportional to the elongation x from the initial state. What work is required to elongate the spring by a length x? (proportionality constant $k = 20\,\text{N/cm} = 2000\,\text{N/m}$)

163. According to *Newton's law of gravitation* (ISAAC NEWTON, 1643–1727) two objects with masses m_1 and m_2 attract each other with the force

$$F = G \cdot \frac{m_1 m_2}{r^2},$$

where r is the distance between the two objects and G is the universal gravitational constant.

a) If one body is at rest, how much work is required to move the other object from $r = a$ to $r - b$, $b > a$?

b) What work (in joules) is required to bring a satellite of mass $m_1 = 1000\,\text{kg}$ vertically into an orbit of altitude $h = 1000\,\text{km}$? The mass ($m_2 = 5.98 \cdot 10^{24}\,\text{kg}$) of Earth can be considered as concentrated at Earth's centre, use $r_E = 6.37 \cdot 10^6\,\text{m}$ as the Earth radius and $G = 6.67 \cdot 10^{-11}\,\text{N} \cdot \text{m}^2/\text{kg}^2$ for the gravitational constant.

c) What amount of work is required to remove a satellite weighing $1000\,\text{kg}$ from Earth's gravitational field? Use the data from part b).

d) Determine the formula for the escape velocity v_0, which is necessary for a rocket of mass m to escape the gravitational field of a planet of mass M and radius R. Use Newton's law of universal gravitation and the fact that the kinetic energy $E_{\text{kin}} = \frac{1}{2}mv_0^2$ supplies the necessary work.

e) Determine the numerical value for the escape velocity at Earth's surface.

f) Lionel is asked by his nephew in which amount of time a rocket with 20'000 km/h would reach the moon. What would be a good response from Lionel?

Medicine/Biology

164. A patient is given a drug to lower blood pressure orally. The concentration $K(t)$ of the drug in the blood after t hours is described by the following model function: $K(t) = A \cdot \left(e^{-bt} - e^{-ct}\right)$ with $b > 0$, $c > 0$, Euler's number e, unit $[\text{ng/ml}]$.

For parts a), b) and c), let: $A = 90$, $b = 0.15$, $c = 0.45$, $0 \le t \le 30$.

a) When is the concentration of the drug in the blood greatest?

b) When is the decrease of the concentration greatest?

c) Calculate $\frac{1}{30}\int_0^{30} K(t)\,\mathrm{d}t$. What does this value mean?

d) Why does the formula for the model function $K(t)$ require that the inequality $c > b$ holds?

165. The growth of wild flowers is described by the family of functions $f_b(x) = \frac{135}{b^2}x^3 - \frac{270}{b}x^2 + 135x$, $1 \leq b \leq 3$, $0 \leq x \leq b$. Here, $f_b(x)$ is the growth rate in centimetres per month at time x and b is a species-specific parameter.

Remark: The growth rate of a wild flower is zero in the beginning (i.e. $f_b(0) = 0$), then increases and finally, towards the end of the growth period, decreases slowly to zero again (i.e. $f_b(b) = 0$ and $f_b'(b) = 0$). This motivates the use of the modelling function $f_b(x) = k \cdot x(x - b)^2$. Here, the coefficient of growth is $k = \frac{135}{b^2}$. The coefficient of growth is therefore smaller as the wild flower has had more time to grow (i.e. the larger b is).

a) How tall does a wild flower get?

b) How long did a wild flower grow if it is approximately 80 cm tall at the end of its growth?

c) In an extended model function, the growth rate can be modified independently from the length of the growth period (parameter b) by an additional parameter a. What might the family of functions look like in this case?

Applications to Economics

> **Economic applications**
>
> Also note the statements in the green box on page 95 and in particular those in the grey box on page 95.
>
> Terms:
> - x: Quantity = number of units sold
> - $p(x)$: Unit price = Sale price per unit of the product when selling x units
> - $E(x)$: Total revenue when selling x units of the product $= x \cdot p(x)$
> - $K(x)$: Total cost for production of x units of the product
> - $G(x)$: Total profit when selling x units of the product $= E(x) - K(x)$
> - $E'(x)$: Marginal revenue = marginal revenue when selling x units of the product
> - $K'(x)$: Marginal cost = marginal cost for production of x units of the product
> - $G'(x)$: Marginal profit = marginal profit from selling x units of the product

166. The marginal cost function for producing x metres of a fabric is modelled as $K'(x) = 5 - 0.008x + 0.000009x^2$ in Fr./m based on data collected. How large are the total costs for the production of the first 2000 metres of fabric?

167. The marginal revenue when selling x units of a product is modelled as $E'(x) = -16x + 64$ in Fr./unit based on data collected. What is the equation giving the total revenue and how large is the maximum total revenue?

168. A manufacturer sells 1000 television sets per week for a unit price of 500 francs. A market research study states that each price reduction of 20 francs increases the number of units sold by 100 units.

   a) What is the unit price function $p(x)$ associated with the number of units x sold per week?

   b) What is the revenue function $E(x)$ associated with the number of units x sold per week?

   c) According to the study, the marginal costs are $K'(x) = 300$. Determine the marginal profit $G'(x)$ when selling x units per week.

   d) How large must the discount be so that the profit is greatest?

   e) The study identified a maximum profit of 662'000 francs per week. What is the equation for the total profit function $G(x)$?

4.7   Differential Equations

Differential equations are considered in chapter 3.3 on page 72 for the first time. Here, this topic is taken up once more.

General Considerations

Many everday activities involve dynamic processes: A physical quantity (e.g. temperature, length, mass) changes with time or position in space. Our aim here, is to find a relationship between a function y, its derivatives y', y'', ... and its argument x. Such a relationship can be written mathematically as $f(\ldots, y'', y', y, x) = 0$. This is called a *differential equation*.

Ideally, the solution $y = y(x)$ can be determined analytically. Otherwise, we need to apply numerical approximation methods.

Differential Equations

A *differential equation*, short *DE*, denotes an equation for an unknown function in which derivatives of this function appear.

The *order* of a differential equation is the order of the highest derivative appearing in it.

A *solution* y of the differential equation $f(\ldots, y'', y', y, x) = 0$ on the interval $[a; b]$ denotes a function $y = y(x)$, which fulfils the differential equation if one evaluates y and all occuring derivatives at any arbitrary point $x \in [a; b]$.

Solutions of differential equations are not unique in general. But in many cases for order 1 there is a unique solution, if one additionally specifies the value of y at one point. This is called the *initial condition*.

Solving an *initial value problem* consists of finding a solution, which fulfils the differential equation as well as the given initial condition or conditions.

169. a) Is $y'(x) = x$ a differential equation?

  b) Is $y'(x) - y(x) = 5$ a differential equation?

  c) Is $y(x) = x$ a solution of the differential equation $y'(x) - 1 = 0$?

  d) Is $y(x) = x$ the only solution of the differential equation $(y'(x))^2 = 1$?

  e) How many solutions does the differential equation $y'(x) = -2 \cdot y'(x)$ have?

  f) Give a differential equation for which $y(x) = 2 \cdot \sin(3x + \frac{\pi}{3})$ is a solution.

  g) Find two solutions of the differential equation $y'(x) = 2 \cdot x$. Is the sum of the two solutions also a solution of the differential equation?

  h) Find two solutions of the differential equation $y'(x) = y(x)$. Is the product of the two solutions also a solution of the differential equation?

170. Determine a differential equation, desribing the following property. In parts b) to d), also answer the following question: What might be the solution of the differential equation? In other words: Which curve might have this property? Give a conjecture.

  a) A curve is sought for which the slope of the tangent in each point $P(x \,|\, y)$ is equal to the negative product of the coordinates of P.

  b) A curve is sought for which the tangent at each point $P(x \,|\, y)$ intersects the x-axis at the point $Q\left(\frac{x}{2} \,\middle|\, 0\right)$.

  c) A curve is sought for which the slope of the tangent in each point $P(x \,|\, y)$ is equal to the slope of the line connecting P and the origin $O(0 \,|\, 0)$.

  d) A curve is sought for which, at each point $P(x \,|\, y)$, the normal passes through the origin.

171. *Cell division.* The more cells there are that can multiply, the faster their number increases. In other words: If $P(t)$ is the size of the population of a cell culture at time t and $P'(t)$ is the instantaneous rate of change, they may, for example, have the relationship that $P'(t)$ is proportional to $P(t)$. The differential equation in this case is given by $P'(t) = k \cdot P(t)$.

  a) For $k = 1$, we have $P'(t) = P(t)$: For which functions does the function coincide with its derivative?

  b) For $k = 2$, we have $P'(t) = 2 \cdot P(t)$: For which functions is the derivative twice the function?

  c) For $k = \ln(2)$, we have $P'(t) = \ln(2) \cdot P(t)$: Which functions solve this differential equation?

  d) Describe the growth of the cell culture from part c).

172. In each of the parts a) to c), verify that the given function is a solution of the given differential equation:

a) The function with the equation $y(x) = x - x^{-1}$ is a solution of $xy' + y = 2x$.

b) The function with the equation $y(x) = \sin(x)\cos(x) - \cos(x)$ is a solution of the initial value problem $y' + \tan(x) \cdot y = \cos^2(x)$, $y(0) = -1$ on the interval $\left]-\frac{\pi}{2}; \frac{\pi}{2}\right[$.

c) Every function of the family of functions with the equation $y(t) = \frac{1 + c \cdot e^t}{1 - c \cdot e^t}$ is a solution of the differential equation $y' = \frac{1}{2}(y^2 - 1)$.

d) Determine a solution of the differential equation $y' = \frac{1}{2}(y^2 - 1)$ from part c) that fulfils the initial condition $y(0) = 2$.

e) For which k does the function $y(t) = \sin(kt)$ solve the differential equation $y'' + 9y = 0$?

f) Show for the values of k found in part e) that every function of the family defined by the equation $y(t) = A\sin(kt) + B\cos(kt)$, $k \neq 0$ also is a solution of the DE $y'' + 9y = 0$.

Slope Fields

We consider the differential equation $y' = x + y$ with the initial condition $y(0) = 1$ and wish to directly sketch the solution without calculating it first.

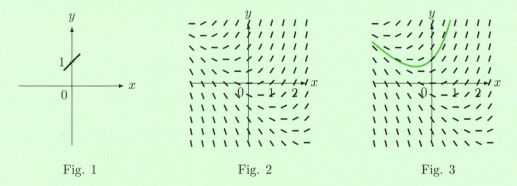

Fig. 1	Fig. 2	Fig. 3

Fig. 1: The slope at the initial point $(0 \mid 1)$ is calculated by $m = 0 + 1 = 1$. The solution thus starts as depicted in the figure.

Fig. 2: To be able to sketch the rest of the curve, we draw many small tangential segments in as many points $(x \mid y)$ as possible with the corresponding slope $m = x + y$. The result of this is called a *slope field* and is depicted here.

Fig. 3: The slope field enables us to visualise the general shape of the solution, by indicating the direction in which the curve must continue at each point. The figure shows a rough sketch of what the solution through the point $(0 \mid 1)$ must look like.

Worked out example:

a) Sketch the slope field for the differential equation $y' = x^2 + y^2 - 1$.

b) Use the slope field from part a) to draw the solution passing through the origin.

Solution:

a) The slopes in multiple points are calculated in the following table:

x	-2	-1	0	1	2	$\ldots$	-2	-1	0	1	2	$\ldots$
y	0	0	0	0	0	$\ldots$	1	1	1	1	1	$\ldots$
$y' = x^2 + y^2 - 1$	3	0	-1	0	3	$\ldots$	4	1	0	1	4	$\ldots$

The slope field is comprised of the tangential segments using the slopes calculated for the corresponding points in the table and is depicted below on the left.

For a)

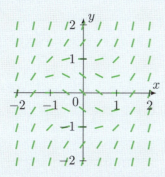

For b)

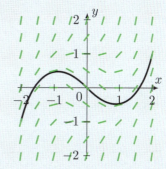

b) We start at the origin and move to the right in the direction of the tangential segment (with slope -1). We continue the curve in sucha way that it always stays parallel to the nearest tangential segment. The resulting curve is depicted above on the right.

As more and more tangential segments are drawn into a slope field, the behaviour of the solutions becomes clearer. Obviously, it is tedious to calculate and draw the slopes for a large number of points. However, computers are ideally suited for this. The adjacent figure shows a denser slope field for the differential equation $y' = x^2 + y^2 - 1$.

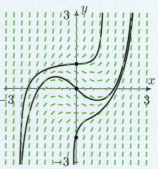

This allows us to draw the solutions with different y-intercepts -2, 0, 1 with reasonable precision.

> **Isoclines and Equilibrium Solutions**
>
> Calculating a slope field for a differential equation can be tedious. Sometimes, it is helpful to first determine the so-called isoclines. *Isoclines* of a differential equation are curves in the xy-plane on which the slope field is constant. The slope of all solution curves of the differential equation is the same at each point of intersection with the isocline. *Equilibrium solutions* are solutions that fulfil $y' = 0$. In other words: The equilibrium solutions of a differential equation are constant functions.

173. The slope field for the differential equation $y' = y(1 - \frac{1}{4}y^2)$ is depicted on the right. Sketch the solutions that fulfil the following initial condition:

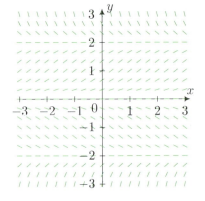

a) $y(0) = 1$ b) $y(0) = -1$

c) $y(0) = -3$ d) $y(0) = 3$

Determine all equilibrium solutions.

174. Which differential equation corresponds to which slope field?

a) $y' = y - 1$ b) $y' = y - x$ c) $y' = y^2 - x^2$ d) $y' = y^3 - x^3$

i)

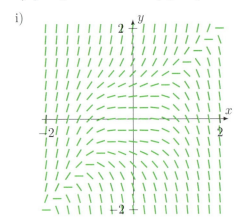

ii)

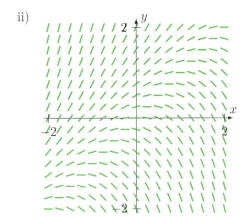

iii)

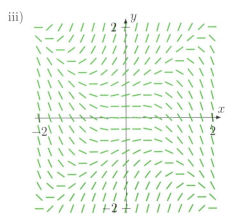

iv)

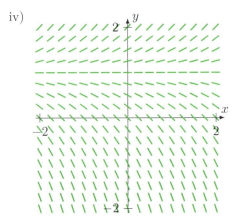

175. Sketch the slope field of the given differential equation. Draw the solution that passes through the given point.

a) $y' = y - 2x$, $(1 \mid 0)$ b) $y' = y + xy$, $(0 \mid 1)$

176. What can be said concerning the solutions of the differential equation $y' = x^2 + y^2$ without calculation? In other words: Determine the isoclines and draw the slope field as well as the solutions for two different initial conditions.

Modelling Exercises

177. *Modelling the cooling of a beverage.* In the following, we wish to model the temperature change of a cup of water that is placed in a room with the help of an experiment and thus derive *Newton's law of cooling* (ISAAC NEWTON, 1643–1727).

a) Bring a cup of hot water into a room. First, measure the room temperature U. Then, measure the temperature of the water every 5 minutes and create a table for the water temperature T (in °C) as a function of time t.

t	T
$\vdots$	$\vdots$

b) In the following table, the measurements from part a) are to be entered in the first two columns. Complete the table by calculating the missing values.

t	T	ΔT	$\Delta T/\Delta t$	$T - U$	$\frac{\Delta T/\Delta t}{T-U}$
$\vdots$	$\vdots$	$\vdots$	$\vdots$	$\vdots$	$\vdots$

Here, we have

- t: time

- T: water temperature in °C

- ΔT: change in water temperature between two measurements

- Δt: time elapsed between two measurements = 5 minutes

- $\Delta T/\Delta t$: average rate of change of the water temperature during the 5 minute interval

- $T - U$: difference between the water temperature and the ambient temperature

- $\frac{\Delta T/\Delta t}{T-U} = k$: approximate proportional constant between the last two columns

c) Fill the gap with one of the words inside the parentheses, based on your own experience or by looking at the table above.

i) The colder the environment, the ... (slower/faster) the water cools.

ii) The beverage initially cools ... (slower/faster) than later on.

iii) The water temperature T is always ... (higher/lower) than the ambient temperature U.

iv) The water temperature T ... (approaches/diverges from) the ambient temperature U over time.

v) The average rate of change of the water temperature $\Delta T/\Delta t$ is... (directly/inversely) proportional to the difference $T - U$ of the water temperature and the ambient temperature.

d) Does the table from part b) confirm the law formulated in item v) of part c)? Write the law as an equation for the temperature T as a function of time t.

e) What stands out in this equation? (In what ways does the unknown function T appear?)

178. *Solutions of important differential equations for growth processes.* Verify that the given function is a solution of the indicated initial value problem.

a) *Exponential growth*

$P'(t) = k \cdot P(t)$ with $P(0) = P_0$ has the solution $P(t) = C \cdot e^{kt}$ with $C = P_0$.

$P(t)$ represents the size of a population at time t with an initial population P_0.

b) *Exponential decay, e.g. law of cooling*

$T'(t) = -k \cdot (T(t) - U)$ with $T(0) = T_0$ has the solution $T(t) = U + C \cdot e^{-kt}$ with $C = T_0 - U$.

$T(t)$ represents the temperature of an object at time t with an initial temperature T_0 and an ambient temperature U.

c) *Bounded growth, e.g. influenza outbreak in an isolated community*

$I'(t) = k \cdot (N - I(t))$ with $I(0) = I_0$ has the solution $I(t) = N - C \cdot e^{-kt}$ with $C = N - I_0$.

$I(t)$ represents the number of infected individuals at time t where I_0 represents the number of infected individuals at the start of the influenza outbreak and N the number of people in the isolated community.

d) *Logistic growth*

$P'(t) = k \cdot P(t) \cdot (S - P(t))$ with $P(0) = P_0$ has the solution $P(t) = \frac{S}{1 + C \cdot e^{-kSt}}$ with $C = \frac{S - P_0}{P_0}$.

As in a), where S represents the carrying capacity (maximum capacity).

179. Newton's law of cooling states that, during the same time interval (e.g. in a minute), the average rate of change of the temperature of an object is proportional to the difference of its instantaneous temperature $T(t)$ and the ambient temperature U.

a) Derive the corresponding differential equation.

b) A cup of tea with an initial temperature of $T(0) = 98\,°C$ is placed in a room with a temperature of $U = 18\,°C$ and cools down to $T(1) = 94\,°C$ after one minute. At which time τ does the cup have a temperature of $75\,°C$?

180. There is an influenza outbreak in an isolated community comprised of N individuals. The number of newly infected individuals on a given day is roughly proportional to the number of uninfected individuals on the preceeding day.

a) Let t be the number of days since the start of the outbreak and $I(t)$ the number of infected individuals on day t. Explain why the following equation gives the number of newly infected individuals: $I(t+1) - I(t) = k \cdot (N - I(t))$.

b) In the following, instead of the time interval of a day, choose a general time interval of length Δt, which can be arbitrarily small. Show that the instantaneous rate of change of infected individuals satisfies the following relationship: $\frac{dI}{dt} = k \cdot (N - I(t))$.

c) A boarding school houses 2500 pupils in dorms on the campus. An influenza breakout starts with 100 infected pupils. The next day there are 220. How many days will it take until 70 % of the pupils on the campus are infected?

181. A better model for the spread of an epidemic compared to the one in exercise 180 is to assume that the instantaneous rate of infection is proportional to the number of infected individuals and also proportional to the number of as yet uninfected individuals. In a city with 5000 inhabitants, isolated from its surroundings, 160 people are infected at the start of the week and 1200 at the end of this week. How long will it approximately take until 80 % of the city's population has been infected?

Remark: A more precise model for the spread of an epidemic is the so-called SIR model, which is comprised of a system of three coupled differential equations and can be found in the pertinent literature.

182. The fishing of halibut in the Pacific in the years from 1970 to 1990 can be described approximately by a logistic growth model. Here, $P(t)$ is the biomass in kilograms, i.e. the total mass of halibut in the population, at time t in years. The carrying capacity is $S = 8 \cdot 10^7$ kg and let $k = 8.875 \cdot 10^{-9}$ kg/year.

a) Determine the biomass after one year, if it was $P(0) = 2 \cdot 10^7$ kg at the start.

b) How long will it take until the biomass reaches $4 \cdot 10^7$ kg?

183. Let $P(t)$ be the size of a population at time t. Given sufficient resources (food and habitat), a population multplies in such a way that its instantaneous rate of change is proportional to the existing population. If the resources are limited, the population can only grow up to a certain maximum value $S \;\widehat{=}\;$ carrying capacity; i.e. the instantaneous rate of change is also proportional to the available resources, which themselves are proportional to $S - P(t)$.

a) Explain the difference equation $P(t+1) - P(t) = k \cdot P(t) \cdot (S - P(t))$.

b) Derive the corresponding differential equation $\frac{\mathrm{d}P}{\mathrm{d}t} = k \cdot P(t) \cdot (S - P(t))$ from it.

c) In an aquarium, the carrying capacity is $S = 500$ and the growth coefficient $k = 0.002$. Calculate and sketch the solutions $P = P(t)$ for different initial population sizes: $P_0 = 0$, $P_0 = 100$, $P_0 = 200$, $P_0 = 300$, $P_0 = 400$, $P_0 = 500$.

184. The population $P(t)$ (at time t) of a country increases for two reasons:

1. The annual average growth rate is 1 % of the population already living in the country. The average growth rate is the average birth rate minus the average death rate.

2. Every year, 20'000 people immigrate.

Derive the corresponding differential equation and show that $P(t) = k \cdot e^{0.01 \cdot t} - 2 \cdot 10^6$ is the solution. What will the population be in 20 years, if it is 5 million now?

4.8   Further Topics

Integration by Substitution

185. Integrals of the form $\int f(g(x)) \cdot g'(x)\,\mathrm{d}x$ can be taken into the form $\int f(u)\,\mathrm{d}u$ by substituting $u = g(x)$ and $\frac{\mathrm{d}u}{\mathrm{d}x} = g'(x)$, where the latter integral may be easier to determine. Determine a suitable substitution and then the integral itself.

a) $\displaystyle\int \cos(x^2) \cdot 2x\,\mathrm{d}x$ b) $\displaystyle\int \sqrt{x^3 + 1} \cdot x^2\,\mathrm{d}x$ c) $\displaystyle\int \frac{x}{1 + x^2}\,\mathrm{d}x$

d) $\displaystyle\int \frac{h'(x)}{h(x)}\,\mathrm{d}x$ e) $\displaystyle\int \frac{-\sin(x)}{\cos(x)}\,\mathrm{d}x$ f) $\displaystyle\int \tan(x)\,\mathrm{d}x$

Method of Integration: Integration by Substitution

If f and g are differentiable functions and $u = g(x)$, then we have:

$$\int_a^b f(g(x))\,g'(x)\,\mathrm{d}x = \int_{g(a)}^{g(b)} f(u)\,\mathrm{d}u.$$

For **186–189**: Determine an antiderivative of the given function.

186. a) $f(x) = (3x - 5)^6$ b) $f(u) = \dfrac{5}{3u - 4}$ c) $g(u) = \dfrac{1}{\sqrt{u + 2}}$ d) $f(u) = \sqrt[4]{au + b}$

187. a) $f(x) = x(x^2 + 1)^3$ b) $g(x) = x\sqrt{x^2 + 1}$ c) $f(x) = \dfrac{4x}{\sqrt[3]{1 - x^2}}$ d) $h(t) = \dfrac{3t^3}{1 + t^4}$

188. a) $f(x) = 2xe^{x^2}$ b) $g(x) = axe^{-bx^2}$ c) $h(t) = \dfrac{1}{t \ln(t)}$ d) $g(u) = \dfrac{a}{u - b}e^{\ln(u - b)}$

189. a) $f(t) = \cos^2(t)\sin(t)$ b) $g(z) = \sin(z)\cos(z)$ c) $f(x) = \dfrac{\sin(\sqrt{x})}{\sqrt{x}}$

For **190–193**: Calculate.

190. a) $\displaystyle\int t^2(3t^3 - 4)\,\mathrm{d}t$ b) $\displaystyle\int \frac{u - 1}{u^2 - 2u + 2}\,\mathrm{d}u$ c) $\displaystyle\int \frac{2x - 1}{\sqrt{x^2 - x - 1}}\,\mathrm{d}x$

191. a) $\displaystyle\int (2x + 1)e^{x^2 + x}\,\mathrm{d}x$ b) $\displaystyle\int \frac{e^{\sqrt{z}-1}}{\sqrt{z}}\,\mathrm{d}z$ c) $\displaystyle\int \frac{\ln(t)}{t}\,\mathrm{d}t$

192. a) $\displaystyle\int_{-1}^0 \frac{1}{(1 - y)^3}\,\mathrm{d}y$ b) $\displaystyle\int_{-1}^0 \frac{3t}{t^2 + 1}\,\mathrm{d}t$ c) $\displaystyle\int_2^6 \sqrt{4x + 1}\,\mathrm{d}x$

193. a) $\displaystyle\int_{\frac{-\pi}{16}}^{\frac{\pi}{12}} \frac{1}{\cos^2(4x)}\,\mathrm{d}x$ b) $\displaystyle\int_0^{-3} e^{2u}\,\mathrm{d}u$ c) $\displaystyle\int_{-1}^1 (2x + 1)e^{-x^2 - x}\,\mathrm{d}x$

194. Let $f(x) = e^{-2x}$ and $g(x) = \sin\left(\frac{x}{2}\right)$ be given. Which of the following statements are true?

i) $\displaystyle\int_0^1 f(x)\,\mathrm{d}x = 1 - \frac{1}{e^2}$ ii) $\displaystyle\int_0^1 f(x)\,\mathrm{d}x = \frac{1}{2e^2}$ iii) $\displaystyle\int_0^\pi g(x)\,\mathrm{d}x = 2$ iv) $\displaystyle\int_0^\pi g(x)\,\mathrm{d}x = -2$

195. Let F be a (non elementary) function that is an antiderivative of f: $y = \frac{\sin(x)}{x}$, $x > 0$. Express $\int\limits_{1}^{3} \frac{\sin(2x)}{x} \,\mathrm{d}x$ using F.

Remark: An *elementary function* is a function defined using the elementary arithmetic operations, compositions and inverse functions of polynomial, exponential, logarithmic and trigonometric functions.

196. Draw a figure illustrating the equation $\int\limits_{a}^{b} f(x)\,\mathrm{d}x = \int\limits_{a-c}^{b-c} f(x+c)\,\mathrm{d}x$.

197. Which of the following regions have equal area? Justify.

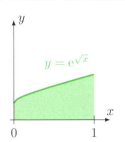
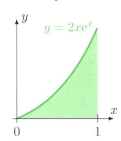
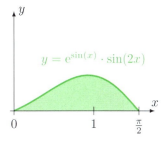

198. Let f, g, F and G be given, where F is an antiderivative of f and G is an antiderivative of g. Are the following statements true or false?

i) $\int f(ax+b)\,\mathrm{d}x = a \cdot F(ax+b)$ ii) $F \circ g$ is an antiderivative of $(f \circ g) \cdot g'$.

199. The functions f and g with $f(x) = x^2 + x + 1$ and $g(x) = ax - 1$ are given. For which values of a does $\int\limits_{0}^{1} f(g(x))\,\mathrm{d}x = \int\limits_{0}^{1} g(f(x))\,\mathrm{d}x$ hold?

Integration by Parts

200. Let f, g, F and G be given functions, where F is an antiderivative of f and G is an antiderivative of g. Is the following statement true or false? «$F \cdot G$ is an antiderivative of $f \cdot g$.»

201. *Integration by application of the product rule.* Recall: $(f(x)g(x))' = f'(x)g(x) + f(x)g'(x)$.
   a) Show that the product rule implies: $\int f'(x)g(x)\,\mathrm{d}x = f(x)g(x) - \int f(x)g'(x)\,\mathrm{d}x$.
   The integral on the right-hand side might be easier to calculate. This method is called *integration by parts*.
   b) We wish to compute $\int \sin(x) \cdot x\,\mathrm{d}x$. Choose $f'(x) = \sin(x)$ and $g(x) = x$ and apply the formula from part a) to determine the integral.
   c) Again, we wish to compute $\int \sin(x) \cdot x\,\mathrm{d}x$. Choose $f'(x) = x$ and $g(x) = \sin(x)$ and apply the formula from part a) to determine the integral. What do you notice?

Method of Integration: Integration by Parts

If f and g are differentiable functions, we have:

$$\int f'(x)\,g(x)\,\mathrm{d}x = f(x)\,g(x) - \int f(x)\,g'(x)\,\mathrm{d}x.$$

202. We consider integrals of the form $\int x^n \ln(x)\,\mathrm{d}x$, $n \in \mathbb{Z}$.

   a) Determine the antiderivative of the functions f in the given order.

   i) $f(x) = x \ln(x)$ ii) $f(x) = x^2 \ln(x)$ iii) $f(x) = x^3 \ln(x)$

   b) What is the antiderivative of $f(x) = x^n \ln(x)$ for $n \in \mathbb{Z}$?

   c) Is this result also valid for $n = 0$ and $n = -1$?

203. Integrals of the form $\int x^n e^x\,\mathrm{d}x$, $n \in \mathbb{N}$: Determine an antiderivative of f in the given order.

   i) $f(x) - x e^x$ ii) $f(x) - x^2 e^x$ iii) $f(x) = x^3 e^x$ iv) $f(x) = x^4 e^x$

204. Integrals of the form $\int e^x \sin(x)\,\mathrm{d}x$ or $\int e^x \cos(x)\,\mathrm{d}x$: Determine an antiderivative of f.

   a) $f(x) = e^x \sin(x)$ b) $f(x) = e^x \cos(x)$

205. Calculate the improper integral, if it exists.

   a) $\displaystyle\int_0^\infty z e^{-z}\,\mathrm{d}z$ b) $\displaystyle\int_0^\infty y^2 e^{-y}\,\mathrm{d}y$ c) $\displaystyle\int_{-\infty}^0 e^u \sin(u)\,\mathrm{d}u$

206. We consider the curve $y = x^3 \ln(x)$ in the interval $I = [0; 1]$, where $\lim\limits_{x \to 0} x^3 \ln(x) = 0$ holds. Calculate the area of the region enclosed by the curve and the x-axis over the interval I.

207. What is the area of the region enclosed by the x-axis, the line $x = 1$ and the curve $k\colon y = x^2 e^x$?

208. Let $y = f(x)$ be a strictly increasing function over $[a; b]$, with inverse function $x = f^{-1}(y)$, see figure:

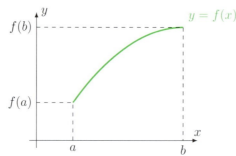

   a) Demonstrate geometrically by use of the figure that: $\displaystyle\int_a^b y\,\mathrm{d}x = [x \cdot y]_a^b - \int_{f(a)}^{f(b)} x\,\mathrm{d}y$.

   b) Use part a) to calculate the two integrals $\displaystyle\int_2^3 \ln(x)\,\mathrm{d}x$ and $\displaystyle\int_0^3 \sqrt{x}\,\mathrm{d}x$.

   c) Is the formula in part a) also valid for any strictly decreasing function g? Justify with a figure.

Arc Length

209. Let the circle with the equation $x^2 + y^2 = r^2$ be given. For $r = 10\,\text{cm}$, draw the circular arc with $\frac{r}{2} \leq x \leq r$.

a) Use a piece of string to measure the length of the arc. This result is called b_1.

b) Create a copy of the arc with a magnification factor of 2. Again, measure its length with a piece of string. This result is called b_2. Due to the magnification, $\frac{b_2}{2}$ is a further measurement for the arc length.

   Which result, b_1 or $\frac{b_2}{2}$, is more accurate? Comment.

c) Describe in words the procedure for a conceived 100-fold magnification.

d) In the following, we derive a formula for the calculation of the arc length L of a general curve defined by the equation $y = f(x)$, $x \in [a; b]$.

   i)   Partition the curve into n segments by means of the points $P_0, P_1, \ldots, P_n$. For $P_k(x_k \,|\, y_k)$, $k = 0, \ldots, n-1$, let the following hold:

   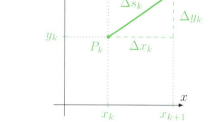

   $\Delta x_k = x_{k+1} - x_k$, $\Delta y_k = y_{k+1} - y_k$ and $\Delta s_k = \overline{P_k P_{k+1}}$.
   Moreover, let Δx_k, Δy_k, Δs_k tend to 0 for $n \to \infty$.

   Explain the equation $(\Delta s_k)^2 = (\Delta x_k)^2 + (\Delta y_k)^2$.

   ii) Explain why $L \approx \sum_{k=0}^{n-1} \sqrt{1 + \left(\frac{\Delta y_k}{\Delta x_k}\right)^2} \cdot \Delta x_k$ holds.

   iii) Take the limit for $n \to \infty$ to obtain an exact formula for L.

e) Use the formula from part d) to calculate the length b of the circular arc. Compare it with b_1 and $\frac{b_2}{2}$ from parts a) and b).

f) The length of a circular arc with central angle α in degrees and radius r is, as we know, equal to $b = \frac{2\pi r}{360°} \cdot \alpha$.

   i)   Determine the central angle α of the circular arc using the fact that $x \in \left[\frac{r}{2}; r\right]$.

   ii) Use this result to determine the length b of the circular arc.

   iii) What do you notice?

Arc Length

If f' is continuous on $[a; b]$, the length l of the curve $y = f(x)$ with $x \in [a; b]$ is given by

$$L = \int_a^b \sqrt{1 + (f'(x))^2}\, dx.$$

210. Draw the graph of the *semicubical parabola* (discovered by William Neile, 1637–1670) $f(x) = \sqrt{x^3}$ for $x \in [0; 2]$ in a coordinate system.

   a) Measure the length of the graph in the interval $[0; 2]$ using a piece of string.

   b) Calculate the length of the graph in the interval $[0; 2]$.

   c) Illustrate the term $\sqrt{1 + (f'(x))^2}\,dx$ by a depiction in the coordinate system.

211. Calculate the length L of the arc of the curve k between $x = a$ and $x = b$.

   a) $k \colon y = \frac{1}{2}\left(\frac{x^3}{3} + \frac{1}{x}\right); \; a = 1, b = 3$

   b) $k \colon y = \frac{1}{2}(e^x + e^{-x}); \; a = -3, b = 3$

212. A corrugated sheet with a sinusoidal profile is to be formed from a rectangular panel. Show that the sine curve is given by the equation $y = 2.5 \cdot \sin\left(\frac{2\pi}{35}x\right)$ in a suitable coordinate system, if the finished corrugated sheet has a width of $70\,\mathrm{cm}$ and a depth of $5\,\mathrm{cm}$. What was the width w of the original piece of panel sheet before it was bent into shape? (Use a calculator to calculate the integral with an accuracy to four significant decimals.)

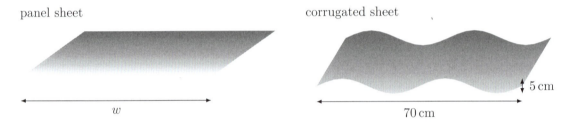

panel sheet   corrugated sheet

w 70 cm   5 cm

4.9 Miscellaneous Exercises

For Chapter 4.1: Integration as the Inverse of Differentiation

213. In each case, determine the formula for $f(x)$.

   a) $f''(x) = 2x + 1$, $f'(1) = 3$ and $f(2) = 7$

   b) $f''(x) = 3x + 5$, the graph G_f has slope 11 at the point $x_0 = 2$ and the point $(6 \mid 90)$ is on G_f.

   c) $f''(x) = 15\sqrt{x} + \frac{3}{\sqrt{x}}$, $f'(1) = 12$ and $f(0) = 5$

   d) $f''(x) = 2x$ and the points $(1 \mid 0)$ and $(0 \mid 5)$ are on G_f.

   e) $f''(x) = \frac{12}{x^4} - \frac{3}{x^3}$ and G_f is tangent to the x-axis at $x_0 = 3$.

214. Computers often calculate the indefinite integral without indicating the constant of integration. Daniel determines $\int \frac{1}{2x+4}\,dx$ with the computer, which yields $\frac{\ln(|2x+4|)}{2}$. Daniela first rewrites $\int \frac{1}{2x+4}\,dx$ as $\frac{1}{2} \cdot \int \frac{1}{x+2}\,dx$ and then determines its result with a computer, which yields $\frac{\ln(|x+2|)}{2}$. These two results are not equivalent. What happened here? Comment.

215. Mirko claims: «Since $\frac{1}{2} \cdot \frac{(1+2x)^4}{4}$ is an antiderivative of $(1+2x)^3$, $\frac{1}{2} \cdot \frac{(1+2x^2)^4}{4}$ is an antiderivative of $(1 + 2x^2)^3$.» Is he right?

216. Where is the error in the following string of arguments?

$$\int 2\sin(x)\cos(x)\,\mathrm{d}x = \int 2\cos(x)\sin(x)\,\mathrm{d}x$$
$$\Leftrightarrow \ \sin^2(x) + C = -\cos^2(x) + C \ \Leftrightarrow \ \sin^2(x) + \cos^2(x) = 0$$

217. Let the function f be given by $f(x) = \frac{1}{x-1}$, $x > 1$. Does the term yield an antiderivative of f?

a) $\ln(x-1)$

b) $\ln(x-1) + \ln(1)$

c) $\displaystyle\int_2^x \frac{1}{t-1}\,\mathrm{d}t$

d) $\ln(x-1) + 1$

218. Let $f(x) = \begin{cases} \frac{1}{2}x, & 0 \le x \le 2 \\ 2x - 3, & x > 2 \end{cases}$ be given.

Determine the antiderivative F of f for which $F(0) = 0$.

219. *Motion described by velocity.* A point mass moves along a straight line with velocity v, which is a function of time. What is the distance of the point mass from its starting point at time t?

a) $v = 3t - 4$ b) $v = at + v_0$ c) $v = 7$ d) $v = t^2 - 3t$

220. *Motion described by acceleration.* A point mass moves along a straight line with acceleration a. How does the velocity v of the point mass vary with time t? How far is the point mass from its starting point at time t_0?

a) $a = 9.81$
   $v(0) = 3$
   $t_0 = 8$

b) $a = 0$
   $v(0) = 4$
   $t_0 = 11$

c) $a = 1.5t$
   $v(2) = 3$
   $t_0 = 20$

d) $a = -\frac{3}{4}t$
   $v(4) = 76$
   $t_0 = 4$

For Chapter 4.2: Calculation of Areas by Means of Lower and Upper Sums

221. Let f be a function that is continuous and monotonically increasing on $[a;b]$. Let $[a;b]$ be partitioned into n subintervals $[x_{k-1}; x_k]$ of equal length $\Delta x = \frac{b-a}{n}$, i.e. $x_k = a + k \cdot \frac{b-a}{n}$, $k = 0, \ldots, n$. The lower sum is then $U_n = \sum_{k=0}^{n-1} f(x_k) \cdot \Delta x$ and the upper sum $O_n = \sum_{k=1}^{n} f(x_k) \cdot \Delta x$.

a) Calculate $O_n - U_n$.

b) Using a), calculate the smallest value of n such that $O_n - U_n \le \frac{1}{2}$ holds.

c) Using a), calculate the limit $\lim\limits_{n \to \infty} (O_n - U_n)$.

222. Let f be a function that is continuous and monotonically increasing on $[a;b]$. Let $[a;b]$ be partitioned into n subintervals $[x_{k-1}; x_k]$ of equal length $\Delta x = \frac{b-a}{n}$, i.e. $x_k = a + k \cdot \frac{b-a}{n}$, $k = 0, \ldots, n$. The lower sum is then $U_n = \sum_{k=0}^{n-1} f(x_k) \cdot \Delta x$ and the upper sum $O_n = \sum_{k=1}^{n} f(x_k) \cdot \Delta x$.

a) Fridolin approximates $\int_a^b f(x)\,\mathrm{d}x$ by using the mean $\frac{U_n + O_n}{2}$. Determine $\lim\limits_{n \to \infty} \frac{U_n + O_n}{2}$.

b) Give an example for f that fulfils $\int_a^b f(x)\,\mathrm{d}x = \frac{U_n + O_n}{2}$ for every n.

For Chapter 4.3: Fundamental Theorem of Calculus

223. True or false? A water pipe delivers water with a rate of $f(t)$ litres per minute. The outflowing water is collected in a tank between the times $t = 2$ and $t = 4$ (t in minutes). The amount of water collected in this way can be specified by the following quantity:

a) $\displaystyle\int_2^4 f(t)\,\mathrm{d}t$ b) $f(4) - f(2)$ c) $(4-2)f(4)$ d) $(4-2)\cdot\frac{f(2)+f(4)}{2}$

224. True or false? The equation $\int\limits_0^x f'(t)\,\mathrm{d}t = f(x)$ is

a) always true. b) sometimes true. c) never true.

225. It is known that for arbitrary numbers $m \in \mathbb{R}^+$ the function f satisfies $\int\limits_{-m}^{m} f(x)\,\mathrm{d}x = 0$. What can be said about this function?

226. Let a curve given by $y = f(x)$ pass through $(0\,|\,0)$ and $(1\,|\,1)$. Calculate $\int\limits_0^1 f'(x)\,\mathrm{d}x$.

227. Which of the following equations describes a version of the fundamental theorem of calculus?

i) $\dfrac{\mathrm{d}}{\mathrm{d}x}\left(\displaystyle\int_a^b f(t)\,\mathrm{d}t\right) = f(b)$ ii) $\dfrac{\mathrm{d}}{\mathrm{d}x}\left(\displaystyle\int_a^x f(t)\,\mathrm{d}t\right) = f(x)$

iii) $\displaystyle\int_a^x \left(\dfrac{\mathrm{d}}{\mathrm{d}t}f(t)\right)\mathrm{d}t = f(x) - f(a)$ iv) $\displaystyle\int_a^t \left(\dfrac{\mathrm{d}}{\mathrm{d}x}f(x)\right)\mathrm{d}x = f(t)$

228. Calculate the parameter k.

a) $\displaystyle\int_{-1}^2 kx^2\,\mathrm{d}x = \frac{2}{3}$ b) $\displaystyle\int_0^k \cos(y)\,\mathrm{d}y = \frac{1}{2}$ c) $\displaystyle\int_0^{\frac{k}{2}} \sin(2x)\,\mathrm{d}x = 1$

d) $\displaystyle\int_1^{k^2} \frac{1}{z}\,\mathrm{d}z = 8$ e) $\displaystyle\int_k^{2k} \mathrm{e}^{-y}\,\mathrm{d}y = \frac{1}{2}\int_0^k \mathrm{e}^{-y}\,\mathrm{d}y$

229. Show that for $p, q \in \mathbb{N}$ we have:

a) $\displaystyle\int_{-\pi}^{\pi} \sin(px)\cos(qx)\,\mathrm{d}x = 0$ b) $\displaystyle\int_{-\pi}^{\pi} \sin(px)\sin(qx)\,\mathrm{d}x = \begin{cases} 0, & \text{if } p \neq q \\ \pi, & \text{if } p = q \end{cases}$

230. Show that for all $a > 0$ and $x > 0$ we have: $\int\limits_1^x \frac{1}{t}\,\mathrm{d}t = \int\limits_a^{ax} \frac{1}{t}\,\mathrm{d}t$. Can this result be interpreted geometrically?

231. Determine the extremum points and the type of extrema of the function f.

a) $f: t \mapsto \displaystyle\int_0^t (x^3 - x)\,\mathrm{d}x$ b) $f: x \mapsto \displaystyle\int_3^x \sqrt{u+1}\,\mathrm{d}u$

c) $f: x \mapsto \displaystyle\int_0^x u(u^2 - x)\,\mathrm{d}u$ d) $f: u \mapsto \displaystyle\int_{-1}^u (t-1)\mathrm{e}^{-t}\,\mathrm{d}t$

For Chapter 4.4: Different Interpretations of the Integral

232. True or false? Justify.

a) $\int_{-5}^{5} (ax^2 + bx + c)\, dx = 2 \int_{0}^{5} (ax^2 + c)\, dx$

b) $\int_{0}^{2} (x - x^3)\, dx$ is the area of the region between the curve $y = x - x^3$ and the x-axis on the interval $[0; 2]$.

233. Let $\int_{0}^{1} x^2\, dx = \frac{1}{3}$ be given. Use this fact and geometric arguments to calculate the integral.

a) $\int_{0}^{1} (1 - x^2)\, dx$ b) $\int_{0}^{1} \sqrt{x}\, dx$ c) $\int_{0}^{1} 3x^2\, dx$ d) $\int_{0}^{1} \left(\frac{x}{3}\right)^2\, dx$

234. Let $\int_{0}^{\frac{\pi}{2}} \sin(x)\, dx = 1$ be given. Use this fact and geometric arguments to calculate the integral.

a) $\int_{0}^{\pi} \sin(x)\, dx$ b) $\int_{0}^{\pi} \cos(x)\, dx$ c) $\int_{0}^{\frac{\pi}{2}} 2\sin(x)\, dx$ d) $\int_{0}^{\pi} \cos\left(\frac{x}{2}\right)\, dx$

235. Calculate.

a) $\int_{0}^{2\pi} |\sin(t)|\, dt$ b) $\left| \int_{0}^{2\pi} \sin(t)\, dt \right|$ c) $\int_{2\pi}^{0} |\sin(t)|\, dt$ d) $\left| \int_{2\pi}^{0} |\sin(t)|\, dt \right|$

236. Prove by calculation that the equation $\int_{r}^{r+3} e^x\, dx = e^r \cdot \int_{0}^{3} e^x\, dx$ holds and describe this relationship geometrically.

237. In many programming languages, the rounding functions are defined as follows:
- Rounding down: floor$(x) = \lfloor x \rfloor \;\widehat{=}\;$ largest integer that is smaller than, or equal to, x.
- Rounding up: ceiling$(x) = \lceil x \rceil \;\widehat{=}\;$ smallest integer that is larger than, or equal to, x.
- Rounding: round$(x) = \lfloor x + 0.5 \rfloor$.

Calculate.

a) $\int_{-3}^{3} \text{floor}(x)\, dx$ b) $\int_{-3}^{3} \text{ceiling}(x)\, dx$ c) $\int_{-3}^{3} \text{round}(x)\, dx$

238. Which terms give the area of the shaded region correctly? Justify.

i) $\left| \int_{0}^{a} f(x)\, dx - \int_{0}^{b} g(x)\, dx \right|$

ii) $\left| \int_{d}^{b} g(x)\, dx - \int_{c}^{a} f(x)\, dx \right|$

iii) $\left| \int_{0}^{a} g(x)\, dx - \int_{0}^{b} f(x)\, dx \right|$

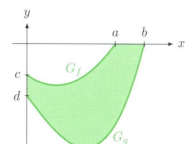

239. Let functions f and g be given by $f(x) = \frac{-3}{16}x^3 + \frac{3}{8}x^2 + 2x$ and $g(x) = mx + q$. $S_1(-2\,|-1)$, $S_2(0\,|\,0)$ and $S_3(4\,|\,2)$ are the intersection points of the graphs G_f and G_g.

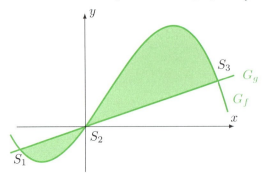

a) Determine the equation of the line g.

b) What is the area of the region enclosed by the two graphs?

240. 1 cubic metre of concrete has a mass of 2.3 (metric) tons.

a) For the construction of a drain, prefabricated concrete segments of length 1 m are used. The figure below on the left shows the cross-section of such a segment (all measurements in metres). The arch is parabolic in shape. Determine the mass of the concrete used in one segment.

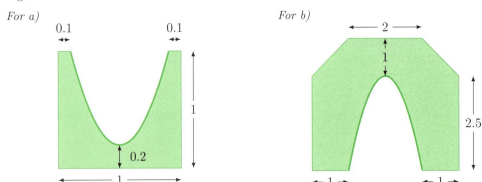

For a)   *For b)*

b) A 10 m long pedestrian tunnel is made using concrete. The figure above on the right shows the cross-section (all measurements in metres) of the tunnel with a parabolic arch. How much concrete is required?

241. The curve defined by the equation $y^2 = x^2(x+3)$ is called the *Tschirnhausen cubic* (EHRENFRIED WALTHER VON TSCHIRNHAUS, 1651–1708). Show that the graph contains a loop. Determine its area.

Hint: $x\sqrt{x+3} = (x+3)\sqrt{x+3} - 3\sqrt{x+3} = (x+3)^{\frac{3}{2}} - 3(x+3)^{\frac{1}{2}}$.

242. a) The graph of $y = \cos(x)$ is approximated by a quadratic parabola p that coincides with the cosine curve at the points at $x = 0$ and $x = \pm\frac{\pi}{2}$. Calculate the area of the region between the two curves.

b) The graph of $y = \cos(x)$ is approximated by a polynomial function p of degree 4 that coincides with the cosine curve at the points $x = 0$ and $x = \pm\frac{\pi}{2}$ and in addition has the same slope as the cosine curve at the points $x = \pm\frac{\pi}{2}$. Calculate the area of the region between the two curves.

243. In a cinema, a screen of 7.5 m height is placed 3 m above the floor (see figure). The first row of chairs is 2.7 m away from the screen and the rows are set 0.9 m apart along the inclined floor. The floor of the seating area is inclined at an angle of $\alpha = 20°$ above the horizontal. Let x be the distance of your seat to the first row along the inclined floor. The cinema has 21 rows, therefore we have $0 \leq x \leq 18$. Your eyes are 1.2 m above the floor.

The best seat is the one from which you see the screen under the largest possible angle of view θ; because the angle θ describes how large the screen appears to you from your seat.

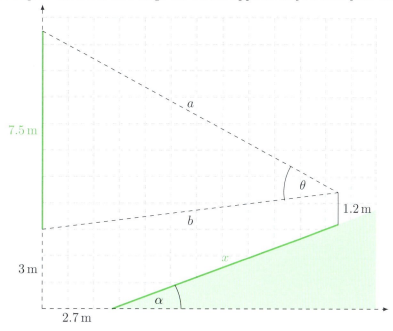

a) Express the angle of view $\theta = \theta(x)$ as a function of the distance x.
Use $a^2 + b^2 - 2ab\cos(\theta) = 7.5^2$ with $a^2 = (2.7 + x\cos(\alpha))^2 + (9.3 - x\sin(\alpha))^2$ and $b^2 = (2.7 + x\cos(\alpha))^2 + (x\sin(\alpha) - 1.8)^2$.

b) Draw the graph of $\theta = \theta(x)$ on the interval $0 \leq x \leq 18$ using a computer and graphically determine the best seat; that is, the value of x, where θ is greatest. Also, determine the worst seat, that is, the value of x, where θ is smallest.

c) Calculate the numerical solution of the equation $\frac{d\theta}{dx} = 0$, i.e. determine the value of x, where θ is a maximum, by purely analytical means. Does this result verify the result of part b)?

d) Consider the graph of θ on the interval $0 \leq x \leq 18$. Geometrically estimate the mean of θ and then calculate this mean value using a computer. Compare this value with the minimum and maximum values of θ.

Remark: The mean value of θ specifies how large the screen appears to a viewer in the cinema auditorium on average.

244. The two well-known formulas for the circumference u of a circle and the area A of the surface of a sphere can be considered as functions of the radius r: $u = u(r) = 2\pi r$ and $A = A(r) = 4\pi r^2$, respectively. For these two functions u and A, calculate the following two definite integrals: $\int_0^r u(t)\,dt$ and $\int_0^r A(t)\,dt$. What do you notice?

245. The segment of the curve $w\colon y = \sqrt{x}$ between the points $A(0\,|\,0)$ and $B(1\,|\,1)$ is first rotated around the x-axis and then around the y-axis. For the calculation of the volume of these two solids, they are cut into n slices, which are replaced by inscribed and circumscribed cylinders.

a) Calculate the lower and upper sums of both solids for $n = 5, 10, 20, 40, 80$.

b) Do the two solids of rotation have the same volume?

246. Let $f(x) = e^x$ and $g(x) = \ln(x)$ be given. The four points $A(0\,|\,0)$, $B(3\,|\,0)$, $C(3\,|\,3)$ and $D(0\,|\,3)$ form a square. Determine the area of the portion of the square that lies between the graphs G_f and G_g.

247. The curve $k_1\colon y = \cos(x)$ is shifted upwards parallel to the axis of the ordinate until it touches the curve $k_2\colon y = \sin(x)$. How far does it need to be shifted and what is the area of the region enclosed by the two touching curves?

248. Show that any tangent to the graph of $f_a\colon y = x^2 + a$, $a > 0$, and the graph of $p\colon y = x^2$ enclose a region of equal area.

249. For which values of b is $\int\limits_{1}^{b}(x^2 - 2x)\,\mathrm{d}x = 0$?

250. Derive the well-known formulas for the following volumes by means of integration:

a) Rotational ellipsoid: $V = \frac{4}{3}\pi ab^2$ if rotated around the x-axis, $V = \frac{4}{3}\pi a^2 b$ if rotated around the y-axis with a, $b \mathrel{\hat{=}}$ semi-axes in the direction of the x- and y-axes

b) Pyramid: $V = \frac{1}{3}GH$, Frustum: $V = \frac{h}{3}(G + \sqrt{GD} + D)$ with $G \mathrel{\hat{=}}$ base area, $D \mathrel{\hat{=}}$ top area, $H \mathrel{\hat{=}}$ total height of pyramid and $h \mathrel{\hat{=}}$ height of frustum

c) Torus: $V = 2\pi^2 R r^2$ with $R \mathrel{\hat{=}}$ major (rotational) radius, $r \mathrel{\hat{=}}$ minor (tube) radius

For Chapter 4.5: Improper Integrals

251. Calculate the improper integral $\int\limits_{1}^{\infty} \dfrac{5}{\sqrt[3]{u^7}}\,\mathrm{d}u$, if it exists.

252. Determine the area of the region extending to infinity, which is enclosed by the curve k with equation $y = k(x) = e^{x-a}$, the tangent t to the curve at point $x = a$ and the x-axis $(a \in \mathbb{R})$. Comment on the result.

253. a) Determining the integral $\int\limits_{0}^{1} \ln(x)\,\mathrm{d}x = [x\ln(x) - x]_0^1$ is non trivial; as $x\ln(x)$ is not defined for

$x = 0$. Therefore, use the geometric relation between $\int\limits_{0}^{1} \ln(x)\,\mathrm{d}x$ and the improper integral

$\int\limits_{-\infty}^{0} e^x\,\mathrm{d}x$ to determine it.

b) Calculate the limit $\lim\limits_{x \to 0^+} x\ln(x)$ using part a).

For Chapter 4.7: Differential Equations

254. To cool down a bottle of mineral water at room temperature ($22\,^\circ$C), it is placed in a refrigerator with an internal temperature of $7\,^\circ$C. After half an hour, the mineral water has cooled to $16\,^\circ$C.

   a) What is the temperature of the mineral water after another half hour?

   b) How long does it take until the mineral water has cooled to $10\,^\circ$C?

255. An intravenous drip uniformly delivers a drug to a patient through the blood. In a hospital, a patient receives 1.5 mg of a drug per minute, that was previously absent in the body, using such a drip infusion. Approximately $7.5\,\%$ of the drug present in the bloodstream is excreted by the kidneys per minute. At the beginning of the treatment, the patient receives 10 mg of the drug by intravenous injection.

   a) Show that this process can be modelled as a bounded growth.

   b) Determine the equation of a function that describes this bounded growth.

   c) What amount of the drug is to be expected in the body of the patient during prolonged treatment?

4.10   Review Exercises

For Chapter 4.1: Integration as the Inverse of Differentiation

256. Determine an antiderivative F for the given function f.

   a) $f(x) = x^2$ b) $f(x) = 4x^3 - 2x + 1$ c) $f(x) = x^n$ d) $f(x) = \frac{2}{x}$

   e) $f(t) = 2 \cdot \cos(t)$ f) $f(t) = \pi \cdot e^t$ g) $f(t) = 3^t$ h) $f(t) = \frac{3}{2}\sqrt{t}$

257. Of a function f, we know its first derivative f' as well as another condition. What is the formula of f?

   a) $f'(x) = 3x^2 - 41,\ f(7) = 65$ b) $f'(x) = 3 + \frac{1}{x^2},\ f(1) = 4$

   c) $f'(t) = 2 \cdot \sin(t),\ f(\frac{\pi}{2}) = 2$ d) $f'(t) = 4 + t,\ f(-4) = -f(4)$

258. Determine the indefinite integral.

   a) $\displaystyle\int 12x^3\,\mathrm{d}x$ b) $\displaystyle\int (14 - 3x^2)\,\mathrm{d}x$ c) $\displaystyle\int (4x - 3)(x^2 + 1)\,\mathrm{d}x$

   d) $\displaystyle\int \frac{2}{x^2}\,\mathrm{d}x$ e) $\displaystyle\int \left(\frac{3}{x^4} - \frac{2}{x^3}\right)\,\mathrm{d}x$ f) $\displaystyle\int \frac{x^3 + 4x - 5}{x}\,\mathrm{d}x$

   g) $\displaystyle\int 3\sqrt{t}\,\mathrm{d}t$ h) $\displaystyle\int (\cos(t) + \sin(t))\,\mathrm{d}t$ i) $\displaystyle\int 12^t\,\mathrm{d}t$

259. True or false?

   a) If F is an antiderivative of f with $f(x) = x^2$ and $F(1) = 1$, then $F(-1) = \frac{1}{3}$.

   b) If F is an antiderivative of f with $f(x) = \sqrt{x}$ and $F(1) = 1$, then $F(-1)$ is not defined.

260. True or false?

    a) If $\int f(x)\,\mathrm{d}x = \int g(x)\,\mathrm{d}x$, then $f(x) = g(x)$.

    b) If $f(x) = g(x)$, then $\int f(x)\,\mathrm{d}x = \int g(x)\,\mathrm{d}x$.

    c) If $f'(x) = g'(x)$, then $f(x) = g(x)$.

261. The shape of a graph G_f of a function f is given. Sketch the graph G_F of the antiderivative F of f that passes through the origin, that is for which $F(0) = 0$ holds.

a)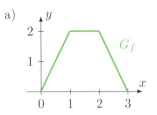

b)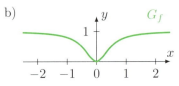

262. Determine an antiderivative F of the given function f by use of the simple substitution rule.

    a) $f(x) = (4x + 3)^2$ b) $f(x) = (3 - \frac{1}{2}x)^4$ c) $f(x) = (5x + 43)^{-2}$ d) $f(x) = (3 - \frac{x}{3})^{-3}$

    e) $f(t) = 3 \cdot \sin(21t)$ f) $f(t) = \cos\left(\frac{-t}{3}\right)$ g) $f(x) = 2 \cdot e^{-2x}$ h) $f(x) = \dfrac{1}{3x - 2}$

For Chapter 4.2: Calculation of Areas by Means of Lower and Upper Sums

263. The graph of the function f defined by the equation $f(x) = \frac{-1}{2} \cdot x^2 + 3x$ and the x-axis enclose a region A over the interval $[0; 6]$. Calculate the lower sums U_6 and U_{12} and the upper sums O_6 and O_{12} using the graph of the function f. What information concerning the area of region A can you deduce from this?

264. Calculate the four following integrals, provided that $\int\limits_{2}^{4} f(x)\,\mathrm{d}x = 3$.

    i) $\displaystyle\int_{2}^{4} f(u)\,\mathrm{d}u$ ii) $\displaystyle\int_{2}^{4} \sqrt{5}f(x)\,\mathrm{d}x$ iii) $\displaystyle\int_{4}^{2} f(t)\,\mathrm{d}t$ iv) $\displaystyle\int_{2}^{4} (-f(x))\,\mathrm{d}x$

265. The functions f and g satisfy $\int\limits_{2}^{7} f(x)\,\mathrm{d}x = -3$, $\int\limits_{4}^{7} f(x)\,\mathrm{d}x = 4$ and $\int\limits_{4}^{7} g(x)\,\mathrm{d}x = 6$. What is the value of each of the following integrals?

    a) $\displaystyle\int_{2}^{7} -4f(x)\,\mathrm{d}x$ b) $\displaystyle\int_{4}^{7} (f(x) + g(x))\,\mathrm{d}x$

    c) $\displaystyle\int_{4}^{7} (2f(x) - 3g(x))\,\mathrm{d}x$ d) $\displaystyle\int_{2}^{4} f(x)\,\mathrm{d}x$

266. Determine the integral function $F_0(x) = \int\limits_{0}^{x} f(t)\,\mathrm{d}t$ of the function f defined by the equation $f(t) = t + 2$, $t \in \mathbb{R}$, using upper and lower sums.

For Chapter 4.3: Fundamental Theorem of Calculus

For **267–269**: Calculate the definite integral.

267. a) $\displaystyle\int_1^6 (3x^2-4x+5)\,\mathrm{d}x$ b) $\displaystyle\int_{-1}^5 (x^2+3x-4)\,\mathrm{d}x$ c) $\displaystyle\int_2^3 \left(x^2 + \frac{1}{x^2}\right)\,\mathrm{d}x$ d) $\displaystyle\int_1^2 \left(3x-12x^{-3}\right)\,\mathrm{d}x$

268. a) $\displaystyle\int_0^{\frac{\pi}{6}} 6\cos(t)\,\mathrm{d}t$ b) $\displaystyle\int_0^\pi (5\sin(t)-4\cos(t))\,\mathrm{d}t$ c) $\displaystyle\int_{\frac{\pi}{6}}^{\frac{\pi}{3}} \frac{3}{\cos^2(t)}\,\mathrm{d}t$

269. a) $\displaystyle\int_0^\pi \mathrm{e}^x\,\mathrm{d}x$ b) $\displaystyle\int_{-1}^1 \mathrm{e}^{x+1}\,\mathrm{d}x$ c) $\displaystyle\int_1^2 3^x\,\mathrm{d}x$ d) $\displaystyle\int_1^e \ln(x)\,\mathrm{d}x$

270. Is the rule correct?

a) $\displaystyle\int_a^b f(x)\,\mathrm{d}x + \int_b^c f(x)\,\mathrm{d}x = \int_a^c f(x)\,\mathrm{d}x$ b) $\displaystyle\int_a^b f(x)\,\mathrm{d}x = -\int_b^a f(x)\,\mathrm{d}x$

c) If $\displaystyle\int_a^b f(x)\,\mathrm{d}x = 0$, then $f(x) = 0$ for $a \le x \le b$.

d) If $f(x) = 0$ for $a \le x \le b$, then $\displaystyle\int_a^b f(x)\,\mathrm{d}x = 0$.

271. Is the integration rule true or false?

a) $\displaystyle\int_a^b \mathrm{d}x = 0$ b) $\displaystyle\int_a^b \mathrm{d}x = b - a$ c) $\displaystyle\int_a^b 0\,\mathrm{d}x = b - a$

272. True or false? Sam Lang walks with velocity $v(t)$ in the time interval $[a; b]$ and his distance to the starting point at time t is given by $s(t)$. Which of the following quantities describes his average velocity in the time interval $[a; b]$?

a) $\displaystyle\frac{1}{b-a}\int_a^b v(t)\,\mathrm{d}t$ b) $\displaystyle\frac{s(b) - s(a)}{b - a}$

c) $v(x)$ for at least one $x \in [a; b]$ d) $\displaystyle\frac{v(a) + v(b)}{2}$

273. The graph G_f of a function f with $y = f(x)$ lies entirely above the x-axis. Let $A(x)$ denote the area of the region in quadrant I that lies below G_f and stretches from the origin to the abscissa x.

a) Describe the relationship between $A(x)$ and an (arbitrary) antiderivative F of the function f.

b) Determine the indefinite integral of $f(x)$, if $f(x) = 3x^2 - 4x + 10$ holds.

For Chapter 4.4: Different Interpretations of the Integral

274. For which values of a is $\displaystyle\int_0^a (x + 1)\,\mathrm{d}x = 12$? Illustrate the result with a figure.

275. Justify geometrically as well as by calculation: $\displaystyle\int_0^1 x^2\,\mathrm{d}x + \int_0^1 \sqrt{x}\,\mathrm{d}x = 1$.

276. Which of the three formulas correctly describe the area of the shaded region A?

i) $\int_0^a f(x)\,dx - \int_0^b g(x)\,dx$

ii) $\int_0^b |f(x) - g(x)|\,dx$

iii) $\int_0^b g(x)\,dx - \int_0^a f(x)\,dx$

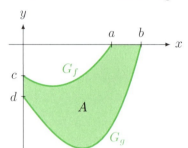

277. Why is the value of the integral $\int_0^{-3} 12x\,dx$ positive, even though the integrand $12x$ is not positive at any point of the considered interval $[-3; 0]$?

278. For which $a \in \mathbb{R}$ does the following equation hold?

a) $\int_{-a}^a \cos(x)\,dx = 0$

b) $\int_{-a}^a \sin(x)\,dx = 0$

279. Justify without calculation why the equation holds.

a) $\int_{-1}^1 \tan(x)\,dx = 0$

b) $\int_{-2}^2 x^3 \cos(x)\,dx = 0$

280. Determine all extrema of $f(x) = \int_0^x (3t^2 - 3)\,dt$.

281. Calculate the area A of the finite region enclosed by the two curves k_1 and k_2.

a) $k_1: y = \frac{1}{4}x^2$
   $k_2: y = 5 - x^2$

b) $k_1: y = x^3 - 3x^2 + 2$
   $k_2: y = 2$

c) $k_1: y = x^3 - 3x^2 - 4x$
   $k_2: y = 0$

282. Assume that Susanne runs faster than Katharina during the whole 1,500-metre run. What is the physical meaning of the area between the velocity curves of the two pupils during the first minute of the race?

283. The region between the two curves is rotated around the specified axis. Calculate the volume of the solid of rotation.

a) $f: y = \frac{1}{2}x$, $g: y = \sqrt{x}$; x-axis

b) $f: y = \sqrt{2x + 2}$, $g: y = 2\sqrt{x - 3}$; x-axis

c) $f: y = 2x$, $g: y = x^2$; y-axis

d) $f: y = \frac{1}{2}x^2 - 1$, $g: y = \frac{1}{4}x^2 + 3$; y-axis

284. The graph of the function f with $f(x) = cx + c$ on the interval $[2; 5]$ is rotated around the x-axis. The volume of the solid of rotation is 7π. Determine c.

285. Lili and Lulu calculate the volume of the solid of rotation formed by rotating the region between the graphs of f and g in the interval $[a; b]$ around the x-axis. Here, $f(x) \geq g(x) \geq 0$ holds for all $x \in [a; b]$. Lili claims that $V = \pi \int_a^b \left((f(x))^2 - (g(x))^2 \right) dx$. Lulu on the other hand claims that $V = \pi \int_a^b (f(x) - g(x))^2\,dx$. Who is right? Justify.

286. Calculate the mean $\bar{f}$ of the function f on the interval $[a; b]$. At which point does this mean occur as the function value?

a) $f(x) = x^2;\ a = 0,\ b = 3$ b) $f(t) = \sin(t);\ a = 0,\ b = \frac{\pi}{2}$

For Chapter 4.5: Improper Integrals

287. Calculate the improper integral.

a) $\displaystyle\int_1^\infty \frac{3}{x^2}\,\mathrm{d}x$ b) $\displaystyle\int_2^\infty \frac{4}{(x-1)^3}\,\mathrm{d}x$ c) $\displaystyle\int_{-\infty}^{-2} \frac{3}{x^4}\,\mathrm{d}x$

288. Examine if the region shaded in the figure and extending to infinity has a finite area A or not. If finite, determine the area.

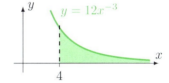

289. The curve $k\colon y = \frac{3}{x}$ is rotated around the x-axis in the domain $x \geq 12$. Calculate the volume V of the solid of rotation that extends to infinity along the x-axis.

For Chapter 4.7: Differential Equations

290. a) What is a differential equation?

b) What is the order of a differential equation?

c) What is an initial condition?

d) What is a solution of a differential equation?

291. What is the slope field of the differential equation $y' = F(x, y)$?

292. a) What is the equation that describes exponential growth? What does it state, expressed in terms of the relative growth rate?

b) Under which conditions is this a suitable model for the growth of a population?

c) What are the solutions of this differential equation?

293. a) What is the equation for logistic growth?

b) Under which conditions is this a suitable model for the growth of a population?

294. True or false? Justify oder disprove by giving a counterexample.

a) All solutions of the differential equation $y' = -1 - y^4$ are monotonically decreasing functions.

b) The function $f(x) = \frac{\ln(x)}{x}$ is a solution of the differential equation $x^2 y' + xy = 1$.

X   Functions

X.1   Fundamental Aspects

Functions

A *function* is an assignment between two sets X and Y, which assigns exactly one element y of Y to each element x of X.

Notation: $f \colon x \mapsto y = f(x)$

In practice, functions are commonly given, or described by, *formulas/equations*, *tables of values* or *graphs* of the functions.

Example: The assignment «$f \colon$ pupil $\mapsto$ score for the soltion of an exercise» is unambiguous and therefore a function. Exactly one score is assigned to each pupil. The inverse assignment «$f \colon$ score $\mapsto$ pupil» for the example given however is not a function, as no name is assigned to the score 3 and multiple names, namely three, are assigned to the score 4.

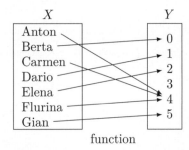
function

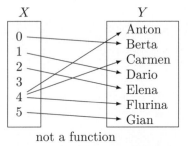
not a function

The element x is called the *argument* and the element y or $f(x)$ the *image* of x under f or the *value* of f at the point x.

Example: For the function given by the equation $y = f(x) = \frac{3}{x-2}$, the image of $x = 1$ under f is $y = f(1) = \frac{3}{1-2} = -3$.

An argument x_0 is called a *root* or *zero*, when $f(x_0) = 0$ holds.

Example: $f \colon x \mapsto x^2 + x - 2$ has a root at point $f \colon x \mapsto x^2 + x - 2$, as $f(1) = 1^2 + 1 - 2 = 0$.

Domain and range

The *domain D* of a function f is the set of all elements either consisting of all possible arguments or a specified subset that is considered as the set of arguments for the function. The *range* or *image W* of f is the set of all function values:

$$D = \{x \mid f(x) \text{ is defined}\}, \ W = \{f(x) \mid x \in D\}$$

Example: For $y = f(x) = \frac{2}{x-5}$, we have $D = \mathbb{R} \setminus \{5\}$ (the fraction is not defined for $x = 5$ as the value of the denominator would be 0) and $W = \mathbb{R} \setminus \{0\}$ (the formula $\frac{2}{x-5}$ can take all values except 0).

Graph of a function

The *graph* G_f of a function f is the set of all points $(x|f(x))$ in the coordinate system, where x is an element of the domain of f.

Notation: $G_f = \{(x|f(x)) \mid x \in D\}$

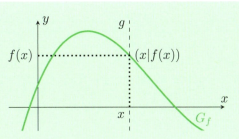

In other words: A graph G_f of a function is a curve or a set of points in the xy-plane, which has exaclty one common point with every vertical line at any arbitrary $x \in D$.

Functions describe the dependence of a quantity on another quantity. Indeed, for a function $f\colon x \mapsto y = f(x)$ the argument x determines the value of the image y. Therefore, y is called the *dependent* and x the *independent* variable.

1. Consider the function given by the equation $y = f(x)$. Determine $f(0)$, $f(5)$, $f\left(-\frac{1}{5}\right)$ and $f(\sqrt{2})$ and simplify as much as possible.

a) $f(x) = x^2$

b) $f(x) = 2 - \frac{1}{4}x$

c) $f(x) = \dfrac{1}{1 - x^4}$

d) $f(x) = \sqrt{3x + 1}$

e) $f(x) = \ln(x + 1)$

f) $f(x) = 3^{x-5}$

2. Let a function be given by the equation $f(x) = \frac{1}{x}$.

a) Determine the value of the function for the argument 10.

b) Determine the argument for which the value of the function is 0.004.

3. Calculate the image of the function or the argument, respectively, for the function f defined by the equation $f(x) = \frac{1}{x}$.

a) $f(2)$

b) $f\left(\frac{1}{3}\right)$

c) $f(0.2)$

d) $f(2.5 \cdot 10^{12})$

e) $f(x) = \frac{2}{3}$

f) $f(x) = 0.25$

g) $f(x) = \frac{-7}{4}$

h) $f(x) = -5 \cdot 10^6$

4. Calculate the image of the function or the argument, respectively, for the function f given by the equation $f(x) = \frac{1}{x}$.

a) $f\left(\frac{p}{q}\right)$

b) $f(x) = \frac{2}{n}$

c) $f(f(8))$

d) $f\left(f\left(-\frac{1}{5}\right)\right)$

5. Determine the value of the function f defined by the equation $f(x) = x^2$.

a) $f(-5)$

b) $f(a)$

c) $f\left(\frac{a}{b}\right)$

d) $f(2a)$

e) $f(n + 2)$

f) $2 \cdot f(4)$

g) $2 \cdot f(x^2)$

h) $f(2) + 3$

i) $f(m) + 3$

j) $f(x - 10) - 100$

k) $2 \cdot f(-7) + 1$

l) $a \cdot f(x - c) + d$

6. Let the function f be defined by the equation

a) $y = f(x) = x - 1$

b) $y = f(x) = x^2$

c) $y = f(x) = x^2 - 1$

d) $y = f(x) = 1 + \frac{1}{x}$

Determine each of the following eight images for each function and simplify as much as possible.

- $f(m)$
- $f(2x)$
- $f(-x)$
- $f(x^2)$
- $f(x + 1)$
- $f(x + h) - f(x)$
- $f\left(\frac{1}{x}\right)$
- $f(\sqrt{x})$

7. 🗎 Complete the table of values and draw the correspponding graph in a coordinate system with $-5 \le x \le 5$, $-5 \le y \le 6$.

a)

x	-2	-1			2.2	10
$f(x)$			0	3		

for $f(x) = 3x$

b)

x		-3	0	1.7		6
$f(x)$	-4				2	

for $f(x) = \frac{2}{3}x$

c)

x	-1	-0.5	0	1	1.5	2
$f(x)$						

for $f(x) = -2x^2 + 7x$

d)

x	-4	-2	0			
$f(x)$				2	4	6

for $f(x) = x + |x|$

e)

x	-1			3		
$f(x)$		1	$\sqrt{3}$		$\sqrt{7}$	3

for $f(x) = \sqrt{x+1}$

8. Draw the graph of the function f defined by the equation $y = \begin{cases} 2x - 4, & x < 4 \\ 4, & 4 \le x \le 8 \\ -2x + 20, & x > 8 \end{cases}$

9. Let the two functions f and g be given by the equations $f(x) = 2x + 3$ and $g(x) = x^2$. For $-3 \le x \le 3$, draw the following five functions in the same coordinate system.

- f
- g
- $f + g$
- $f - g$
- $f \cdot g$

10. At which points does the graph of the function defined by $y = f(x)$ intersect the coordinate axes?

a) $y = -9x - 2$ b) $y = x^2 - 25$ c) $y = x^2 + x + 1$

d) $y = \sqrt{x - 3} - x + 5$ e) $y = \ln(x + 2)$ f) $y = 2 \cdot e^x$

11. Determine the intersections of the graphs of the functions f and g.

a) $f(x) = 2x - 1$, $g(x) = \frac{1}{5}x + 2$ b) $f(x) = 3x + 1$, $g(x) = -5x^2 - x + 10$

c) $f(x) = 4x + 8$, $g(x) = 0.1x^2 + 4x + 8$ d) $f(x) = -8x^2 - x + 10$, $g(x) = 2x^2 + 1$

e) $f(x) = x - \frac{3}{2}$, $g(x) = \sqrt{x + 1}$ f) $f(x) = 5^{2x}$, $g(x) = 4 \cdot 2^x$

12. Determine the value of the parameter a such that the function f has a root at the point $x = 4$.

a) $f(x) = a \cdot x - 1$ b) $f(x) = x^2 + x + a$

c) $f(x) = -x^3 + a \cdot x^2 - 4a \cdot x + 64$ d) $f(x) = a^{x-2} - a^2$

e) $f(x) = \frac{4}{3} \cdot \sqrt{x + a} - a + 1$, $x \ge -a$ f) $f(x) = \mathrm{lb}(2x + a) - 3$, $x > \frac{-a}{2}$

13. The graph of the function f defined by the equation $f(x) = x \cdot \sqrt{x - \frac{1}{2}a}$ passes through the point $\left(\frac{1}{2} \mid 3\right)$. Determine the value of the parameter a.

14. The functions p and q are given by the equations $p(x) = \frac{1}{4}x^2 - ax + 7$ and $q(x) = 3x - 9a$. Determine the value of the parameter a such that the graphs of the functions p and q share exactly one point.

15. Determine the largest possible domain D for the function g given by the equation.

    a) $g(x) = \sqrt{x^2}$ b) $g(x) = \sqrt{2x - 6}$ c) $g(x) = \dfrac{9}{x^2 - 9}$ d) $g(x) = \dfrac{-4}{x^2 + 4}$

16. Let the function f be defined by the equation $y = -(x - 1)^2 + 4$ and the domain D. The graph of the function is depicted below on the left. Give the range W of the function f.

    a) $D = \mathbb{R}$ b) $D = [-2; 2]$ c) $D = [5; \infty[$

For exercise 16

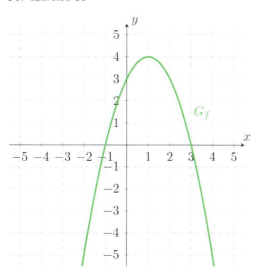

For exercise 17

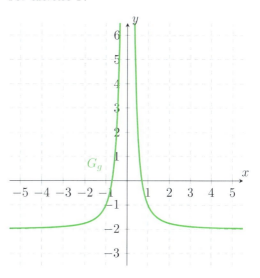

17. Determine the range W of the function g given by the equation $y = \frac{1}{x^2} - 2$ and depicted above on the right.

    a) $D = \mathbb{R}\backslash\{0\}$ b) $D =]-3; -0.5]$ c) $D = [1; \infty[$

18. Determine the range of the function $f \colon x \mapsto x^2$ for the domain given.

    a) $[2; 5]$ b) $[-1; 4]$ c) $[-4; 1]$ d) $[-3; 3]$ e) $]-3; 3[$ f) $\mathbb{R}$

19. Determine the domain and range of the functions from exercise 1. Also, draw the corresponding graphs, if necessary with the aid of a computer.

20. Give an equation for a function that has the given domain D.

    a) $D = \mathbb{R}_0^+$ b) $D = [3; \infty[$ c) $D = \mathbb{R}$ d) $D = \mathbb{R}\backslash\{-1, 2\}$

21. In number theory, the function that assigns the number of divisors to each natural number n is denoted by σ. Determine $\sigma(n)$ or n, respectively. *Note:* 1 and n are divisors of n.

a) $\sigma(15)$ b) $\sigma(36)$ c) $\sigma(n) = 2$ d) $\sigma(n) = 1$

e) $\sigma(64)$ f) $\sigma(11^3)$ g) $\sigma(13 \cdot 19)$ h) $\sigma(n) = 3$

i) What is the domain D of the function σ?

j) Draw the graph of the function σ for the subdomain $1 \leq n \leq 16$.

22. The function d assigns the remainder after divison by 4 to each natural number n.

a) Determine the images $d(15)$, $d(32)$, $d(102)$ and $d(1291)$.

b) Sketch the graph G_d for the subdomain $1 \leq n \leq 14$ and give the range W of d.

c) Simplify the images as much as possible: $d(n + 12)$, $d(4n + 11)$, $d(8n - 3)$; $n \in \mathbb{N}$.

d) Graphically represent the function s defined by $s(n) = d(n) + d(n+1) + d(n+2) + d(n+3)$ for $n \in \mathbb{N}$.

23. Scales of temperature: A temperature C in degrees Celsius is converted into the corresponding temperature F in degrees Fahrenheit using the formula $F = \frac{9}{5} \cdot C + 32$.

a) What is the domain of the temperature C in degrees Celsius?

b) Determine the range of F.

c) What is the meaning of the slope of the line that represents the dependence of F on C?

d) What is the meaning of the y-intercept of the line that represents the dependence of F on C?

e) Determine the formula with which, conversely, a temperature F in degrees Fahrenheit can be converted into C in degrees Celsius.

24. A coin is tossed vertically into the air. The coin is let go at a height of 1 m above the floor and has an initial velocity of 9.5 m/s. The corresponding function is given by $h(t) = 1 + 9.5t - \frac{1}{2} \cdot 10t^2$ (t: time in seconds, h: height above the floor in metres).

Remark: The air resistance is disregarded and gravitational acceleration g is set to $10 \, \text{m/s}^2$ for simplicity's sake.

a) Transfer the table of values to your notebook and complete it.

t	0	0.5	1	1.5	2
$h(t)$					

b) Draw the graph of the function using the table from part a).

c) How many seconds after the coin is let go will it hit the floor?

d) What values can t and $h(t)$, respectively, take (domain and range)?

25. *Temperature profile.* The following temperature profile was observed on a cloudless day in September. At 5:30 the temperature was minimal, at 16:00 maximal.

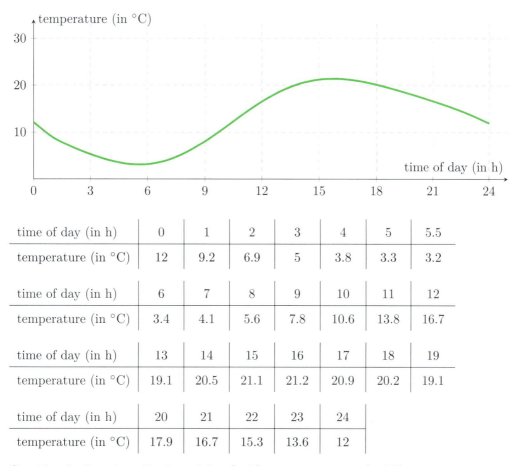

time of day (in h)	0	1	2	3	4	5	5.5
temperature (in °C)	12	9.2	6.9	5	3.8	3.3	3.2

time of day (in h)	6	7	8	9	10	11	12
temperature (in °C)	3.4	4.1	5.6	7.8	10.6	13.8	16.7

time of day (in h)	13	14	15	16	17	18	19
temperature (in °C)	19.1	20.5	21.1	21.2	20.9	20.2	19.1

time of day (in h)	20	21	22	23	24
temperature (in °C)	17.9	16.7	15.3	13.6	12

Consider the function «T: time of day (in h) $\longmapsto$ temperature (in °C)».

a) List the following temperatures: $T(2\,\mathrm{h})$, $T(11\,\mathrm{h})$, $T(17\,\mathrm{h})$, $T(22\,\mathrm{h})$.

b) At what time t is $T(t) = 19.1\,°\mathrm{C}$?

c) For which time interval is $T(t) \geq 16.7\,°\mathrm{C}$?

d) For which time interval is $T(t) \leq 16.7\,°\mathrm{C}$?

e) Determine the domain and the corresponding range of the function T.

26. A newly delivered metal box containing 120 cans of food and weighing 95 kg arrives in a grocery store on Monday morning. After 50 cans of food are taken out and sold in the first couple of days, the box with the remaining cans has a weight of 60 kg.

a) What is the weight $G(d)$ of the box including the remaining cans, if $1 \leq d \leq 120$ cans have been removed? Derive a formula.

b) Calculate the weight of the metal box including its content, if half of the initial number of cans are removed.

c) How many cans are still in the box, if the total weight is 32 kg?

d) When the total weight falls under 20 kg, a new box is ordered. How many cans have been removed when this is the case?

27. The figure depicts four different shapes of flower vases and below them four different graphs, which show the volume V of water as a function of the fill height h. Which graph fits to which shape of vase?

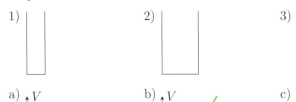 

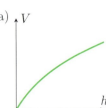 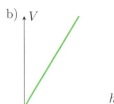 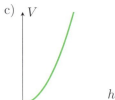 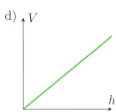

28. The figure depicts the cross-sections of two different flower vases. Draw a graph that depicts the volume V of water as a function of the fill height h for each vase (see exercise 27).

 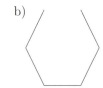

29. The empty vases from exercise 27 are filled at a water tap, whose water stream does not vary during the filling stage, i.e. the flow rate per minute is constant. Sketch the graph of the function «fill time t [min] $\longmapsto$ fill height h [cm]» for each shape of vase.

30. Two people march from the same point, one of them to the north with a constant velocity of $5\,\mathrm{km/h}$ and the other to the east with a constant velocity of $6\,\mathrm{km/h}$. Determine a formula for the distance d inbetween the two of them at time $t \geq 0$ (in hours).

31. A funnel-shaped water tank has the shape of an inverted circular cone with a base radius of $R = 2\,\mathrm{m}$ and height $H = 4\,\mathrm{m}$. The tank is filled up to a height h with water. Determine the volume V of water as a function of h for $0 \leq h \leq H$.

32. In parts a) to d), determine a linear function, using one of the two forms of formulas for linear functions, that maps the interval I_1 onto the interval I_2.

a) $I_1 = [0; 1]$, $I_2 = [0; 5]$ b) $I_1 = [0; 1]$, $I_2 = [2; 8]$

c) $I_1 = [3; 9]$, $I_2 = [0; 5]$ d) $I_1 = [-3; 3]$, $I_2 = [-8; 8]$

e) Determine the formula of a quadratic function that maps the interval I_1 onto the interval I_2 from part d). *Note:* There are infinitely many possible solutions.

33. Do p and q describe the same function? Reason with reference to the equation.

Remark: Two functions are equal if and only if their terms are equivalent and they have the same domain.

a) $p(x) = x - 1$, $q(x) = \dfrac{x^2 - 1}{x + 1}$ b) $p(x) = \ln(x^4)$, $q(x) = 4 \cdot \ln(x)$

Monotonicity and Boundedness

A function f is called *monotonically increasing* (or *monotonically decreasing*) on an interval I, if for all x_1, $x_2 \in I$ one has:

$$x_1 \leq x_2 \;\Rightarrow\; f(x_1) \leq f(x_2) \quad (\text{or } x_1 \leq x_2 \;\Rightarrow\; f(x_1) \geq f(x_2)).$$

A function f is called *strictly increasing* (or *strictly decreasing*) on an interval I, if for all x_1, $x_2 \in I$ one has:

$$x_1 < x_2 \;\Rightarrow\; f(x_1) < f(x_2) \quad (\text{or } x_1 < x_2 \;\Rightarrow\; f(x_1) > f(x_2)).$$

A function f is *bounded (from) below and (from) above*, if there are two numbers m and M such that all images of the function lie between them:

$$m \leq f(x) \leq M \quad \text{for all } x \in I.$$

Examples:
- f with $f(x) = x^3$ is strictly increasing on the whole of $\mathbb{R}$.
- f with $f(x) = \ln(x)$ is strictly increasing for $x > 0$.
- f with $f(x) = e^{-x}$ is strictly decreasing on the whole of $\mathbb{R}$.
- f with $f(x) = \sin(x)$ is bounded below by -1 and above by 1.
- f with $f(x) = \cos(x)$ is bounded below by -1 and above by 1 also.
- f with $f(x) = e^x$ only has positive images and so is bounded below by 0.

For **34–37**: Proceed similarly as in the following example.

Show that the function f with $f(x) = x^2$ is stricly increasing on $[0; \infty[$.

Argumentation: $\quad 0 \leq x_1 < x_2 \Leftrightarrow x_1 - x_2 < 0$. As $x_1 + x_2 > 0$ it follows that

$$(x_1 - x_2)(x_1 + x_2) < 0 \;\Leftrightarrow\; x_1^2 - x_2^2 < 0 \;\Leftrightarrow\; x_1^2 < x_2^2 \;\Leftrightarrow\; f(x) < f(y).$$

34. Show that the function f with $f(x) = x^2$ is strictly decreasing on $]-\infty; 0]$.

35. Show that the function f with $f(x) = \sqrt{x}$ is strictly increasing on $[0; \infty[$.

36. Show that the function f with $f(x) = \frac{1}{x}$ is strictly decreasing for $x < 0$ as well as for $x > 0$.

37. Show that the function f with $f(x) = \frac{1}{x^2}$ is strictly increasing for $x < 0$ and strictly decreasing for $x > 0$.

38. Answer the question using the graph of the function f. In which intervals
  a) is $f : x \mapsto y = \sin(x)$ strictly increasing or strictly decreasing, respectively?
  b) is $f : x \mapsto y = \frac{1}{2}x^2 - 2x + 1$ strictly increasing or strictly decreasing, respectively?
  c) is $f : x \mapsto y = x^3$ monotonically increasing?

39. a) The function $f\colon x \mapsto y = \frac{1}{x^2}$ is bounded below by 0 and unbounded above. Is the function $g\colon x \mapsto y = \frac{1}{1+x^2}$ also bounded below by 0 and unbounded above?

   b) The function $f\colon x \mapsto y = \mathrm{e}^{-x}$ is bounded below by 0 and unbounded above. Is the function $g\colon x \mapsto y = \mathrm{e}^{-x^2}$ also bounded below by 0 and unbounded above?

   c) Give the equation of a function f, which is bounded above by 1 and below by 0.

Continuity

A function f is called *continuous* at the point $a \in D$, if the following condition is fulfilled: The value $f(x)$ tends to the value $f(a)$, when the argument x tends to the value a, both from the left and the right.

If a function f is continuous at every point of an interval I, the function is called continuous on the whole interval I. For arguments on the boundary of the interval, it suffices to consider the left- or right-sided limit, respectively.

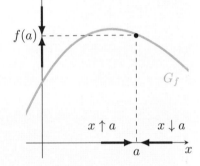

Notation: $f(x) \to f(a)$ for $x \to a$ or
$$\lim_{x \to a} f(x) = f\left(\lim_{x \to a} x\right) = f(a)$$

Note: More on limits is found in chapter 2 on page 30.

In all of the intervals where f is continuous, its graph G_f can be drawn as a contiguous curve.

A point x_0, where a function is *not continuous* is called a *point of discontinuity*.

To show the continuity of a function f at a point a, one often simply verifies the condition
$$\lim_{x \uparrow a} f(x) = \lim_{x \downarrow a} f(x) = f(a).$$

Example: Let the graph of the function f be given. Here, green points are points on the curve while white points are points that are not part of the curve, i.e. excluded.

- f is discontinuous at the point 0, as $f(0)$ is not defined (singularity).

- $f(2) = 3$, but $\lim_{x \uparrow 2} f(x) = 3$, $\lim_{x \downarrow 2} f(x) = 1$,
  i.e. $\lim_{x \to 2} f(x)$ does not exist.
  Thus, f is discontinuous at the point 2.

- $\lim_{x \uparrow 5} f(x) = 2$ and $\lim_{x \downarrow 5} f(x) = 2$,
  i.e $\lim_{x \to 5} f(x) = 2$, but $f(5) = 1.5 \neq 2$.
  Thus, f is discontinuous at the point 5.

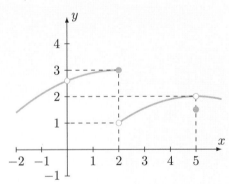

40. Determine all intervals in the region $-5 \leq x \leq 8$, where the depicted function is continuous. *Note:* Filled green points are points on the curve while white points are points that are not part of the curve, i.e. excluded.

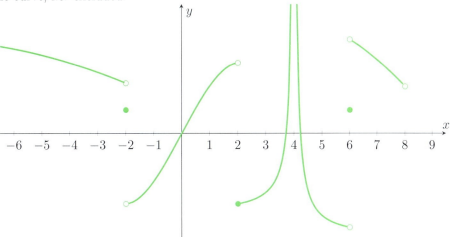

41. Is the function f continuous or discontinuous at the point $x = 0$?

a) $f(x) = x^2$ b) $f(x) = |x|$ c) $f(x) = \frac{1}{x}$ d) $f(x) = \frac{x}{x}$

42. Is the function f continuous or discontinuous at the point a?

a) $f(x) = x^3 - 9x$, $a = 4$ b) $f(x) = (x+5)^{-2}$, $a = -5$ c) $f(x) = |x-3|$, $a = 3$

43. True or false? If a function f is continuous at the point 5 with $f(5) = 2$, then we have $\lim\limits_{x \to 2} f(4x^2 - 11) = 2$.

44. Is the function f continuous or discontinuous at the point $x = 0$?

a) $f(x) = \begin{cases} x - 1, & x \leq 0 \\ x + 1, & x > 0 \end{cases}$ b) $f(x) = \begin{cases} 1, & x \leq 0 \\ 2^x, & x > 0 \end{cases}$

c) $f(x) = \begin{cases} 1, & x \leq 0 \\ 2^{-x}, & x > 0 \end{cases}$ d) $f(x) = \begin{cases} 1, & x \leq 0 \\ 2^{\frac{1}{x}}, & x > 0 \end{cases}$

45. Determine the points of disconinuity of the function f given.

a) $f(x) = \dfrac{x-1}{x^2}$ b) $f(x) = \dfrac{x^2-1}{x^2+1}$ c) $f(x) = \dfrac{x}{(x^2-1)(x^2+1)}$

46. An object of mass $1\,\mathrm{kg}$ has the distance r to the centre of Earth. The gravitational force F the Earth exerts on this object is given by $F(r) = \begin{cases} \frac{G \cdot M}{R^3} \cdot r, & 0 \leq r < R \\ \frac{G \cdot M}{r^2}, & r \geq R \end{cases}$ ($M \, \hat{=} \,$ Earth's mass, $R \, \hat{=} \,$ Earth's radius, $G \, \hat{=} \,$ gravitational constant). Is F a continous function for $r \geq 0$?

47. For which values of c is the function f with $f(x) = \begin{cases} cx^2 + 2x, & x < 2 \\ x^3 - cx, & x \geq 2 \end{cases}$ continuous on the whole of $\mathbb{R}$?

X.2   Further Aspects Concerning Functions

Composition

48. *Typical questions.* Let the following functions be given:

- $f_1(x) = 2x$
- $f_2(x) = x + 2$
- $f_3(x) = \frac{1}{x}$
- $f_4(x) = x^2$

Determine the equation of the formula and simplify the formula as much as possible.

a) $m(x) = f_1(x) + f_2(x)$ b) $m(x) = f_1(x) - f_4(x)$ c) $m(x) = f_2(x) \cdot f_3(x)$

d) $m(x) = f_4(x) : f_3(x)$ e) $m(x) = f_1(f_4(x))$ f) $m(x) = f_4(f_1(x))$

Composition of functions

Let the two functions u and v be given. The new function $u \circ v$ defined by the formula $(u \circ v)(x) = u(v(x))$ is called the *composition* of u and v.

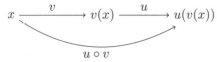

The function u is called the *outer function* and v the *inner function*.

Example: Let the functions $u(x) = \tan(x)$ and $v(x) = 5x^3 - 9$ be given. Then, we have:

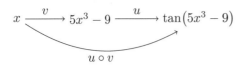 

$$u(v(x)) = u(5x^3 - 9) = \tan(5x^3 - 9) \qquad v(u(x)) = v(\tan(x)) = 5\tan^3(x) - 9$$

This example shows that the compostion of two functions does not have to be *commutative*. This means that generally $u \circ v \neq v \circ u$.

Naturally, functions can be multiply composed.

Example: Let $u(x) = \sqrt{x}$, $v(x) = 6x - 1$ and $w(x) = \sin(x)$. Then, we have:

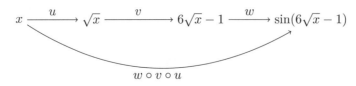

$$w(v(u(x))) = w(v(\sqrt{x})) = w(6\sqrt{x} - 1) = \sin(6\sqrt{x} - 1)$$

On the other hand, functions can also be written as the composition of two or more functions.

Example: For example, the function $h(x) = (x^2 + 3x)^4$ can be written as the composition of two functions:

$$\left. \begin{array}{l} f(x) = x^4 \\ g(x) = x^2 + 3x \end{array} \right\} \quad \Rightarrow \quad h(x) = f(g(x)) = f(x^2 + 3x) = (x^2 + 3x)^4$$

49. Determine $f(g(x))$ and $g(f(x))$.

   a) $f(x) = x^2$, $g(x) = 2x + 7$ b) $f(x) = \sqrt{x}$, $g(x) = 3 - 4x$ c) $f(x) = \frac{2}{x}$, $g(x) = x^2 + 3$

50. The three functions $u = u(x) = \sqrt{x}$, $v = v(x) = 1 - x^2$ and $w = w(x) = \cos(x)$ with $0 \le x \le \frac{\pi}{2}$ are composed into a new function in different ways. Determine the simplified formula of the function.

   a) $v \circ u$ b) $v \circ v$ c) $v \circ w$ d) $w \circ v$

   e) $u \circ v \circ w$ f) $w \circ v \circ u$ g) $v \circ v \circ u$ h) $v \circ v \circ w$

51. Determine $f(x)$ and $g(x)$ such that $w(x) = f(g(x))$ as given.

   Remark: The trivial composition with $g(x) = x$ and $f(x) = w(x)$ is not allowed.

   a) $w(x) = (3x + 10)^3$ b) $w(x) = \dfrac{1}{2x + 4}$ c) $w(x) = \sqrt{x^2 - 3x}$

   d) $w(x) = 6 \cdot \sin(1 - x)$ e) $w(x) = \dfrac{10}{(3x - x^2)^3}$ f) $w(x) = \cos(\sqrt{x} + 1)$

   g) $w(x) = \tan^5(x)$ h) $w(x) = 8 \cdot e^{\sqrt{x}}$ i) $w(x) = 5 \cdot \ln(\tan(x)) - 14$

52. Let the composite functions F_1 and F_2 be defined as follows: $F_1(x) = f(g(x)) = \frac{g(x)}{4} - 3$, with $g(x) = 4x + 12$, and $F_2(x) = u(v(x)) = \frac{1}{v(x)} + 2$, with $v(x) = \frac{1}{x-2}$.

   Show that F_1 and F_2 are the same function and, for $x \ne 2$, one therefore has $F_1(x) = F_2(x)$.

53. Let the function f be given. Find a function g such that $f(g(x)) = x$.

   a) $f(x) = \frac{1}{4}x - 1$ b) $f(x) = \sqrt{2x} + 1$

   c) $f(x) = -4 \cdot (x - 9)^2$ d) $f(x) = 10 \cdot \ln(6x + 13) + 1$

54. Write $f(x) = (x^2 - 1)^3$ as a composition of

   a) two functions.   b) three functions.   c) four functions.

55. ▯ Let two functions be given by the equations $t_1(x) = x - \frac{3}{2}$ and $t_2(x) = \frac{1}{2}x$. Sketch the graph of the functions $f \circ t_1$ and $f \circ t_2$, if f is the given function. Describe the influence of the composition on the graph of the function f.

   a) $f(x) = x^2$ b) $f(x) = x(x + 2)(x - 2)$ c) $f(x) = \sin(2\pi \cdot x)$

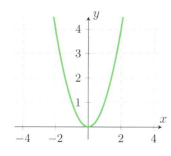 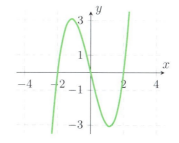 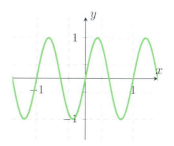

56. Let the function f be given by the equation $y = \sqrt{x}$ and g by $y = \frac{1}{x}$. Determine the domain D and the range W of

   a) f b) g c) $f \circ g$

57. Verify that the function given by $y = f(x)$ is *involutory*. This means that for every x in its domain it fulfils: $f(f(x)) = x$.

   a) $y = f(x) = -x$ b) $y = f(x) = \frac{1}{x}$

   c) $y = f(x) = \frac{-1}{x}$ d) $y = f(x) = \begin{cases} 3 - \frac{x}{2}, & x \in [0; 2] \\ 6 - 2x, & x \in \,]2; 3] \end{cases}$

58. Show that the function h given by $y = h(x)$, for every x, fulfils: $h(h(x)) = h(x)$. Functions with this property are called *idempotent*.

   a) $h(x) = x$ b) $h(x) = k \,\hat{=}\, \text{konst.}$ c) $h(x) = |x|$ d) $h(x) = \begin{cases} 0, & x \leq 0 \\ 1, & x > 0 \end{cases}$

59. Let the three functions u with $u(x) = 3x$, v with $v(x) = \frac{x}{3}$ and w with $w(x) = x - 12$ be given. Simplify the triple compositions as much as possible.

   a) $u(v(w(x)))$ and $v(w(u(x)))$ b) $u(w(u(x)))$ and $w(u(u(x)))$

   c) $v \circ v \circ w$ and $v \circ w \circ v$ d) $(u \circ w \circ w) - 2 \cdot (w \circ u \circ w) + (w \circ w \circ u)$

60. The figure shows the graphs of the functions f and g defined by $f(x) = \frac{1}{1+x}$ and $g(x) = x$ as well as the points $P_1, \ldots, P_5$ on these graphs.

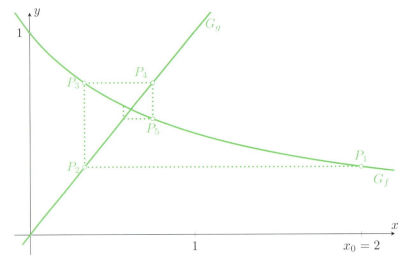

   a) Calculate the coordinates of the points P_1 to P_5.

   b) Start with an arbitrary x_0 and express the coordinates of the points this yields using the compositions f, $f \circ f$, $f \circ f \circ f, \ldots$.

   c) Describe in words what happens, if one continues to iterate the compositions $(f \circ f \circ \ldots \circ f)(x_0)$ endlessly.

61. Let the function f be given by $f(x) = \begin{cases} |x|, & x \geq -2 \\ x+4, & x < -2 \end{cases}$

Determine all solutions of the equation $f(f(f(x))) = 0$.

62. Let the function f be given by the equation $y = f(x) = \frac{1}{1-x}$. Calculate the 1999th composition for $x = 2000$, i.e. $f^{\langle 1999 \rangle}(2000) = f(f(f(\ldots f(2000)\ldots)))$.

Transformations

63. 🗋 *Changes in the equation of a function and its effects on the graph.* We consider the function f given by $y = f(x) = 2^x$. Changing the equation of a function effects the graph of the function. Sketch the graph of the new function. Which geometric transformation causes the new graph to result from the graph G_f?

a) $y = 2^x - 3$, $y = 2^x + 3$ b) $y = 2^{x-3}$, $y = 2^{x+3}$

c) $y = -2^x$, $y = 2^{-x}$ d) $y = 3 \cdot 2^x$, $y = 2^{3x}$

Transformations

Overview of the most important transformations of a function f ($a \in \mathbb{R}\backslash\{0\}$):

The equation $y = f(x)$ transitions into …	Transformation of the graph G_f
$y = f(x-a)$, i.e., x is replaced by $x - a$.	Translation by a in directionof the x-axis
$y = f(x) + a$, i.e., y is replaced by $y - a$.	Translation by a in directionof the y-axis
$y = f(-x)$, i.e., x is replaced by $-x$.	Reflection across the y-axis
$y = -f(x)$, i.e., y is replaced by $-y$.	Reflection across the x-axis
$y = f(ax)$, i.e., x is replaced by ax.	Scaling by the factor $\frac{1}{a}$ in direction of the x-axis (horizontal stretching)
$y = af(x)$, i.e., y is replaced by $\frac{1}{a} \cdot y$.	Scaling by the factor a in direction of the y-axis (vertical stretching)

---

64. Explain how the graph of the function g emerges from the graph of f.

a) $f(x) = x^4$, $g(x) = \frac{1}{3}x^4$ b) $f(x) = \sqrt{x}$, $g(x) = \sqrt{x+5}$

c) $f(x) = 2^x$, $g(x) = 2^x + 3$ d) $f(x) = \lg(x)$, $g(x) = \lg(\frac{1}{3}x)$

e) $f(x) = \frac{1}{x^3}$, $g(x) = \frac{1}{(-x)^3}$ f) $f(x) = \sin(x)$, $g(x) = -\sin(2x)$

65. The function $y = k(x) = -3(x-2)^3 + 7$ emerges from the function given by the equation $y = x^3$ using a few transformations. List the transformations and sketch the graph of the function k.

66. Sketch G_f by hand.

a) $f(x) = \sqrt{x-4} - 1$ b) $g(x) = 4x^2 + 3$ c) $h(x) = -(x+5)^2$

d) $f(x) = 3 \cdot 10^{-x} + 3$ e) $g(x) = 3 \cdot \lg(x-1) - 1$ f) $h(x) = 2\sin(\pi \cdot x)$

67. We consider the function f defined by the equation $w(x) = \sqrt{x-8}$. What is the equation of the new function, if the following transformation is applied to the graph of the function w?

    a) The graph is shifted 6 units to the left.

    b) The graph is reflected across the y-axis and shifted up 1 unit.

    c) The graph is stretched by the factor 2.5 in direction of the y-axis and shifted down 3 units.

    d) The graph is reflected across the x-axis and stretched by the factor $\frac{1}{8}$ in direction of the x-axis.

68. Let the function g be given by $g(x) = \sqrt{x-3}$. What is the equation of the new function, if the following transformations are applied to the graph of the function g in the given order? The graph is reflected across the x-axis, then stretched by the factor 2 in direction of the x-axis and finally shifted 6 units to the left.

69. The function $p(x) = \sqrt{4x+1}$ emerges from $q(x) = \sqrt{x}$ by applying elementary transformations. List the individual transformations in order.

70. Give the transformation that maps G_g onto G_f and draw the graphs of g and f.

    a) $g(x) = x^2$, $f(x) = -2(x+3)^2 + 5$ b) $g(x) = \sqrt{x}$, $f(x) = -\sqrt{\frac{1}{3}x - 2}$

71. 📄 In the coordinate system below on the left, a part of the graph of the function $f(x) = x^2$ is shown. Draw the graph of the function $g(x) = (x-2)^2 + 3$ in the same coordinate system.

For exercise 71

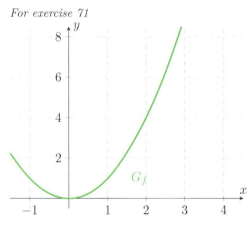

For exercise 72

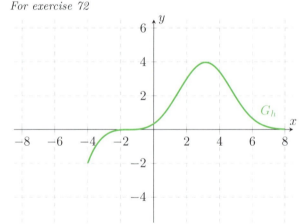

72. 📄 In the coordinate system above on the right, a part of the graph of the function h is shown.

    a) Draw the graph of the function a defined by $a(x) = \frac{1}{2} \cdot h(x)$ in the same coordinate system.

    b) Draw the graph of the function b defined by $b(x) = h(-x)$ in the same coordinate system.

73. The graph of the power function $y = x^3$ is reflected across

    a) the origin $(0\,|\,0)$.   b) the x-axis.   c) the y-axis.   d) the line $y = x$.

    What is the equation of the reflected graph?

74. The depicted graph G_f is the result of translating, reflecting or scaling the function g. What is the equation for the function f?

a) $g(x) = x^2$

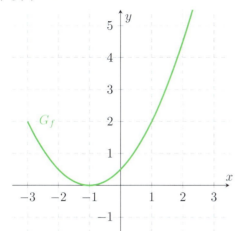

b) $g(x) = x^3$

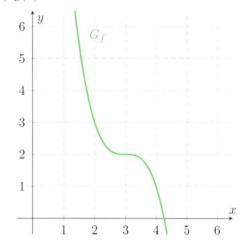

c) $g(x) = x^4$

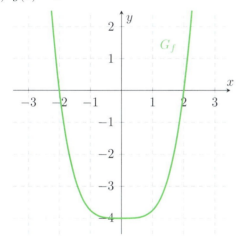

d) $g(x) = \sqrt{x}$

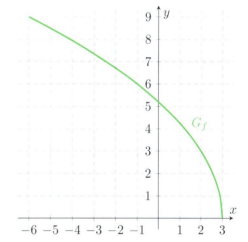

e) $g(x) = 2^x$

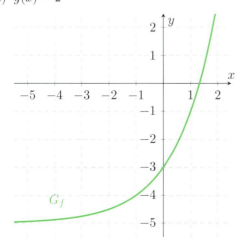

f) $g(x) = \ln(x)$

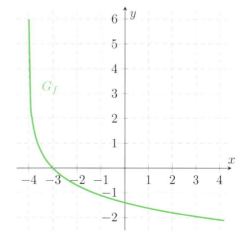

75. 📄 Let the graph G_f be given. Draw the graphs of the following function.

a) $g(x) = f(-x)$ b) $h(x) = -f(x)$ c) $k(x) = 1 + f(x+1)$

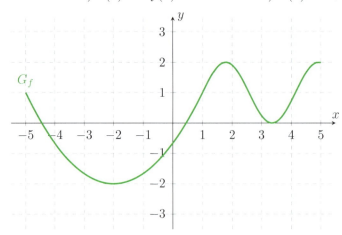

Symmetry and Periodicity

Symmetry

G_f is *symmetric* with respect to the y-axis, if and only if for all x in the domain we have: $f(-x) = f(x)$.

G_f is *symmetric* with respect to the origin O, if and only if for all x in the domain we have: $f(-x) = -f(x)$.

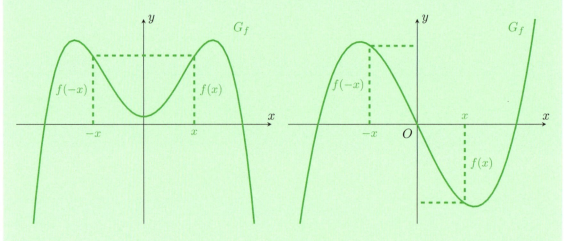

Functions f with this property are called *even*.   Functions f with this property are called *odd*.

76. Is the function given even, odd or neither?

a) $f(x) = -3 + 4x^8$ b) $g(x) = x^4(x^2 - 4)$ c) $h(x) = x^2 + 3x - 7$

d) $f(t) = \dfrac{t}{t^2 + 1}$ e) $g(t) = t^2 + \frac{1}{t}$ f) $h(t) = t + \frac{1}{t}$

g) $f(z) = |z|$ h) $g(z) = e^{\frac{-z^2}{2}}$ i) $h(z) = \sin(z)$

77. 📄 Complete the incomplete graph of the function f (see below) such that f is

    a) an odd function.   b) an even function.

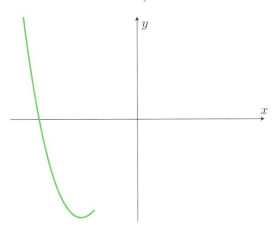

78. 📄 Complete the graph of the function $y = \sqrt{x}$ in the coordinate system such that the compounded graph is part of an

    a) even function.   b) odd function.

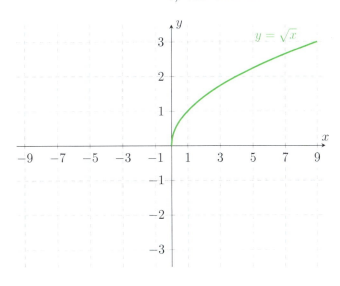

79. Give the equation of the function for the graph of the compounded function from exercise 78.

80. Let g be an even function and u an odd function. Is the compounded function even, odd or neither?

    a) $f(x) = (g \circ g)(x)$ b) $f(x) = (g \circ u)(x)$ c) $f(x) = u(g(x))$
    d) $f(x) = (u \circ u)(x)$ e) $f(x) = u(x) + g(x)$ f) $f(x) = u(x) \cdot g(x)$

81. Let the two functions $f_1(x) = x^4 + 2x^2 - 10$ and $f_2(x) = x^7 + x^3 - 6x$ be given. Calculate $f_1(-1)$, $f_1(1)$, $f_1(-2)$, $f_1(2)$, $f_1(-x)$ and $f_2(-1)$, $f_2(1)$, $f_2(-2)$, $f_2(2)$, $f_2(-x)$.

What do you notice? Can the symmetry be seen directly from the formula of the function?

82. Give a formula of a function f which is both even and odd.

83. Discuss the statements concerning a function with finitely many roots.

   a) An odd function always has an odd number of roots. An even function however can have an even or odd number of roots.

   b) A strictly increasing function on $\mathbb{R}$ cannot be even.

84. Mia claims that the function f given by $f(x) = \sqrt{1 - \cos^2(x)}$ is even, as the cosine function is even and therefore $f(-x) = \sqrt{1 - \cos^2(-x)} = \sqrt{1 - \cos^2(x)} = f(x)$ must hold. Jan disagrees: f is odd, as the Pythagorean trigonometric identity yields $f(x) = \sqrt{1 - \cos^2(x)} = \sin(x)$ and the sine function is odd. Comment on the claims of Mia and Jan.

85. Let the function f be given by the formula $f(x)$ with $x \in \mathbb{R}$.

   a) Show: The function g defined by $g(x) = \frac{f(x) + f(-x)}{2}$ is even.

   b) Show: The function u defined by $u(x) - \frac{f(x) - f(-x)}{2}$ is odd.

   c) Show using parts a) and b): Every function f can be written as the sum of an even and an odd function. Illustrate this fact using the example $f(x) = e^x$.

86. Let the function defined by the equation $y = f(x) = \frac{x^2}{x-1}$ be given. Express the formula $f(x)$ of the function as a sum $g(x) + u(x)$, where $g(x)$ and $u(x)$ are the formulas of an even and an odd function, respectively. If necessary, use the insight from exercise 85.

Periodic Functions

A function f is called *periodic with period T*, if its graph coincides with itself when shifted by T units in the direction of the x-axis, i.e. if one has:

$$f(x + T) = f(x) \text{ for all } x \text{ in the domain of } f.$$

Remarks and examples:

- If T is a period of f, then $-T$ is also a period of f, which is why one often assumes $T > 0$.
- If T is a period of f, then $2T$, $3T$, $4T$, $5T$, ... are also periods of f: $f(x + k \cdot T) = f(x)$, $k \in \mathbb{Z}$.

   Note: Usually, one is only interessted in the smallest positive period of f.

- $y = \sin(x)$ and $y = \cos(x)$ are both periodic with period $T = 2\pi$; $y = \tan(x)$ is periodic with period $T = \pi$.

---

87. State whether the function is even or odd, periodic or aperiodic. For periodic functions, also give the smallest positive period T.

   a) $f(x) = \sin^3(x)$ b) $f(x) = \sin(x^3)$ c) $f(x) = \cos(2x) + \cos(4x)$

   d) $f(x) = \sin(4x) + \cos(2x)$ e) $f(x) = \sin(x) \cdot \cos(2x)$

88. Determine the smallest positive period T of the function f.

   a) $f(x) = -2\sin\left(\frac{2}{3}x + 1\right)$ b) $f(x) = \sin(3x) + \cos(6x)$

   c) $f(x) = \sin^2(x)$ d) $f(x) = \sin^2(x) + \cos^2(x)$

89. The function f is given by the formula $f(x) = x^3$ in the region $-4 \leq x < 3$. On the rest of $\mathbb{R}$, f is periodically extended in such a way that the extended function has the period $T = 7$. Calculate $f(2021)$.

90. The varying water level h in a sea port can be described by $h(t) = a \cdot \sin(b(t - c)) + d$. The maximal value is $h = 9\,\text{m}$ and the minimal value is $h = 6\,\text{m}$. The length of the period is 12 hours. The water level is at $7.5\,\text{m}$ 3 hours after the start of observation and is decreasing. Determine a, b, c and d.

91. A Ferris wheel has a diameter of $50\,\text{m}$ and its axle is $30\,\text{m}$ above the ground. It completes a revolution once every 8 minutes. A point on the circumference of the wheel is at the very bottom at time $t = 0$. How high above the ground is this point at time t (in minutes)?

92. The mass bob of an ideal spring-mass system is lifted up $5\,\text{cm}$ from the resting position, at time $t = 0$, and then let go. The mass bob begins to oscillate with a frequency of $f = \frac{1}{2}\,\text{Hz}$. *Note*: $f = \frac{1}{T}$.

   a) Determine the displacement of the mass bob at time t (in seconds) compared to the resting postion.

   b) After how many seconds will the mass bob traverse the resting position a second time?

   c) Is the distance between the mass bob and its resting position after 1.75 seconds more than half the amplitude?

93. The equation $\cos(x) = \frac{-1}{2}$ has two solutions in the interval $[0; 2\pi[$.

   a) How many solutions does the equation $\cos(7x) = \frac{-1}{2}$ have in the interval $[0; 2\pi[$?

   b) How many solutions does the equation $\cos\left(3x + \frac{\pi}{2}\right) = \frac{-1}{2}$ have in the interval $[0; 2\pi[$?

   c) How many solutions does the equation $3\cos(12x - \pi) = \frac{-1}{2}$ have in the interval $[0; 2\pi[$?

   d) How many solutions does the equation $\tan(9x) = \frac{-1}{2}$ have in the interval $[0; 2\pi[$?

94. Let the sine function f be given by $y = f(x) = 2\sin(3x - 4)$. Determine a cosine function g, which «cancels» the sine curve, i.e. determine g such that $f + g = 0$.

95. Show: If f is periodic with period T, then g with $g(x) = f(ax)$ is periodic with period $\frac{T}{a}$.

96. Show:

   a) A strictly increasing function on $\mathbb{R}$ has at most one root.

   b) A strictly decreasing function on $\mathbb{R}$ cannot be even.

   c) A strictly increasing function on $\mathbb{R}$ cannot be periodic.

   d) The constant function $f \colon x \mapsto y = c$, $c \in \mathbb{R}$ is monotonically decreasing.

Inverse Functions

97. *When formulas become functions.* The volume of a right circular cylinder of radius r and height h is given by the equation: $V = \pi r^2 h$, $(r, h > 0)$.

a) Let $r = \frac{1}{2}$ and solve the equation once for V and once for h. The two equations for V and h can be interpreted as functions. Sketch the approximate shape of the functions given by the equations $V = V(h)$ and $h = h(V)$ in quadrant I of a coordinate system.

b) Let $h = \frac{1}{3}$ and solve the equation once for V and once for r. The two equations for V and r can be interpreted as functions. Sketch the approximate shape of the functions given by the equations $V = V(r)$ and $r = r(V)$ in quadrant I of a coordinate system.

c) Let $V = 100$ and solve the equation once for h and once for r. The two equations for h and r can be interpreted as functions. Sketch the approximate shape of the functions given by the equations $h = h(r)$ and $r = r(h)$ in quadrant I of a coordinate system.

Inverse Functions

A function $f: x \mapsto y = f(x)$ is *invertible* on an interval $I \subset D$, if for all $x \in I$, we have: Different x-values are assigned to different function values $f(x)$, i.e. $x_1 \neq x_2$ always implies $f(x_1) \neq f(x_2)$ for all $x_1, x_2 \in I$.

Expressed graphically: Invertible on the interval I signifies that the graph of the function on the interval I has at most one intersection point with any horizontal line.

Example: $f: x \mapsto y = f(x) = 3x + 1$ is invertible on the whole of $\mathbb{R}$, as different points x also have different images under f.

If f is invertible on the interval I, the formula for the *inverse function* f^{-1} can be determined by solving the equation of f for x and then swapping the variables x and y. By swapping x and y, the inverse function is presented as usual (argument x on the x-axis and images $y = f^{-1}(x)$ on the y-axis). *Warning:* $f^{-1}(x) \neq \frac{1}{f(x)}$.

Example: $f: x \mapsto y = f(x) = 3x + 1$. Solving $y = 3x + 1$ for x gives $x = \frac{y-1}{3}$. Swapping x and y gives $y = \frac{x-1}{3} = \frac{1}{3}x - \frac{1}{3}$. Therefore, $f^{-1}: x \mapsto y = f^{-1}(x) = \frac{1}{3}x - \frac{1}{3}$ is the inverse function of f.

Because of the swapping of x and y, the graph of the inverse function f^{-1} is the reflection of the graph of f across the angle bisector $y = x$.

The domain of f becomes the range of f^{-1} and inversely, the range of f is the domain of f^{-1}.

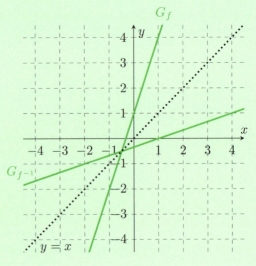

98. 🗋 On the basis of the graph, assess whether the function is invertible on the whole of $\mathbb{R}$ or not, and sketch the graph of the inverse function, if applicable.

a) $y = f(x) = -\frac{3}{2}x$

b) $y = f(x) = x^2$

c) $y = f(x) = x^3$

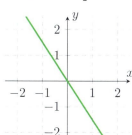
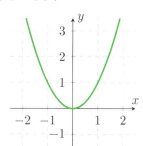
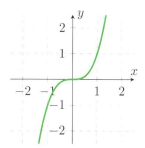

d) $y = f(x) = \sqrt{x}$

e) $y = f(x) = x^3 - x$

f) $y = f(x) = \frac{1}{3x}$

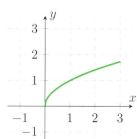
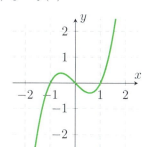
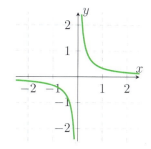

99. Determine the formula of the inverse function f^{-1} of f, if f is invertible.

a) $f(x) = \frac{1}{4}x$

b) $f(x) = x^3,\ x \geq 0$

c) $f(x) = 3x + 2$

d) $f(x) = 1 - x$

e) $f(x) = 4$

f) $f(x) = x^2,\ x \leq 0$

100. 🗋 The graph G_f is depicted in the coordinate system. Draw the graph of the inverse function f^{-1} in the same coordinate system.

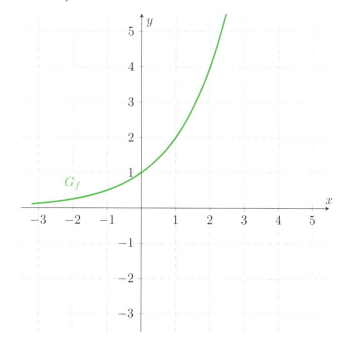

101. Determine a domain D_f, where the function f given is invertible and derive the formula of its inverse function f^{-1}.

a) $f(x) = x^2$ b) $f(x) = (x-2)^2$ c) $f(x) = \dfrac{2x-1}{3}$ d) $f(x) = \sqrt{x-4}$

e) $f(x) = \ln(x)$ f) $f(x) = 2^x$ g) $f(x) = |x|$ h) $f(x) = \sin(x)$

102. The function d assigns the remainder of the division of n by 4 to each natural number n (see exercise 22 on page 180). Give two different domains of maximum size, on which d is invertible.

103. How might the following statements be justified?

a) A strictly increasing function is invertible.

b) A strictly decreasing function is invertible.

c) The inverse function of an odd function is odd, when it exists.

104. Let the functions u and v be defined by the equations $u(x) = 3x - 2$ and $v(x) = 4x + 1$.

a) Determine the formula of the function w with $w(x) = u(v(x))$.

b) Give the formula for w^{-1}.

c) How is w^{-1} related to the inverse functions u^{-1} and v^{-1}? Find an equation that describes how w^{-1} can be determined using u^{-1} and v^{-1}.

105. Express $(u \circ v \circ w)^{-1}$ using an equation comprised of u^{-1}, v^{-1} and w^{-1}. In other words: How is $(u \circ v \circ w)^{-1}$ related to u^{-1}, v^{-1} and w^{-1}?

106. Find some functions that are their own inverse function.

107. a) Determine the intersection point of G_f and $G_{f^{-1}}$ in the interval $[0; \infty[$ for the function f with $f(x) = x^2 - 2$.

b) Use part a) to determine all intersection points of G_f with the parabola that arises through reflection of the graph of f at the angle bisector $y = x$.

c) Determine the intersection point of G_g and $G_{g^{-1}}$ in the interval $]0; \infty[$ with $g(x) = \ln(x) + x$.

108. True or false? The graphs of a function f and its inverse function f^{-1}

a) only have finitely many points of intersection.

b) always have a single common point.

c) only intersect on the angle bisector $y = x$.

X.3   Polynomial Functions

Fundamentals and Symmetry

> **Polynomial functions**
>
> A function given by the equation $y = f(x) = a_n x^n + a_{n-1} x^{n-1} + \ldots + a_2 x^2 + a_1 x + a_0$ with $n \in \mathbb{N}$, real coefficients $a_i \in \mathbb{R}$ and $a_n \neq 0$ is called a *polynomial function*.
>
> The largest occuring exponent n is called the *degree* of the polynomial function. For $n = 1$, this is a linear function (the graph is a line) and for $n = 2$. it is a quadratic function (the graph is a parabola).

> **Symmetry of polynomial functions**
>
> If the terms of a polynomial function f only have even exponents, then f is an even function and the graph of f is symmetric with respect to the y-axis.
>
> If the terms of a polynomial function f only have odd exponents, then f is an odd function and the graph of f is symmetric with respect to the origin $(0 \,|\, 0)$.

109. Is f a polynomial function?   If yes, write the equation of the function f in the form $y = a_n x^n + a_{n-1} x^{n-1} + \ldots + a_1 x + a_0$.

a) $f(x) = x - 2x^2 - \frac{9}{7} - \frac{1}{2} x^4$ b) $f(x) = \dfrac{5x^2 + 2x}{2}$ c) $f(x) = \frac{1}{x}$

d) $f(x) = 4$ e) $f(x) = \left(2x + \sqrt{2}\right)^2$ f) $f(x) = x + 10^x$

g) $f(x) = 2x + \sqrt{x}$ h) $f(x) = (x - 1)(x + 1)$ i) $f(x) = \dfrac{x^3 - 2x^2 + 1}{x}$

110. Let the function $P(x) = x^3 + x^2 + x + 1$ be given. Determine $Q(x)$ and simplify it.

a) $Q(x) = P(2x)$ b) $Q(x) = P(-x)$ c) $Q(x) = P(x^2)$ d) $Q(x) = P(x - 1)$

111. Let $p(x) = (3 + 2x - x^2)^3$.   What is the degree of the polynomial function defined by this expression?

a) $x \cdot p(x)$ b) $x + p(x)$ c) $x^8 \cdot p(x)$ d) $x^8 + p(x)$ e) $x^6 + p(x)$

112. $P(x) = (1 - 2x^3)^4$ and $Q(x) = (1 - 2x^4)^3$. Determine the degree of the polynomial function R.

a) $R(x) = P(x) \cdot Q(x)$ b) $R(x) = P(x) + Q(x)$

c) $R(x) = P(x) - Q(x)$ d) $R(x) = P(x) + 2 \cdot Q(x)$

113. Let P and Q be two polynomial functions of degree m and n, respectively. What is the degree of the polynomial function $P \cdot Q$ and of the polynomial function $P + Q$?

114. Consider the four polynomial functions given by the equations $P_1(x) = (x^2 - 0.5)^3$, $P_2(x) = (x^3 - 0.5)^2$, $P_3(x) = (x+2)^2 - (x-2)^2$ and $P_4(x) = (x+2)^3 \cdot (x-2)^3$. For which indices $n \in \{1, 2, 3, 4\}$ is the graph of the polynomial function $y = P_n(x)$ symmetric with respect to

a) the y-axis?   b) the origin $O(0\,|\,0)$?

115. Choose a value for the parameter a such that the graph of the polynomial function f is symmetric with respect to either the y-axis or the origin.

a) $f(x) = 4(x + a)^6$ b) $f(x) = x^7 - x^5 - ax^4$

c) $f(x) = 5(x + a)(x + 8)$ d) $f(x) = x^a - x$

116. The polynomial $2x^5 - 0.5x^4 + x^2 - x$ has coefficients 2, -0.5, 0, 1, -1, 0.

a) What are the coefficients of the polynomial $x^6 - x^4 + 3x^2$?

b) Give the polynomial that has coefficients 1, 0, 0, 0, -1.

117. Let the polynomial $P(x)$ have coefficients 1, -1, -3, 0 and the polynomial $Q(x)$ have coefficients 1, 0, 0. What are the coefficients of the polynomial $R(x)$?

a) $R(x) = P(x) + Q(x)$ b) $R(x) = P(x) \cdot Q(x)$

c) $R(x) = P(Q(x))$ d) $R(x) = Q(P(x))$

118. Determine the equation of a polynomial function p, whose graph passes through the given points.

a) $(-1\,|\,10)$, $(3\,|-2)$, $(4\,|\,5)$ b) $(-3\,|-12)$, $(1\,|\,2)$, $(5\,|\,0)$

c) $(-2\,|-8)$, $(0\,|\,0)$, $(2\,|\,0)$, $(3\,|\,\frac{9}{2})$ d) $(-4\,|-30)$, $(-1\,|\,6)$, $(0\,|\,2)$, $(3\,|\,26)$

Global Behaviour

For a polynomial function f with $f(x) = a_n x^n + a_{n-1} x^{n-1} + \ldots + a_2 x^2 + a_1 x + a_0$, the term $a_n x^n$ with the largest exponent n is responsible for the behavior of the function f for $x \to \pm\infty$ $(a_n \neq 0)$.

The *global behaviour* of a polynomial function, i.e. the shape of the graph for $x \to \pm\infty$, can be reduced to the following four basic types:

Type 1:   Type 2:   Type 3:   Type 4:
n even, $a_n > 0$ n even, $a_n < 0$ n odd, $a_n > 0$ n odd, $a_n < 0$

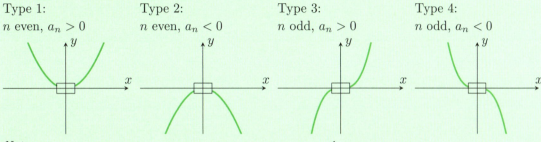

Note:
- $x \to \infty$, then $x^n \to \infty$
- $x \to -\infty$, then $x^n \to \begin{cases} +\infty, & \text{if } n \text{ even} \\ -\infty, & \text{if } n \text{ odd} \end{cases}$

Example: Consider the two polynomial functions $f(x) = 3x^4 - 14x^3 - 20$ and $g(x) = 3x^4$. We compare the global shape of these two functions using their graphs for two different regions.

Region 1:   *Region 2:*
$-4 \leq x \leq 6$; $-200 \leq y \leq 200$ $-30 \leq x \leq 30$; $-50'000 \leq y \leq 1'200'000$

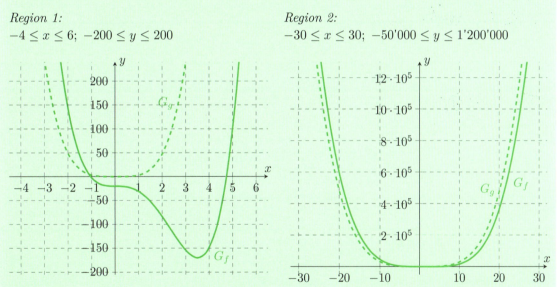

In this region, the graphs of f and g are easily distinguished.

The larger a region is chosen, the more the graph of f looks like the graph of g.

119. From the preceeding box, we know that the behaviour of polynomials for $x \to \pm\infty$ can be categorised by the four basic types. Give a concrete example for each basic type.

120. Assign the behaviour of the polynomial function p for $x \to \pm\infty$ to the appropriate basic type that corresponds to its global behaviour.

a) $p(x) = -3x^5 + 12x^3 - 8$

b) $p(x) = \frac{1}{2}x^4 - 28x^3 + 3x - 8$

c) $p(x) = 4x^3 + 2x^2 - 7x + 12$

d) $p(x) = -2x^4 + 10^9 x^3$

121. Let the polynomial function $p(x) = 5x^3 - x^2 + 1$ be given. Justify by calculation, why the behaviour of p for $x \to \pm\infty$ is like that of the monomial $q(x) = 5x^3$. *Hint:* Factor out the term $5x^3$.

Polynomial Division and Roots

Polynomial Long Division

The *division of polynomials* is done analogously to the long division of numbers.

Examples:

Long division without remainder:

```
286 : 22 = 13
-22
 66
-66
  0
```

Polynomial division without remainder:

$$\left(x^3 + 6x^2 + 3x - 10\right) : \left(x + 5\right) = x^2 + x - 2$$

$$\underline{- x^3 - 5x^2}$$
$$x^2 + 3x$$
$$\underline{- x^2 - 5x}$$
$$- 2x - 10$$
$$\underline{2x + 10}$$
$$0$$

The powers of the variable for polynomials correspond to the powers of ten for numbers. For the division of polynomials, one first sorts the dividend and the divisor so that their exponents of the variable are in descending order. After that, one consecutively divides, multiplies back and forms the new rest. The calculation terminates if the rest becomes 0 or if the largest exponent of the polynomial rest is smaller than the largest exponent of the polynomial divisor.

Example: Polynomial division with remainder:

$$\left(4x^3 + 2x^2 + 6x - 12\right) : \left(2x + 5\right) = 2x^2 - 4x + 13 + \frac{-77}{2x + 5}$$

$$\underline{- 4x^3 - 10x^2}$$
$$- 8x^2 + 6x$$
$$\underline{8x^2 + 20x}$$
$$26x - 12$$
$$\underline{- 26x - 65}$$
$$- 77$$

122. Carry out the polynomial long division.

a) $(4x^3 + 17x^2 + 14x - 3) : (x + 3)$

b) $(12x^3 + 9x^2 - 34x + 5) : (3x^2 + 6x - 1)$

c) $(3x^3 + 4x^2 + 8) : (x + 2)$

d) $(x^4 + x^2 + 1) : (x^2 + x + 1)$

e) $(12x^3 - 19x^2 + 23x - 3) : (3x - 1)$

f) $(x^5 + 1) : (x + 1)$

g) $(5x^3 - 11x^2 - 14x - 10) : (x - 3)$

h) $(x^7 - 1) : (x^3 - 1)$

Roots and Splitting Off a Factor

If f is a polynomial function of degree n, the following properties hold:

1) If x_0 is a root of f, the formula of the function can be written as a product of the linear factor $(x - x_0)$ with a polynomial $g(x)$:

$$f(x) = (x - x_0) \cdot g(x)$$

Here $g(x)$ has degree $n - 1$. This is called *splitting off the factor* $(x - x_0)$.

2) If x_0 is a root of f and the linear factor $(x - x_0)$ can be split off exactly k times, then x_0 is a *root of multiplicity k* $(1 \le k \le n)$ of f. In this case, we have: $f(x) = (x - x_0)^k \cdot g(x)$, where $g(x)$ has degree $(n - k)$.

3) If x_0 is a root of multiplicity k of the function f, the shape of the graph passing through the x-axis near the root x_0 is described by one of the following cases:

$k = 1$ 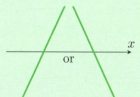 $k > 1$ *and k is even* 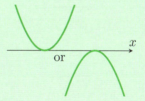 $k > 1$ *and k is odd*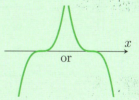

4) The function f has at most n roots and at least one when n is odd.

---

123. Which of the following four polynomial functions given by the formula $p_i(x)$ has the roots $x_1 = -1$, $x_2 = 2$, $x_3 = x_4 = 5$ (root of multiplicity 2)?

• $p_1(x) = x^3 - 6x^2 + 3x + 10$

• $p_2(x) = x^4 + 21x^3 + 33x^2 + 5x - 50$

• $p_3(x) = x^4 - 11x^3 + 33x^2 - 5x - 50$

• $p_4(x) = x^4 + 11x^3 - 33x^2 + 5x + 50$

124. One root x_1 of the polynomial function f is known. Determine all the remaining roots.

a) $f(x) = 2x^3 + 12x^2 - 2x - 60$, $x_1 = 2$

b) $f(x) = x^3 - 39x + 70$, $x_1 = 5$

125. Let the polynomial functions f and g be given by the equations $f(x) = x^3 - 2x^2 - 5x + 6$ and $g(x) = (x + 2)(x - 1)(x - 3)$. What are the roots of f and g? What do you notice? What is the relationship between f and g?

126. Show that $x_0 = 2$ is a root of the polynomial $f(x) = 20x^3 - 36x^2 - 11x + 6$ and determine the factor $g(x)$ in the equation $f(x) = (x - 2) \cdot g(x)$.

127. Determine the roots of the polynomial function f given by $f(x) = x^4 - 15x^2 + 10x + 24$ and write the formula of the function as a product.

128. Use the roots given to develop the formula of a polynomial function p with these roots and integer coefficients, and of minimal degree.

a) $0, -5$ b) $1 \pm \sqrt{2}, 3$ c) $\pm\frac{\sqrt{3}}{3}, 2$ d) $-1, 4$ and double root 2

129. Determine all the roots and write the formula of the function as a product of linear factors.

a) $f(x) = x^3 + 5x^2 + 2x - 8$ b) $g(x) = x^4 - x^3 - 6x^2 + 4x + 8$

130. Determine all the roots of the polynomial function f. Furthermore, for each root, state whether it is a simple, double, triple etc. root and sketch the graph G_f.

a) $f(x) = (x - 2)(x + 5)^4$ b) $f(x) = x^3(x - 1)(x + 1)^3$

c) $f(x) = x^2 + 8x + 16$ d) $f(x) = x^6 - x^5$

e) $f(x) = (x + 5)(x^2 - 2x + 1)$ f) $f(x) = -x^5 + x$

g) $f(x) = x^3 - 3x + 2$ h) $f(x) = (x^2 - 3x - 10)(x^3 - 3x + 2)$

131. Let the graph of a polynomial function of degree n be given. Give the formula of a suitable function that fits the graph.

a) $n = 5$ b) $n = 4$ c) $n = 4$

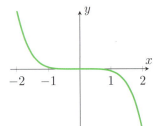 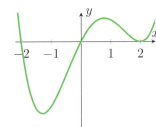

d) $n = 4$ e) $n = 7$ f) $n = 4$

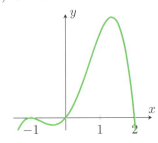 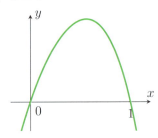

132. True or false? The polynomial function p with $p(x) = x^3 + 2x - 3$ has, appart from $x = 1$,

a) 3 further roots.   b) another double root.

c) no further roots.   d) only one further root.

e) a simple and a double root.

133. Determine a polynomial function of the smallest possible degree with the given roots $x_1, x_2, \ldots$ and whose graph passes through the point P.

a) $x_1 = 1, x_2 = -6, \ P(2\,|\,8)$ b) $x_1 = 0, x_2 = 3, \ P(-1\,|\,2)$

c) $x_1 = 3, x_2 = 4, x_3 = 7, \ P(-1\,|\,{-160})$ d) $x_1 = -1, x_2 = 0, x_3 = 2, x_4 = 3, \ P(1\,|\,8)$

e) $x_1 = -5, x_2 = x_3 = x_4 = 2$ (triple root), $x_5 = 3, \ P(4\,|\,{-144})$

134. The polynomial function f of degree 3 has roots at $x_1 = -6$, $x_2 = 1$ and $x_3 = 5$. What roots and what degree n does the polynomial function g have, if

a) $g(x) = f(x - 4)$? \hspace{4cm} b) $g(x) = f(x^2 + 1)$?

X.4 Rational Functions (Poles, Singularity, Asymptotes)

A *rational function* f is a function, whose formula can be described by a fraction where the enumerator p and the denominator q are polynomial functions: $f(x) = \frac{p(x)}{q(x)}$.

Examples: $f(x) = \dfrac{x^2 + 2}{2x}$, $g(x) = \dfrac{1}{x^2 + 2x + 4}$, $h(x) = \dfrac{3x^3 + 2x^2 + x - 10}{x^2 + 1}$

Singularities, Poles, Asymptotes

Rational functions f can have the following distinctive features:

1) The function f has a *restricted domain D*, i.e. it might have *singularities*. A domain might be restricted in one of two ways:

   a) The singularity x_h is a *removable singularity*, if the limit $\lim_{x \to x_h} f(x)$ exists. In this case, one can define $f(x_h) = \lim_{x \to x_h} f(x)$.

   b) The singularity x_p is a *pole*, i.e. the graph of the function f approaches the line parallel to the y-axis that passes through the point x_p.

   This line orthogonal to the x-axis is called a *vertical asymptote* of the function f.

   Let x_p be a pole of the function f, then two cases of how f approaches the vertical asymptote are distinguished:

   - x_p is a *pole without change of sign*.
   - x_p is a *pole with change of sign*.

Example for singularities: $f(x) = \frac{x}{x(x-1)}$ has two singularities, one at $x_1 = 0$ and one at $x_2 = 1$. Since the factor x appearing in the fraction can be cancelled for $x \neq 0$, $x_1 = 0$ is a removable singularity. In contrast to this, $x_2 = 1$ is a pole with change of sign.

2) *Asymptotic behaviour:* The graph of the rational function f approaches a graph of a polynomial function g for $x \to \pm\infty$. g is then called the *polynomial asymptote*.

   If the polynomial numerator $p(x) = a_n x^n + a_{n-1} x^{n-1} + \ldots + a_1 x + a_0$ has degree n and the polynomial denominator $q(x) = b_m x^m + b_{m-1} x^{m-1} + \ldots + b_1 x + b_0$ has degree m, then the following three cases are distinguished for the function $f(x) = \frac{p(x)}{q(x)}$:

   - For $n < m$, the function f has the horizontal asymptote $y = g(x) = 0$.
   - For $n = m$, the function f has the horizontal asymptote $y = g(x) = \frac{a_n}{b_m}$.
   - For $n > m$, g is a non constant polynomial function. The formula for the function g can be determined using the polynomial division $p(x) : q(x)$.

     For $n = m + 1$, this yields a linear polynomial asymptote g, whose graph is an oblique asymptote of f.

135. Assign the equation of the function to the corresponding graph 1), 2), ... 6).

a) $y = \dfrac{1}{x^2 - 4}$

b) $y = \dfrac{1}{(x - 2)^2} + 1$

c) $y = \dfrac{1}{(x - 4)^2}$

d) $y = \dfrac{1}{x - 4}$

e) $y = x + \dfrac{1}{x - 2}$

f) $y = x + \dfrac{1}{(x - 2)^2}$

1)
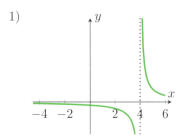

2)
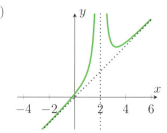

3)
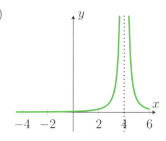

4)
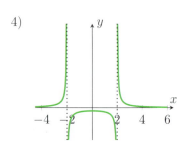

5)
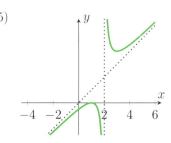

6)
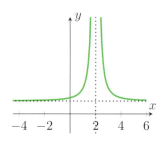

136. Determine the domain, poles (with/without change of sign) and removable singularities of the given function.

a) $y = \dfrac{1}{(x + 9)^2}$

b) $y = \dfrac{x}{x^3 - x}$

c) $y = \dfrac{3x + 11}{x^2 - x - 6}$

d) $y = \dfrac{x + 3}{(x - 3)^2}$

e) $y = \dfrac{(2x + 6)(x - 2)}{(x - 2)(x + 8)}$

f) $y = \dfrac{1}{x^4 - 16}$

g) $y = \dfrac{(2x - 7)(x - 3)}{x^2 - 6x + 9}$

h) $y = \dfrac{x^3 - x^2 - 12x}{x^3 - 5x^2 - 8x + 48}$

137. Examine the behaviour of the function f for $x \to \pm\infty$ and give the equation of its horizontal or oblique asymptote.

a) $f(x) = \dfrac{x - 1}{3 - 2x}$

b) $f(x) = \dfrac{1 - x}{(x - 2)(x + 1)}$

c) $f(x) = \dfrac{4x}{x^2 + 5} + x - 1$

d) $f(x) = \dfrac{x^2 - 6}{2x}$

e) $f(x) = \dfrac{2x^2 + 6x + 1}{x + 3}$

f) $f(x) = \dfrac{1 - x^4}{x^3 - 9x}$

138. Examine the behaviour of the function f for $x \to \pm\infty$ and give the equation of its polynomial asymptote.

a) $f(x) = \dfrac{4x - 5x^3}{\frac{1}{2}x^3 + x^2 + 5x - 100}$

b) $f(x) = \dfrac{5x^4 + x - 5}{x^2(1 + 2x + 3x^2)}$

c) $f(x) = \dfrac{x^3 - 4x^2 + 4x + 1}{x - 1}$

d) $f(x) = \dfrac{2x^3 + 3x^2 - 5x + 1}{x^2 + x + 2}$

139. a) Determine the formula of a rational function, which has the oblique asymptote $y = 4x + 5$ and a single pole (without change of sign) at $x = 3$.

b) Determine the formula of a rational function with the following properties: Root at $x = 1$, double root at $x = -1$, only poles at $x = 0$ and $x = 2$. For $x \to \pm\infty$, the equation of the asymptote is $y = 1$.

140. Let the formula $f(x)$ of a function, the parameters a and b and the graph of a function be given. Determine the parameters such that the formula of the function fits the graph.

a) $y = \dfrac{1}{(x-a)^2} + b$

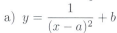

b) $y = \dfrac{1}{x+a} + b$

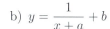

c) $y = x + a + \dfrac{1}{4(x-b)^2}$

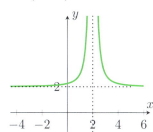
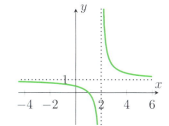
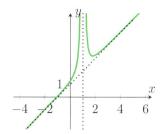

141. Let the formula $f(x)$ of a function, the parameters a and b and the graph of a function be given. Examine if the parameters need to be chosen as odd or even natural numbers such that the formula of the function fits the graph.

a) $f(x) = \dfrac{1}{(x-1)^a}$

b) $y = \dfrac{x}{(x-2)^a(x+2)^b}$

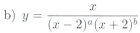

c) $y = x + \dfrac{1}{(x+2)^a(x-1)^b}$

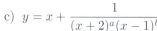

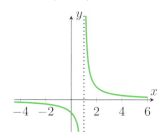
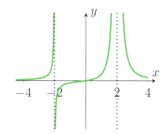
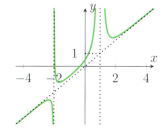

142. The graph of the function $f(x) = \dfrac{x^3 + ax^2}{bx^2 + 3}$ has the oblique asymptote $y = x - 1$. Determine the values of the parameters a and b.

143. Show: All functions f of the form $f(x) = \dfrac{ax+b}{cx+d}$ have a graph symmetric with respect to a point.

Hint: Determine the pole (vertical asymptote) and the (horizontal) asymptote.

144. Let the graph of a rational function be given. Determine a possible formula for the function.

a)

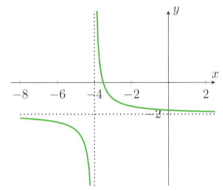

b)

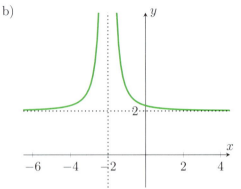

c)

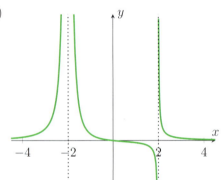

d)

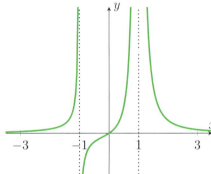

e)

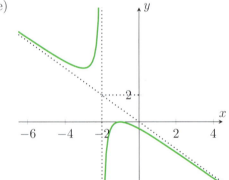

f)

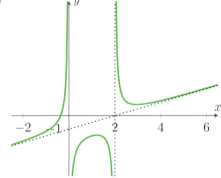

X.5   Trigonometric, Exponential and Logarithmic Functions

145. What symmetry properties does the graph of the function f have?

a) $f(x) = \sin(x)$ b) $f(x) = \cos(x)$ c) $f(x) = \tan(x)$

d) $f(x) = \sin(x) + \cos(x)$ e) $f(x) = \sin(x^3)$ f) $f(x) = \cos^3(x)$

146. Which function is even, which is odd?

a) $t \mapsto \sin(t) - \tan(t)$ b) $t \mapsto \sin^2(t)$ c) $t \mapsto \sin(t^2)$ d) $t \mapsto \tan^3(t) + \sin^3(t)$

147. Sketch the graphs of the functions f and g in the same coordinate system.

a) $f(x) = 3\sin(x),\ g(x) = \sin(3x)$ b) $f(x) = \cos\left(x - \frac{3\pi}{2}\right),\ g(x) = \cos(x) - \frac{3\pi}{2}$

For **148–150**: Sketch the graph of the given function.

148. a) $f(t) = 3 + \sin(3t)$ b) $f(t) = 2 - 3\tan(2t)$ c) $f(t) = 3 + 2\cos\left(\frac{t}{2}\right)$

149. a) $f(t) = \frac{1}{2}\tan\left(t - \frac{\pi}{4}\right)$ b) $f(t) = -\cos\left(\frac{t}{3} - \frac{\pi}{2}\right)$ c) $f(t) = -3\sin\left(2t + \frac{3\pi}{2}\right)$

150. a) $f(t) = \sin(t) - \sin(2t)$ b) $f(t) = \sin\left(\frac{t}{3}\right) + \cos\left(\frac{t}{4}\right)$

151. Give the formula of a function that fits the graph.

a)

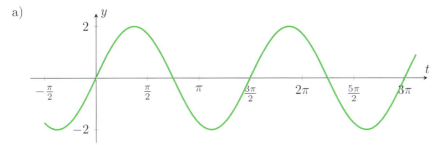

b)

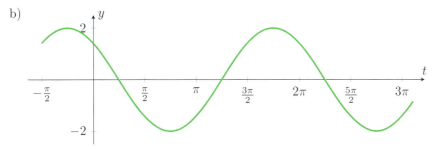

c)

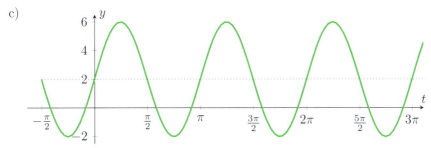

152. Describe the graph drawn

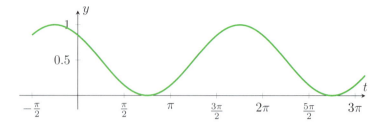

    a) as a sine function.

    b) as a cosine function.

153. The parts of this exercise examine the behaviour of the function $f(x) = \sin\left(\frac{1}{x}\right)$ near $x = 0$.

    a) For which values of x is $\sin(x) = 0$, $\sin(x) = 1$, $\sin(x) = -1$?

    b) Using part a) determine the values of x, where $\sin\left(\frac{1}{x}\right) = 0$, $\sin\left(\frac{1}{x}\right) = 1$, $\sin\left(\frac{1}{x}\right) = -1$.

    c) Using part b) describe the behaviour of $f(x) = \sin\left(\frac{1}{x}\right)$ near the point $x = 0$.

154. Assign each of the five graphs to one of the six given function formulas.

    i) $f_1(x) = 3 \cdot 3^x$

    ii) $f_2(x) = \left(\frac{1}{4}\right)^x$

    iii) $f_3(x) = 3^x$

    iv) $f_4(x) = 3 \cdot 5^x$

    v) $f_5(x) = 3 \cdot \left(\frac{1}{4}\right)^x$

    vi) $f_6(x) = 1 + \frac{1}{2}x$

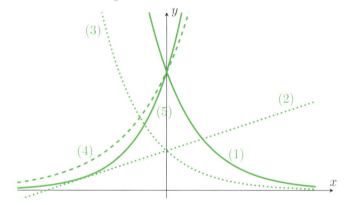

155. At which points do the graphs of f and g intersect?

    a) $f: x \mapsto 2^x$, $g: x \mapsto 2^{4-x}$
                                               b) $f: x \mapsto e^{x-1}$, $g: x \mapsto e^{-x}$

    c) $f: u \mapsto e^{u-2}$, $g: u \mapsto e^u - 2$
                                             d) $f: t \mapsto \ln(t + 3)$, $g: t \mapsto \ln(t) + 3$

156. Give the largest possible domain D and corresponding range W for the following function.

    a) $f(x) = e^{\sqrt{x}}$ b) $f(t) = \sqrt{\ln(t)}$ c) $f(u) = \ln(\sqrt{u})$

    d) $g(x) = \ln(x^2)$ e) $g(t) = \ln^2(t)$ f) $g(x) = \ln(\ln(x))$

157. Express the equation of the function f in the form $f(x) = a \cdot b^x$.

    a) $f(x) = 2^{x-1}$ b) $f(x) = 2^{2x+5}$ c) $f(x) = 3^{1-x}$

158. Express the equation of the function f in the form $f(t) = a \cdot e^{\lambda t}$.

    a) $f(t) = 2^t$ b) $f(t) = 2^{t-5}$ c) $f(t) = 2^t - 5$

159. What is the equation of the inverse function f^{-1} of f?

    a) $f: x \mapsto \left(\frac{1}{2}\right)^{x-2}$ b) $f: x \mapsto \frac{1}{2} \cdot 5^x$

    c) $f: t \mapsto 4 \cdot 2^{3t}$ d) $f: t \mapsto 2\ln(t^2 - 1)$ for $t > 1$

Answers

1 Sequences and Series

1. Possible continuations:

a) 100'000, 1'000'000, ...

b) 19, 21, ...

c) 48, 63, 80, ...

d) 3, 5, ...

e) 0, 0, 0, 5, ...

f) 21, 34, ...

g) $\frac{1}{42}$, $\frac{1}{56}$, ...

h) Lucky number 1 in Swiss lottery

i) 22, 29, ...

j) 15, 26, ... or 16, 32, ...

k) 70, 92, ...

l) 30, 31, ...

m) 0, 7 (date)

n) 60, 120, ...

o) 19, 23, ...

p) 7, 10, ...

q) 40, 57, ...

r) 20, 50, ...

s) 4, 3, ...

t) 1, 1 (phone number)

u) 15'622, 78'122, ...

v) 1, 0, ...

2. a) 1, 3, 5, 7, 9, 11
12, 12, 14, 18, 24, 32
−1, 1, −1, 1, −1, 1

b) 13, 22, 40, 76, 148
4, 9, −6, 39, −96
2, 1, 3, 4, 7
7, 8, 10, 13, 17

3. a) $a_k = 2k$ b) $a_k = 2k - 1$ c) $a_k = 2^k$ d) $a_k = k^2$

e) $a_k = 7k$ f) $a_k = 7k + 1$ g) $a_k = 10k - 23$ h) $a_k = 3^k$

4. a) $a_k = \frac{1}{k}$ b) $a_k = \frac{k}{k+1}$ c) $a_k = \frac{(-2)^k}{2k+1}$ d) $a_k = \frac{8k-4}{8k-1}$

5. a) $a_{50} = \frac{99}{200}$, $a_{51} = \frac{101}{204}$ b) $a_{50} = \frac{52}{151}$, $a_{51} = \frac{53}{154}$

6. a) $a_1 = 3$, $a_{k+1} = a_k + 4$, $k \in \mathbb{N}$ b) $a_1 = 6$, $a_{k+1} = 2a_k$, $k \in \mathbb{N}$

c) $a_1 = 6$, $a_{k+1} = 2a_k + 1$, $k \in \mathbb{N}$ d) $a_1 = 4$, $a_{k+1} = 3a_k - 1$, $k \in \mathbb{N}$

7. a) 1, 3, 2, −1, −3, −2, 1, 3, 2, −1 $\Rightarrow$ $a_{100} = -1$, $a_{101} = -3$, $a_{102} = -2$ and $a_{107} = a_{101} = -3$

b) 2, 1, $\frac{1}{2}$, $\frac{1}{2}$, 1, 2, 2, 1, $\frac{1}{2}$, $\frac{1}{2}$ $\Rightarrow$ $a_{100} = \frac{1}{2}$, $a_{101} = 1$, $a_{102} = 2$ and $a_{107} = a_{101} = 1$

8. Note that $k \in \mathbb{N}$ holds for each of the following answers.

a) 11, 15, 19, 23, 27 and $a_k = 4k - 1$ b) −1, −5, −9, −13, −17 and $a_k = 11 - 4k$

c) 2, 3, 4, 5, 6 and $a_k = k - 1$ d) −3, −2, −1, 0, 1 and $a_k = k - 6$

9. a) 16, 32, 64, 128, 256 and $a_k = 2 \cdot 2^k = 2^{k+1}$, $k \in \mathbb{N}$

b) 9, 3, 1, $\frac{1}{3}$, $\frac{1}{9}$ and $a_k = 243 : 3^k = 3^{5-k}$, $k \in \mathbb{N}$

c) 36, −54, 81, $-\frac{243}{2}$, $\frac{729}{4}$ and $a_k = 16 \cdot \left(\frac{-3}{2}\right)^{k-1} = \frac{-32}{3} \cdot \left(\frac{-3}{2}\right)^k$, $k \in \mathbb{N}$

d) −2, −2, −2, −2, −2 and $a_k = -2$, $k \in \mathbb{N}$

10. a) $a_1 = 36$, $a_{k+1} = a_k + 2$, $k \in \mathbb{N}$ b) $a_1 = -1$, $a_{k+1} = a_k - 2$, $k \in \mathbb{N}$

c) $a_1 = -3$, $a_{k+1} = -3a_k$, $k \in \mathbb{N}$ d) $a_1 = 1$, $a_{k+1} = a_k + 2k + 1$ or
$a_{k+1} = (\sqrt{a_k} + 1)^2$, $k \in \mathbb{N}$

11. a) explicit: $a_k = (-1)^{k+1}$; recursive: $a_1 = 1$, $a_{k+1} = -a_k$, $k \in \mathbb{N}$

    b) explicit: $a_k = (-1)^k$; recursive: $a_1 = -1$, $a_{k+1} = -a_k$, $k \in \mathbb{N}$

    c) explicit: $a_k = \frac{1}{2} - (-1)^k \cdot \frac{1}{2}$; recursive: $a_1 = 1$, $a_2 = 0$, $a_{k+2} = a_k$, $k \in \mathbb{N}$

    d) explicit: $a_k = \sin\left((k-1) \cdot \frac{\pi}{2}\right)$; recursive: $a_1 = 0$, $a_2 = 1$, $a_{k+2} = -a_k$, $k \in \mathbb{N}$

12. a) $1, \frac{1}{2}, \frac{1}{6}, \frac{1}{24}, \frac{1}{120}, \frac{1}{720}$ and $a_k = \frac{1}{k!}$, $k \in \mathbb{N}$

    b) $1, \frac{3}{2}, \frac{7}{4}, \frac{15}{8}, \frac{31}{16}, \frac{63}{32}$ and $a_k = \frac{2^k - 1}{2^{k-1}} = 2 - 2^{1-k}$, $k \in \mathbb{N}$

    c) $1, -4, 9, -16, 25, -36$ and $a_k = (-1)^{k-1} \cdot k^2$, $k \in \mathbb{N}$

    d) $2, 1, \frac{-1}{2}, \frac{-1}{4}, \frac{1}{8}, \frac{1}{16}, \frac{-1}{32}, \frac{-1}{64}, \frac{1}{128}, \ldots$ and $a_k = \sqrt{2} \cdot 2^{2-k} \cdot \sin\left((2k-1)\frac{\pi}{4}\right)$, $k \in \mathbb{N}$

13. a) $(a_k) = (b_k) = (c_k)$: $1, 11, 111, 1111, 11'111, \ldots$

    b) $(a_k) = (b_k) = (c_k)$: $1, \frac{1}{2}, \frac{1}{3}, \frac{1}{4}, \frac{1}{5}, \ldots$

14. a) $1, 4, 9, 16, 25, 36$; $s_{100} = 10'000$

    b) $1, -1, 2, -2, 3, -3$; $s_{100} = -50$

    c) $1, 3, 7, 15, 31, 63$; $s_{100} = 2^{100} - 1$

15. a) 60 b) 242 c) 350 d) 290 e) $n^2 + 11n$

16. a) $\displaystyle\sum_{k=1}^{7} 5k$ b) $\displaystyle\sum_{k=1}^{50} 5k$

    c) $\displaystyle\sum_{k=5}^{21} 5k = \sum_{k=1}^{17} (20 + 5k)$ d) $\displaystyle\sum_{k=1}^{5} 3^k$

    e) $\displaystyle\sum_{k=1}^{20} k^2$ f) $\displaystyle\sum_{k=1}^{116} 3(k-1) = \sum_{k=1}^{116} (3k - 3)$

17. a) $s_3 = 1$, $s_8 = 2$ b) $s_4 = \frac{24}{25}$, $s_9 = \frac{99}{100}$ c) $s_{15} = \frac{3}{4}$

    $s_n = \sqrt{n+1} - 1$ $s_n = \frac{n(n+2)}{(n+1)^2}$ $s_n = \frac{\sqrt{n+1} - 1}{\sqrt{n+1}}$

18. a) recursive: $a_1 = 1$, $a_{k+1} = a_k + 2$, $k \in \mathbb{N}$;   explicit: $a_k = 2k - 1$, $k \in \mathbb{N}$

    b) $s_n = n^2$ c) –   d) $999'900$

19. a) $\frac{1}{6}, \frac{1}{12}, \frac{1}{20}, \frac{1}{30}, \frac{1}{42}$ b) $\frac{1}{2}, \frac{2}{3}, \frac{3}{4}, \frac{4}{5}, \frac{5}{6}$; $s_n = \frac{n}{n+1}$ c) $a_k = \frac{1}{k(k+1)}$

20. $s_n = \ln(n+1)$

21. Check by evaluating the difference for every pair of successive terms: $a_{k+1} - a_k$.

    a) AP. Recursive: $a_1 = 1$, $a_{k+1} = a_k + 3$;   explicit: $a_k = 1 + (k-1) \cdot 3 = 3k - 2$.

    b) AP. Recursive: $a_1 = 1$, $a_{k+1} = a_k + 1$;   explicit: $a_k = 1 + (k-1) \cdot 1 = k$.

    c) Not an AP, as $18 - 13 = 5 \neq 7 = 25 - 18$.

    d) AP. Recursive: $a_1 = 12$, $a_{k+1} = a_k + 8$;   explicit: $a_k = 12 + (k-1) \cdot 8 = 8k + 4$.

    e) Not an AP, as $11 - 14 = -3 \neq -2 = 9 - 11$.

    f) AP. Recursive: $a_1 = 50$, $a_{k+1} = a_k - 10$;   explicit: $a_k = 50 + (k-1) \cdot (-10) = 60 - 10k$.

22. a) $a_5 = 38$ b) $a_5 = 19$ c) $a_5 = 33$ d) $a_5 = 54$

23. a) $a_1 = 2$, $a_2 = 191$ and $a_3 = 380$

b) recursive: $a_1 = 2$, $a_{k+1} = a_k + 189$, $k \in \mathbb{N}$; explicit: $a_k = 2 + (k-1) \cdot 189$, $k \in \mathbb{N}$

24. a) $a_{20} = \frac{27}{2} = 13.5$ b) $a_4 = \frac{42}{5} = 8.4$

25. $a_1 = 831$, $a_{k+1} = a_k + 31$, $k \in \{1, 2, \ldots, 24\}$

26. 6 terms

27. $m_1 = -\frac{3}{2}$ and $m_2 = 2$

28. The AP is either 7, 10, 13 or 13, 10, 7.

29. 159'305

30. –

31. a) 5050   b) 2500

c) 18'430   d) $s_{50} = 2550$

e) $s_{45} = 220.5$ f) $a_1 = 50, d = -1.2, s_{50} = 1030$

32. a) 8, 15, 22, 29, 36, 43, 50   b) 351   c) 7485

33. a) 25'002'550   b) 24'750'999   c) 4'323'627

34. a) $s_{111} = 12'987$; $s_n = n^2 + 6n = n(n+6)$ b) $s_{111} = 13'542$; $s_n = n^2 + 11n = n(n+11)$

35. a) 19'160   b) 26   c) 2255   d) -25

36. a) $s_7 = \sum\limits_{k=1}^{7} 3k = 84$ b) $s_8 = \sum\limits_{k=1}^{8} (50 - 5k) = 220$

c) $s_6 = \sum\limits_{k=1}^{6} \left(\frac{1}{2}k + \frac{13}{2}\right) = 49.5$ d) $s_{14} = \sum\limits_{k=1}^{14} (17 - 5k) = -287$

e) $s_{45} = \sum\limits_{k=1}^{45} \left(\frac{1}{3}k\right) = 345$ f) $s_{11} = \sum\limits_{k=1}^{11} (8k - 39) = 99$

37. $a_1 = 3$, $d = 4$ or $a_1 = -14$, $d = 12.5$

38. $13 \, \text{cm}$

39. after 14 years; Fr. 3000.–

40. $270 \, \text{m}$; $20 \, \text{s}$

41. Scheme B is better than scheme A.

42. a) GP. Recursive: $a_1 = 1$, $a_k = 4 \cdot a_{k-1}$; explicit: $a_k = 4^{k-1}$.

b) Not a GP, as $\frac{6.75}{4.5} = \frac{3}{2} \neq \frac{4}{3} = \frac{9}{6.75}$.

c) GP. Recursive: $a_1 = 2$, $a_k = 3 \cdot a_{k-1}$; explicit: $a_k = 2 \cdot 3^{k-1}$.

d) Not a GP, as $\frac{54}{-18} = -3 \neq -3.5 = \frac{-189}{54}$.

e) GP. Recursive: $a_1 = 12$, $a_k = 0.5 \cdot a_{k-1}$; explicit: $a_k = 12 \cdot 0.5^{k-1}$.

f) GP. Recursive: $a_1 = 10$, $a_k = -2 \cdot a_{k-1}$; explicit: $a_k = 10 \cdot (-2)^{k-1}$.

43. a) $q = 1.5$, $a_8 = 1093.5$ b) $q = 3$, $a_8 = 5832$

c) $q = -0.5$, $a_8 = -50$ d) $q = \pm\sqrt{2}$, $a_8 = 18$

44. a) Yes; $a_8 = 1.9487\ldots$ b) Yes; $a_8 = 12.8$ c) No

d) No   e) Yes; $a_8 = 0.43047\ldots$ f) Yes; $a_8 = -205.03125$

45. 17, 34 and 68

46. The GP is either 3, −6, 12 or 12, −6, 3.

47. 217 terms are smaller.

48. 907 terms are larger.

49. There are 24 terms $(a_{74}, \ldots, a_{97})$ which lie inbetween.

50. a) 7.18 %   b) 14.2 years

51. a) $\approx 2.56\,\%$ b) calculated energy comsumption:
$113.8 \cdot 10^9$ kWh

52. 2 and 5; denominations of Swiss coins and banknotes

53. $\frac{2047}{512} = 3.998\ldots$

54. –

55. a) 665   b) −3280   c) 2730   d) $-2'391'484.\overline{4}$

56. a) $\frac{5115}{4}$ b) $\frac{245'745}{16}$ c) 531'440   d) $\frac{58'025}{13'122}$ e) $1093\sqrt{3}+3280$

57. a) 838'861   b) 6560   c) $\frac{129'009'091}{24}$ d) 13'107

58. At least 238 terms need to be added.

59. ≈ 14.491

60. a) $A_2 = 3.75$, $A_3 = 2.8125$ b) $A_k = 5 \cdot \left(\frac{3}{4}\right)^{k-1}$ c) $k = 39$

d) $a_k = 3^{k-1}$, $k \in \mathbb{N}$ e) $b_k = \frac{3^{k-1}-1}{2}$

61. a) 490   b) approx. 0.9 mm

62. a) 9   b) $\frac{2}{3}$ c) 500   d) $3+3\sqrt{3}$ e) $\frac{1}{3}$ f) $\frac{8}{15}$ g) $\frac{1}{18'000}$ h) 54

63. a) $10\left(1 - 0.9^n\right)$ b) $9.999734\ldots$ c) 10

64. $\frac{120}{7} \approx 17.14$

65. a) $q = \frac{3}{4}$; $s = 16'384$ b) $q = \frac{-2}{3}$; $s = 36$

66. a) $q = \frac{5}{6}$ b) $q_1 = \frac{1}{3}$ and $q_2 = \frac{2}{3}$

67. $a_3 = 2 \cdot \sqrt{2} - 2$, $a_4 = \sqrt{2} - 2$, $a_5 = \sqrt{2} - 1$ and $s = 12 \cdot \sqrt{2} - 16 \approx 0.97$

68. a) $\frac{4}{9}$ b) $\frac{17}{99}$ c) 1 d) $\frac{167}{37}$

69. a) $0 < |x| < \frac{1}{5}$ b) $1 < x < 3$, $x \neq 2$

70. $16\,\text{cm}$; $585.14\,\text{cm}^3$

71. $s = \frac{3}{2}\,\text{m}$

72. $s = 25\pi\,\text{cm} \approx 78.5\,\text{cm}$

73. $s = 117$; $Z(27\,|\,18)$

74. Both paths have length $\pi \cdot r$.

75. They need to decrease by at least $70.4\,\%$.

76. –

77. –

78. $42.92\,\%$

79. a) 1st DS: 1, 4, 9, 16, 25, $\ldots$ $\Rightarrow$ 36, 49   b) 1st DS: 7, 5, 3, 1, -1, -3, $\ldots$ $\Rightarrow$ -5, -7
   c) 1st DS: 2, 3, 5, 7, 11, 13, $\ldots$ $\Rightarrow$ 17, 19   d) 1st DS: 1, -2, 3, 1, -2, 3, 1, $\ldots$ $\Rightarrow$ -2, 3

80. a) 102, 111   b) $\frac{161}{60}$, $\frac{191}{60}$ c) 133, 138   d) 14, 12

81. No one.

82. They differ by a constant.

83. a) 1st DS: -1, 3, -1, 3, -1, 3, -1, $\ldots$ 2nd DS: 4, -4, 4, -4, 4, -4, 4, $\ldots$
      3rd DS: -8, 8, -8, 8, -8, 8, -8, $\ldots$ 4th DS: 16, -16, 16, -16, 16, -16, 16, $\ldots$
   b) 6th DS: 64, -64, 64, -64, 64, -64, 64, $\ldots$ 7th DS: -128, 128, -128, 128, -128, $\ldots$
   c) $d_k = (-1)^{n+1+k} \cdot 2^n$ für $n \in \mathbb{N}$

84. a) –   b) $d_k = 2k$

85. $d_k = 2k - 1$ for $k \in \mathbb{N}$

86. a) $a_k = k^2 - k + 1$

$\ a_{101} = 10'101$

b) $a_k = -k^2 + 11$

$\ a_{101} = -10'190$

c) $a_k = \frac{1}{2}k^2 - \frac{3}{2}k + 4$

$\ a_{101} = 4953$

d) $a_k = -\frac{3}{2}k^2 + \frac{7}{2}k + 10$

$\ a_{101} = -14'938$

e) $a_k = k^2 - 5k + 6$

$\ a_{101} = 9702$

f) $a_k = -k^2 + 3k + 1$

$\ a_{101} = -9897$

87. a) $a_{22} = 10'626$ b) $a_{22} = 10'649$ c) $a_{22} = 1938$ d) $a_{22} = -8000$

88. a) 4, 7, 9, 12, 14, 17, 19, 22, ...   b) 1, 8, 27, 64, 125, 216, ...

89. $-10, -5, 15, 53, 112, 195$

90. No

91. a) The 1st DS is also a GP.   b) No

92. Yes

93. –

94. –

95. The order does not change in all three cases.

96. f is linear.

97. explicit: $a_k = k^2 - k + 2$; recursive: $a_1 = 2$, $a_{k+1} = a_k + 2k$, $k \in \mathbb{N}$

98. a) 35, 51, 70

b) $F_{k+1} = F_k + 3k + 4$, $F_1 = 5$

c) $F_k = 1.5k^2 + 2.5k + 1$

d) 15'251

99. a) $a_1 = 4$, $a_2 = \frac{16}{3}$ b) $a_k = 3 \cdot \left(\frac{4}{3}\right)^k$ c) The sequence diverges.

100. a) $u_k = 3s \cdot \left(\frac{4}{3}\right)^k$, $k \in \mathbb{N}_0$; (u_k) is divergent.

b) $A_k = A_0 \cdot \left(\frac{1}{9}\right)^k = \frac{\sqrt{3}}{4}s^2 \cdot \left(\frac{1}{9}\right)^k$, $k \in \mathbb{N}_0$

c) $\frac{A_0}{3} \cdot \left(\frac{4}{9}\right)^{k-1}$, $k \in \mathbb{N}$

d) $F_k = A_0 \left(1 + \frac{3}{5}\left(1 - \left(\frac{4}{9}\right)^k\right)\right)$, $k \in \mathbb{N}_0$

e) $F = \frac{8}{5}A_0 = \frac{2\sqrt{3}}{5}s^2$

101. a) Fr. 1628.90   b) Fr. 613.90

102. a) Fr. 3760.90   b) Fr. 3761.15

103. Fr. 14'284.–

104. a) Fr. 6643.20   b) Fr. 315.10

105. Fr. 6862.80

106. Fr. 7497.75

107. 14 times

108. Fr. 313.55

109. 269'347.80 francs

110. a) 1, 1, 2, 3, 5, 8, 13, 21, 34, 55, 89   b) $F_0 = F_1 = 1$ and $F_{k+2} = F_{k+1} + F_k$, $k \in \mathbb{N}_0$

111. $W_k = W_{k-2} + W_{k-1}$ for $k \geq 3$ with $W_1 = 1$ and $W_2 = 2$

112. $M_k = M_{k-2} + M_{k-1}$ for $k \geq 3$ with $M_1 = 1$ and $M_2 = 2$

113. $T_k = T_{k-2} + T_{k-1}$ for $k \geq 3$ with $T_1 = 1$ and $T_2 = 2$

114. –

115. a) 1, 2, 3, 5, 8, 13, 21, 34, 55, 89   b) 2, 1, 3, 4, 7, 11, 18, 29, 47, 76
     c) 3, 4, 1, −3, −4, −1, 3, 4, 1, −3   d) 2, −1, 1, 0, 1, 1, 2, 3, 5, 8

116. a) Both: $1, \frac{1}{2}, \frac{2}{3}, \frac{3}{5}, \frac{5}{8}, \frac{8}{13}$ b) –   c) $g = \frac{\sqrt{5}-1}{2}$

117. a) true   b) $F_{31} + F_{32} = 2'178'309 + 3'524'578 = 5'702'887 = F_{33}$ c) –

118. a) 0, 1, 1, 2, 3, 5, 8, 13, 21, ...
     b) $T(4) = T(6) = T(10) = +1$ and $T(5) = T(13) = -1$
     c) $s_4 = 7$, $s_7 = 33$, $s_8 = 54$, $s_{11} = 232$; $s_n = u_{n+2} - 1$
     d) $U(2) = 3$, $U(3) = 8$, $U(4) = 21$, $U(8) = 987$, $U(11) = 17'711$; $U(n) = u_{2n}$

119. a) $2 + 4 + 6 + \ldots + 2n = n \cdot (n+1)$
     b) $1 + 2 + 3 + 4 + \ldots + n = \frac{n \cdot (n+1)}{2}$
     c) $1 + \frac{1}{2} + \frac{1}{4} + \frac{1}{8} + \ldots + \frac{1}{2^{n-1}} = 2 - \frac{1}{2^{n-1}}$

120. –

121. a) $2 + 6 + 18 + 54 + \ldots + 2 \cdot 3^{n-1} = 3^n - 1$
     b) $1 \cdot 2 + 2 \cdot 3 + 3 \cdot 4 + \ldots + n \cdot (n+1) = \frac{n(n+1)(n+2)}{3}$

122. a) $s_n = n^2$ b) $s_n = \frac{n}{n+1}$
     c) $s_n = \frac{n}{2n+1}$ d) $s_n = \frac{2n}{n+1}$
     e) $s_n = (n+1)! - 1$ f) $s_n = \left(\frac{n(n+1)}{2}\right)^2$

123. a) $a_k = k^2 + 1$ b) $a_k = \frac{k}{k+1}$

124. $P_n = \frac{1}{n}$

125. –

126. –

127. –

128. –

129. –

130. –

131. –

132. –

133. –

134. –

135. –

136. a) neither
b) GP: $a_1 = 40$, $a_{k+1} = 0.75a_k$;
$a_k = 40 \cdot 0.75^{k-1}$

   c) AP: $a_1 = 3$, $a_{k+1} = a_k + 4$; $a_k = 4k - 1$
d) GP: $a_1 = 5$, $a_{k+1} = -3 \cdot a_k$; $a_k = 5 \cdot (-3)^{k-1}$

137. recursive: $a_1 = 3$, $a_{k+1} = 10a_k + 3$; explicit: $a_k = \frac{1}{3} \cdot 10^k - \frac{1}{3}$, $k \in \mathbb{N}$

138. $a_3 = 2a_2 - a_1 = 2m - 1$, $a_4 = 2a_3 - a_2 = 3m - 2$, $a_5 = 2a_4 - a_3 = 4m - 3$, $a_6 = 2a_5 - a_4 = 5m - 4$

   $\Rightarrow$ $a_k = (k - 1)m - (k - 2) = (m - 1)k + (2 - m)$ $\Rightarrow$ It is an AP.

139. a) $\frac{1}{5}, \frac{1}{6}, \frac{1}{7}, \frac{1}{8}, \frac{1}{9}$; $a_k = \frac{1}{k+2}$, $k \in \mathbb{N}$
b) $\frac{1}{4}, \frac{1}{3}, \frac{1}{2}, \frac{1}{1}$, a_7 not def.; $a_k = \frac{1}{7-k}$, $k \leq 6$

   c) $\frac{1}{13}, \frac{1}{15}, \frac{1}{17}, \frac{1}{19}, \frac{1}{21}$; $a_k = \frac{1}{2k+7}$, $k \in \mathbb{N}$
d) $\frac{-1}{8}, \frac{-1}{13}, \frac{-1}{18}, \frac{-1}{23}, \frac{-1}{28}$; $a_k = \frac{1}{7-5k}$, $k \in \mathbb{N}$

140. $d_{\max} = 9$

141. AP: a), d), e), h), j); GP: b), c), f), g), k); neither: i), l)

142. a) (b_k) is a GP.
b) (b_k) is an AP.

143. –

144. 100

145. a) –
b) approximately 5.3 million francs

146. a) converges; $\frac{3}{2}\,\mathrm{m}^2$
b) diverges

147. –

148. $11.9747\ldots \approx 12\,\ell$

149. $80\,\mathrm{km}$ from U and $39\,\mathrm{km}$ from V, respectively

150. Yes

151. At the latest in 2044

152. 720 francs

153. $u = 4$, $v = 1$, $w = -2$

154. –

155. a) $5 + 8 + 11 + 14 + 17 + \ldots$

    b) $2 + 2 + 4 + 8 + 16 + 32 + \ldots$

    c) $1 + 0 + 1 + 0 + 1 + 0 + 1 + 0 + 1 + \ldots$

    d) $1 + (-1) + 1 + (-1) + 1 + (-1) + 1 + \ldots = 1 - 1 + 1 - 1 + 1 - 1 + 1 \mp \ldots$

    e) $32 + (-16) + 8 + (-4) + (-1) + \ldots = 32 - 16 + 8 - 4 - 1 \mp \ldots$

    f) $1 + (-2) + 3 + (-4) + 5 + (-6) + 7 + \ldots = 1 - 2 + 3 - 4 + 5 - 6 + 7 \mp \ldots$

156. The given series converges and has limit 4.

157. a) $\frac{2+5x}{1-x^2}$ b) $\frac{2-3x}{1-x^2}$ c) $\frac{x+x^2}{1+x^2}$

158. $a_{12} = -69$

159. a) a_7 b) b_5

160. a) 61.46   b) 916.82   c) 56.05   d) 30.44   e) 804.54   f) 45.87

161. a) $w_2(x) = 4x - 3$, $w_3(x) = 8x - 7$, $w_4(x) = 16x - 15$

    b) –   c) Yes   d) –

162. –

163. a) They lie: i) on a line, ii) on an exponential curve and iii) on a parabola.

    b) –

164. a) $17, 33, 65$ b) $\frac{1}{2}, -\frac{1}{4}, \frac{1}{8}$ c) $3, 8, 63$ d) $8, 19, 46$

165. a) $a_3 = 6$, $a_4 = 8$, $a_5 = 10$; $a_k = 2k$ b) $a_3 = -1$, $a_4 = -3$, $a_5 = -5$; $a_k = 5 - 2k$

166. 900 numbers

167. Let $k \in \mathbb{N}$ for all answers.

    a) GP; $a_{15} = 49'152$; rec.: $a_1 = 3$, $a_{k+1} = a_k \cdot (-2)$; expl.: $a_k = 3 \cdot (-2)^{k-1} = \frac{-3}{2} \cdot (-2)^k$

    b) neither

    c) neither

    d) AP; $a_{15} = -789$; rec.: $a_1 = 765$, $a_{k+1} = a_k - 111$; expl.: $a_k = 876 - 111k$

    e) GP; $a_{15} = 9'565'938$; rec.: $a_1 = 2$, $a_{k+1} = a_k \cdot (-3)$; expl.: $a_k = 2 \cdot (-3)^{k-1} = -\frac{2}{3} \cdot (-3)^k$

    f) neither

    g) neither

    h) AP; $a_{15} = 789$; rec.: $a_1 = -765$, $a_{k+1} = a_k + 111$; expl.: $a_k = -876 + 111k$

168. a) expl.: $a_k = 3k - 2$ b) expl.: $a_k = 7k - 1$ c) expl.: $a_k = 2^{k+1}$

       rec.: $a_1 = 1$, $a_{k+1} = a_k + 3$ rec.: $a_1 = 6$, $a_{k+1} = a_k + 7$ rec.: $a_1 = 4$, $a_{k+1} = 2a_k$

169. a) $s_n = 37n - \frac{7}{2}n(n-1)$ b) $s_n = 64 \cdot \left(1 - \left(\frac{3}{4}\right)^n\right)$

170. a) $s_{24} = \sum\limits_{k=1}^{24} (12 + 3k) = 1188$ b) $s_6 = \sum\limits_{k=1}^{6} 4 \cdot 3^{k-1} = 1456$

     c) $s_{12} = \sum\limits_{k=1}^{12} 15'360 \cdot \left(-\frac{3}{2}\right)^{k-1} = -791'017.5$ d) $s_{10} = \sum\limits_{k=1}^{10} (23 - 6k) = -100$

171. a) $a_{48} = 103$, $s_{48} = 2688$ b) $i = 20$ c) $a_1 = -8.125$, $s_{24} - s_7 = 85$

172. a) 124 terms   b) 7997 terms

173. 103, 115, 127

174. The sequence consists of 20 terms.

175. The sequence consists of 100 terms.

176. The AP consists of 55 terms at most and we have : $a_n = u_{55} = 3$.

177. a) $a_1 = 1000$, $s_9 = 1428.54\ldots$ b) $i = 9$
     c) $q = \pm\sqrt{3}$, $a_1 = \frac{2}{3}$ d) $i = 17$

178. a) 6905 terms   b) 14 terms   c) at least 8 terms

179. $a_2 = 30$

180. $s = 512$

181. $2 \cdot \sqrt{6} + 4 \cdot \sqrt{3}$

182. a) $2500 \, \text{cm}^2$; $625 \, \text{cm}^2$; $156.25 \, \text{cm}^2$ b) $Q_n = \frac{10'000}{4^n}$
     c) from the 64th square   d) $\approx 3333.28 \, \text{cm}^2$
     e) $3333.\overline{3} \, \text{cm}^2$

183. a) 56.7   b) 13.5   c) 1   d) $\frac{30}{7}$

184. $0 < |m| < \frac{2}{3}$

2 Limits

1. The number 1

2. a) The room temperature of the kitchen.   b) The room temperature of the kitchen.

3. a) $\frac{1}{3}$ b) 2   c) $\frac{2}{3}$ d) 3

4. a) 0   b) 2   c) 0   d) 1

5. a) $0.12345678910111213\ldots$ b) $0.10100100010000100000\ldots$
     c) no limit   d) 2

6. a) 2   b) 4   c) 3   d) 5   e) 6   f) 0

7. a) 0   b) 0   c) ∞; prop. divergent

8. a) 0   b) 3   c) ∞; prop. divergent

9. a) $a_k = 70 \cdot \left(\frac{2}{3}\right)^k$ b) ambient temperature, i.e. $24\,°\mathrm{C}$

10. a) $a = 0$; $n = 10$ b) $a = 0$; $n = 3$

11. a) $n = 1001$; $n = 10^{10} + 1$ b) $n = 10^6 + 1$; $n = 10^{20} + 1$ c) $n = 501$; $n = 5 \cdot 10^9 + 1$

12. $n = 51$

13. a) $n = 10$ b) $n = 10^{(10^6)-2}$ c) $n = 5 \cdot 10^{10} + 1$

14. a) Example 1: $a_k = 2$; Example 2: $a_k = 2 + \frac{1}{k} = \frac{2k+1}{k}$; Example 3: $a_k = 2 + \left(\frac{1}{4}\right)^k$

   b) Examples: $a_k = \frac{1}{2} + \frac{1}{3k}$; $a_k = \frac{1}{2} + \left(-\frac{1}{4}\right)^k$

15. a) No   b) Yes   c) No   d) Yes   e) No

16. a) No   b) No

17. a) 2   b) 0   c) $\frac{-1}{5}$
   d) 0   e) 2   f) 5
   g) $\frac{7}{4}$ h) 8   i) -16

18. a) $\frac{-4}{5}$ b) $-\infty$; properly divergent   c) oscillating

19. a) 0   b) 1   c) 0

20. a) 1   b) 0   c) 0
   d) properly divergent to ∞ e) properly divergent to ∞ f) properly divergent to $-\infty$

21. a) No, oscillating   b) Yes, 0   c) Yes, 1
   d) No, prop. divergent to ∞ e) Yes, 0   f) No, prop. divergent to $-\infty$

22. a) 0   b) 0   c) 1000

23. $\lim\limits_{k \to \infty} \frac{4 \cdot k^m}{7 \cdot k^\ell} = \begin{cases} \infty, & m > \ell \\ \frac{4}{7}, & m = \ell \\ 0, & m < \ell \end{cases}$

24. a) $a = 0$: limit 0; $a \neq 0$: limit 1   b) $a = 0$: limit 0; $a \neq 0$: limit $\frac{1}{a}$

25. $\lim\limits_{k \to \infty} a_k = 2 \;\Rightarrow\;$ e.g. $b_k = a_k + 2 = \frac{4k+5}{k+3}$; $b_k = 2 \cdot a_k = \frac{4k-2}{k+3}$; $b_k = (a_k)^2 = \frac{4k^2-4k+1}{k^2+6k+9}$

26. a) e^2 b) $\frac{1}{e} = e^{-1}$ c) e   d) e^2

27. a) 4   b) $\frac{4}{5}$ c) $1 + \sqrt{2}$ d) 1

28. $\frac{1}{2}$

29. a) No   b) Yes

30. a) 0; 0   b) 0; 0   c) 0; 0

31. a) 0   b) 1   c) oscillating   d) properly divergent

32. a) False   b) True   c) False

33. a) $\frac{1}{3}$ b) $-\infty$, prop. div.   c) $\frac{3}{2}$ d) $-\frac{1}{2}$
     e) 3   f) ∞, prop. div.   g) -1 h) $\frac{1}{3}$

34. a) $\sqrt{2}$ b) 2   c) 0   d) 5

35. $\lim\limits_{x\to\infty}\left(\sqrt{2x+1}-\sqrt{2x}\right) = \lim\limits_{x\to\infty}\left(\frac{\sqrt{2x+1}-\sqrt{2x}}{1}\cdot\frac{\sqrt{2x+1}+\sqrt{2x}}{\sqrt{2x+1}+\sqrt{2x}}\right) = \lim\limits_{x\to\infty}\frac{1}{\sqrt{2x+1}+\sqrt{2x}} = 0$

36. a) $\frac{3}{4}$ b) $\frac{3}{8}$

37. a) 1   b) 0   c) $\frac{3}{2}$

38. a) $\frac{2}{5}$, $\frac{2}{5}$ b) 0, 0   c) 0, does not exist   d) 0, does not exist

39. a) 4   b) 2   c) -128 d) $\frac{1}{2}$

40. Answers for $x \to \infty$ and $x \to -\infty$, respectively
     a) 0, divergent   b) 0, divergent   c) 0, divergent   d) 0, 0
     e) divergent, 0   f) divergent, 0   g) divergent, 0   h) both divergent

41. a) divergent   b) 0   c) divergent   d) 0

42. Answers for $x \to \infty$ and $x \to -\infty$, respectively
     a) -1, -1 b) 1, -1 c) $\frac{2}{3}$, $\frac{2}{3}$ d) 0, does not exist

43. $a = -4$, $b = 5$, $c = 6$

44. a) $m = 4$ b) $m \geq 5$ c) no m d) $m = 0, 1, 3$ e) $m = 2$

45. a) 0   b) 7   c) 0   d) -3 e) a f) 3

46. a) 2   b) 6   c) oscillating
     d) 0   e) oscillating   f) oscillating

47. a) 4   b) -20 c) $\frac{4}{3}$ d) $\frac{1}{2}$

48. a) $\frac{1}{8}$, oscillating   b) oscillating, $\frac{1}{10}$
     c) 0, 14   d) 26, 0

49. a) 3   b) oscillating   c) $\frac{11}{18}$
    d) -4 e) -6 f) $\frac{1}{2}$

50. a) 2   b) 12   c) $\frac{1}{4}$

51. a) 1   b) 2

52. a) 4   b) 12   c) -8 d) $-4p$

53. a) $-\frac{1}{2}$ b) $\frac{1}{6}$ c) $-\frac{4}{5}$

54. The first equation is not true and the second is true.

55. a) -2 b) $\frac{1}{2}$ c) 4   d) $-\frac{1}{2}$ e) $\frac{1}{3}$ f) 0

56. a) $\frac{1}{2}$ b) $\frac{1}{72}$ c) oscillating
    d) $-\infty$, properly divergent   e) $-\frac{1}{4}$ f) 48

57. a) 16   b) 100   c) $4m^2$

58. a) $\frac{\sqrt{2}}{4}$, $\frac{\sqrt{2}}{4}$ b) $\frac{\sqrt{m}}{2m}$, $\frac{\sqrt{m}}{2m}$

59. a) 10   b) 4, 4   c) $\frac{5}{4}$

60. a) $\frac{1}{8} = 0.125$ b) 0   c) $-$ d) $f(x) = \frac{1}{\sqrt{x^2+16}+4}$

61. a) $-$ b) (i) 0   (ii) 2   (iii) 1

62. a) Yes   b) No   c) Yes   d) No   e) Yes

63. a) Yes   b) Yes

64. a) (strictly) monotonically decreasing   b) (strictly) monotonically increasing
    c) (strictly) monotonically decreasing   d) (strictly) monotonically increasing

65. a) (a_k) is strictly decreasing and bounded by 0 and $\frac{1}{7}$.
    b) (a_k) is strictly decreasing and bounded by 0 and $\frac{1}{2}$.

66. a) (a_k) is strictly increasing and bounded by $\frac{4}{5}$ and 1.
    b) (a_k) is strictly increasing and bounded by $\frac{\sqrt{2}}{2}$ and 1.

67. a) (strictly) monotonically decreasing   b) (strictly) monotonically decreasing

68. a) False   b) False   c) True   d) True

69. a) False; counterexample: $a_k = (-1)^k$ b) False; counterexample: $a_k = (-1)^k$
    c) True by definition of monotonicity

70. a) e.g. $a_k = 5 + \frac{1}{k}$ b) e.g. $a_k = -1 + \left(-\frac{1}{2}\right)^k$

71. a) False   b) False   c) False   d) False   e) False   f) False   g) False   h) False

72. a) Limit 6   b) Limit 1

73. –

74. a) No   b) No   c) No   d) No

75. a) Yes; 8   b) Yes; $\frac{1}{2}$ c) Yes; $\frac{1}{2}$

76. a) $\lim\limits_{x \to 4^+} f(x) = 3$ and $\lim\limits_{x \to 4^-} f(x) = 3$; $\lim\limits_{x \to 4} f(x) = 3$ exists.

     b) $\lim\limits_{x \to 2^+} f(x) = 2$ and $\lim\limits_{x \to 2^-} f(x) = 1$; $\lim\limits_{x \to 2} f(x)$ does not exist.

77. a) No; -1, 1   b) No; 0, ∞ (divergent)

78. a) ∞ or $-\infty$, resp., (divergent) b) 6   c) ∞ (divergent) or 0, resp.

79. a) $-\infty$ or ∞, resp., (divergent) b) 0   c) ∞ (divergent) or 0, resp.

     d) 0   e) 1   f) $\frac{1}{2}$

80. a) True   b) True   c) True   d) True   e) True   f) False

81. a) Yes, 2   b) Yes, $-\frac{1}{4}$ c) No   d) Yes, 1   e) No   f) No

82. a) convergent, $\ln(2)$ b) divergent   c) divergent

83. a) 1   b) -1 c) no limit   d) 0

84. oscillating

85. b

86. $a_1 = 0$; $a_n = \frac{2}{n^2+n}$ for $n \geq 2$; $\lim\limits_{n \to \infty} s_n = 1$

87. a) $a_5 = \frac{1}{30}$, $a_7 = \frac{1}{56}$, $a_{15} = \frac{1}{240}$, $a_{99} = \frac{1}{9900}$ and $a_{110} = \frac{1}{12'210}$, $a_{1110} = \frac{1}{1'233'210}$,

       $a_{11110} = \frac{1}{123'443'210}$, $a_{111110} = \frac{1}{12'345'543'210}$

     b) $s_1 = \frac{1}{2}$, $s_2 = \frac{2}{3}$, $s_3 = \frac{3}{4}$, $s_4 = \frac{4}{5}$, $s_5 = \frac{5}{6}$ and $s_n = \frac{n}{n+1}$

     c) $\lim\limits_{k \to \infty} a_k = 0$ and $\lim\limits_{n \to \infty} s_n = 1$ d) –

88. –

89. a) $\frac{\pi^2}{24}$ b) $\frac{\pi^2}{8}$

90. a) (f_n): 1, 1, 2, 3, 5, 8, 13, 21, 34, 55, 89, 144, ...

     b) (q_n): $\frac{1}{1}$, $\frac{2}{1}$, $\frac{3}{2}$, $\frac{5}{3}$, $\frac{8}{5}$, $\frac{13}{8}$, $\frac{21}{13}$, $\frac{34}{21}$, $\frac{55}{34}$, $\frac{89}{55}$, ...

     c) (q_n): 1, 2, 1.5, 1.666..., 1.6, 1.625, 1.615..., 1.619..., 1.617..., 1.618..., ...

     d) $q_n = 1 + \frac{1}{q_{n-1}}$, $n \geq 2$ and $q_1 = 1$ e) $\frac{1+\sqrt{5}}{2} = 1.618...$ f) –

     g) No   h) 1, $1 + \frac{1}{1} = 2$, $1 + \frac{1}{1+\frac{1}{1}} = \frac{3}{2}$, $1 + \frac{1}{1+\frac{1}{1+\frac{1}{1}}} = \frac{5}{3}$

91. a) False   b) False   c) False   d) True   e) False   f) False   g) False

92. a) $\lim\limits_{x \to -2} f(x) = 4$, $\lim\limits_{x \to -2} \frac{f(x)}{x} = -2$ b) $\lim\limits_{x \to 0} g(x) = 0$, $\lim\limits_{x \to 0} \frac{g(x)}{x} = 0$

93. a) 4   b) $-\frac{1}{2}$

94. An overview of the three cases. In all cases $x_0 = 0$ (see chapter 2.3 «Further topics» for right-handed and left-handed limits):

$\lim\limits_{x \uparrow 0} 2^{\frac{1}{x}} = 0$ $\lim\limits_{x \uparrow 0} \frac{1}{1 + 2^{\frac{1}{x}}} = 1$ $\lim\limits_{x \uparrow 0} 2^{-\frac{1}{x^2}} = 0$

$\lim\limits_{x \downarrow 0} 2^{\frac{1}{x}} = \infty$ $\lim\limits_{x \downarrow 0} \frac{1}{1 + 2^{\frac{1}{x}}} = 0$ $\lim\limits_{x \downarrow 0} 2^{-\frac{1}{x^2}} = 0$

$\lim\limits_{x \to 0} 2^{\frac{1}{x}}$ does not exist $\lim\limits_{x \to 0} \frac{1}{1 + 2^{\frac{1}{x}}}$ does not exist $\lim\limits_{x \to 0} 2^{-\frac{1}{x^2}} = 0$ (removable discontinuity)

asymptote $y = 1$ asymptote $y = \frac{1}{2}$ asymptote $y = 1$

95. $\lim\limits_{g \to \infty} \left(\frac{1}{b} + \frac{1}{g} \right) = \frac{1}{b} = \frac{1}{f}$ $\Leftrightarrow$ $b = f$, i.e. the image lies in the focal plane of the lens.

96. a) $\lim\limits_{m_2 \to \infty} \frac{x_1 \cdot m_1 + x_2 \cdot m_2}{m_1 + m_2} = x_2$ b) $\lim\limits_{m_1 \to \infty} \frac{x_1 \cdot m_1 + x_2 \cdot m_2}{m_1 + m_2} = x_1$

97. a) 0.488   b) 2.120028   c) 0.73908

98. a) Limit 0   b) Limit -2

99. $n = 1001$

100. a) No   b) Yes   c) e.g. $a_k = k$

101. a) 10   b) $\frac{-5}{4}$ c) prop. divergent
  d) 3   e) oscillatory   f) $\frac{-1}{2}$

102. a) 1   b) 0   c) 0   d) 1   e) 1   f) 1

103. a) Yes; limit 0 (GP with $q = \frac{1}{3}$)   b) No; prop. div. to $-\infty$ (AP with $d < 0$)

104. $\frac{1 + \sqrt{5}}{4}$ (GP with $q = \frac{1}{\sqrt{5}}$)

105. a) $\lim\limits_{x \to \infty} \frac{2x+3}{x+1} = 2$, $\lim\limits_{x \to -\infty} \frac{2x+3}{x+1} = 2$ b) $\lim\limits_{x \to -\infty} \frac{7x}{x-7} = 7$, $\lim\limits_{x \to \infty} \frac{7x}{x-7} = 7$

106. a) $\frac{1}{3}$ b) 1   c) 1

107. a) prop. divergent, ∞ b) prop. divergent, ∞ c) convergent, $-\frac{1}{2}$ d) convergent, $\frac{1}{3}$

108. a) $x \to \infty$: divergent; $x \to -\infty$: limit 0   b) $x \to \infty$: divergent; $x \to -\infty$: limit 0
  c) both divergent   d) both convergent with limit 0

109. a) 9   b) $\frac{9}{4}$

110. a) 6   b) -10

111. a) $2x$ b) h

112. a) 27   b) 27

113. a) $\frac{8}{9}$ b) properly divergent, ∞

3 Differential Calculus

1. a) $v_1 = 0.5\,\text{m/s}$ b) $v_2 = 0.25\,\text{m/s};\ v_3 = 0.75\,\text{m/s}$
   c) $v_4 \approx 0.6\,\text{m/s};\ v_5 \approx 0.7\,\text{m/s};\ v_6 \approx 0.8\,\text{m/s};\ v_7 = 1\,\text{m/s}$ d) $v = v(6) \approx 1.2\,\text{m/s}$

2. a) $\frac{f(4)-f(1)}{4-1} > 0$ b) $\frac{f(2)-f(1)}{2-1} < 0$
   c) $\lim\limits_{x\to 4} \frac{f(x)-f(4)}{x-4} > 0$ d) $\lim\limits_{x\to 6} \frac{f(x)-f(6)}{x-6} < 0$
   e) $\frac{f(2)-f(1)}{2-1} > \frac{f(6)-f(5)}{6-5}$ f) $\lim\limits_{x\to 3} \frac{f(x)-f(3)}{x-3} < \lim\limits_{x\to 4} \frac{f(x)-f(4)}{x-4}$
   g) $\lim\limits_{x\to 0.5} \frac{f(x)-f(0.5)}{x-0.5} = \lim\limits_{x\to 5} \frac{f(x)-f(5)}{x-5}$

3. a) –   b) $v_{\text{average}} = 1\,\text{m/s}$ c) –
   d) –   e) $v_{\text{inst}} = 0.5\,\text{m/s}$ f) $v_{\text{inst}} = \lim\limits_{t\to 1} \frac{s(t)-s(1)}{t-1} = 0.5$
   g) –   h) $v(t_0) = 0.5 t_0$

t_0	0	2	3	4
$v(t_0)$	0	1	1.5	2

4. a) i) $m_{AB} = \frac{2}{25}$,   ii) $m_{CD} = \frac{3}{65}$,   iii) $m_{DE} = \frac{-1}{10}$,   iv) $m_{FG} = \frac{-1}{25}$
   b) greatest: AB; smallest: DE c) increasing on $[0; 380]$, decreasing on $[380; 600]$
   d) D and F e) $\approx 0.3\,\%$; No

5. a) $D = \mathbb{R};\ m_1 = -2;\ m_2 = 9$ b) $D = \mathbb{R};\ m_1 = 3;\ m_2 = -8$
   c) $D = [-2; \infty[;\ m_1 = \sqrt{2};\ m_2 = \frac{2}{5}$ d) $D = \mathbb{R}\backslash\{-3\};\ m_1 = -2;\ m_2 = \frac{-3}{25}$

6. a) $-1.4\,^\circ\text{C/min}$ and $-0.8\,^\circ\text{C/min}$ b) –   c) by their slopes

7. a) growth of bacteria in the time interval $[t_0; t_1]$
   b) average rate of growth of bacteria during the time interval $[t_0; t_1]$
   c) instantaneous rate of growth of bacteria at time t_0

8. a) $L(x) = 0.25^x,\ x \geq 0$
   b) It is the effective change of luminosity from depth x_1 to depth x_2.
      $L(5) - L(2) = -0.061523\ldots$
   c) $\frac{\Delta L}{\Delta x} = -0.02050\ldots$
   d) It is the instantaneous rate of change in luminosity at a depth of 5 metres.

9. a) $I_{[t_1;t_2]} = \frac{W(t_2)-W(t_1)}{t_2-t_1}$ b) $I(t_1) = \lim\limits_{t_2\to t_1} \frac{W(t_2)-W(t_1)}{t_2-t_1}$

10. a) $\nu_R = \frac{M(t_0+\Delta t)-M(t_0)}{\Delta t}$ b) $\nu_R(t_0) = \lim\limits_{\Delta t \to 0} \frac{M(t_0+\Delta t)-M(t_0)}{\Delta t}$

11. a) $\rho_{[z_0;z]} = \frac{m(z)-m(z_0)}{(z-z_0)\cdot Q}$ b) $\rho_{z_0} = \frac{1}{Q} \cdot \lim\limits_{z \to z_0} \frac{m(z)-m(z_0)}{z-z_0}$

12. a) True b) True c) True

13. a) $55\,\mathrm{m/s}$; $50.5\,\mathrm{m/s}$; $50.25\,\mathrm{m/s}$; $50.005\,\mathrm{m/s}$ b) $50\,\mathrm{m/s}$ c) $v_0 = 10t_0$

14. a) $136\,\mathrm{m}$ b) $52\,\mathrm{m/s}$ c) $92\,\mathrm{m/s}$

15. a) i) 10 ii) $8+h$ iii) $2x_0 + h$

 b) i) 10 ii) $x_1 + 4$ iii) $x_1 + x_0$

 c) i) 8 ii) 8 iii) $2x_0$

16. a) $f'(1) = -1$ b) $f'(2) = \frac{-1}{4}$ c) $f'(5.5) = \frac{-4}{121} \approx -0.033$

17. a) $f'(2) = 4$ b) $f'(5) = \frac{-2}{125}$ c) $f'(-1) = 3$

18. a) $f'(5) = 7$ b) $f'(-1) = -2$ c) $f'(0) = 0$ d) $f'(3) = m$

19. $\frac{\mathrm{d}}{\mathrm{d}x} f(1) = \frac{1}{4}$

20. a) $f(x) = \sqrt{x}$ and $x_0 = 49$ b) $f(x) = 2^x$ and $x_0 = 5$ c) $f(x) = \lg(x)$ and $x_0 = 100$
 d) $f(x) = \sin(x)$ and $x_0 = \frac{5\pi}{6}$ e) $f(x) = x^4$ and $x_0 = 3$ f) $f(x) = \frac{1}{x}$ and $x_0 = \frac{1}{5}$

21. a) $f'(x) = 1$ b) $f'(x) = 2x$ c) $f'(x) = 3x^2$
 d) $f'(x) = 6x$ e) $f'(x) = \frac{1}{2}$ f) $f'(x) = 6x^2 - 2x$
 g) $f'(x) = \frac{1}{2\sqrt{x}}$ h) $f'(x) = 2ax + b$ i) $f'(x) = \frac{-c}{x^2}$

22. a) $2x$; $2x$ b) $3x^2 + h^2$; $3x^2$
 c) $\frac{-1}{x^2-h^2}$; $\frac{-1}{x^2}$ d) $2x - 1$; $2x - 1$

23. slopes of secants: $\frac{1}{2}$, $\frac{2}{3}$, $\frac{3}{4}$, $\frac{4}{5}$, $\frac{5}{6}$; $m_0 = 1$

24. a) $F(a + \Delta a) - F(a) = 2a\Delta a + (\Delta a)^2$ b) $\frac{\Delta F}{\Delta a} = 2a + \Delta a$ c) $\frac{\mathrm{d}F}{\mathrm{d}a} - 2a$

25. $\frac{\mathrm{d}F}{\mathrm{d}r} = 2\pi \cdot r$

26. No

27. i) True ii) False iii) False

28. I–c), II–a), III–b), IV–d)

29. –

30. –

31. –

32. –

33. a) No   b) No   c) Yes   d) No   e) No

34. Graph 2

35. Graph 2

36. $t = 4$

37. a) $f'(x) = 0$ b) $f'(x) = -9$ c) $f'(x) = 16x^{15}$
   d) $f'(x) = 3$ e) $f'(x) = 3x^2$ f) $f'(t) = 14x + 2$
   g) $f'(x) = \frac{-1}{2}x^4$ h) $f'(x) = \frac{3}{2}x^2 - 2$ i) $f'(x) = \frac{1}{3}x^3 - \frac{1}{3}x^2$
   j) $f'(x) = \frac{-5}{x^6}$ k) $f'(x) = \frac{1}{4}x^{-\frac{3}{4}} = \frac{1}{4\sqrt[4]{x^3}}$ l) $f'(x) = -3 \cdot \frac{1}{5}x^{-\frac{4}{5}} = \frac{-3}{5\sqrt[5]{x^4}}$

38. a) $g'(t) = -4t + 3$ b) $g'(t) = 2t + 15t^2$ c) $g'(t) = -12t^3 - 3 + 6t^{-3}$
   d) $g'(t) = 4at^3 + 2bt$ e) $g'(t) = \frac{a}{\sqrt{t}}$ f) $g'(t) = 9x^2t^2$

39. a) $\frac{dy}{dx} = -20x^4 + 2$ b) $\frac{dy}{dx} = -\frac{18}{x^3}$ c) $\frac{dy}{dx} = \frac{3}{2}x^2$
   d) $\frac{dy}{dx} = x^{-\frac{1}{5}} - x^{-\frac{1}{3}}$ e) $\frac{dy}{dx} = \frac{1}{3}x^{-\frac{2}{3}} + \frac{1}{5}x^{-\frac{4}{5}}$ f) $\frac{dy}{dx} = \frac{1}{2\sqrt{x}} - \frac{2}{\sqrt[3]{x^2}}$

40. a) $\frac{dy}{dx} = 10x + 15$ b) $\frac{dy}{dx} = 6x + 2$ c) $\frac{dy}{dx} = \frac{7}{4}$

41. a) $y' = 3x^2 + 1$ b) $y' = 4x^3 - 6x^2$ c) $y' = 18x - 6$
   d) $y' = 8x + 3$ e) $y' = 2 - \frac{1}{x^2}$ f) $y' = 1 + \frac{12}{x^3}$

42. a) $V'(a) = 3a^2$ b) $O'(h) = 2\pi r$ c) $A'(r) = 2\pi r$ d) $V'(r) = 4\pi r^2$

43. a) 4 b) $8x - 4$ c) 4
   d) -2 e) $4x$ f) $2x + 2$

44. For the following formulas, $c \in \mathbb{R}$ may be an arbitrary constant.
   a) $f(x) = x^2 + c$ b) $f(x) = 2x^4 + 3x + c$ c) $f(x) = \frac{1}{6}x^6 - \frac{1}{4}x^4 + \frac{1}{2}x^2 + c$

45. Yes

46. $q = 1$

47. a) $f'(x) = -12x^3 + 2x$ b) $g'(z) = -3z^2 + 4z - 2$
   c) $h'(t) = -60t^4 - 48t^2 + 42t$ d) $f'(x) = \frac{15}{2}x\sqrt{x}$
   e) $h'(x) = \frac{5x^2 - 1}{2\sqrt{x}}$ f) $g'(u) = \frac{14u + 2}{5\sqrt[5]{u^3}}$
   g) $f'(z) = 16z^3 + \frac{7}{2}z^2\sqrt{z} - \frac{1}{\sqrt{z}} - 8$ h) $f'(x) = 16x^3 + 60x^2 + 74x + 30$

48. a) $f'(x) = a - 2x$ b) $f'(x) = -a$ c) $f'(x) = -4a^2x^3 + 3ax^2$
   d) $g'(t) = \frac{3t + a}{2\sqrt{t}}$ e) $g'(t) = 2at + 1$ f) $g'(t) = 4t^3 + 4a^2t$

49. In both cases: $f'(x) = 9x^2 - 8x + 3$

50. –

51. –

52. –

53. a) $P' = f' \cdot g \cdot h + f \cdot g' \cdot h + f \cdot g \cdot h'$; $P'(x) = 3x^2(x+1)(x^2-1) + x^3(x^2-1) + 2x^4(x+1)$

b) $P' = 3f^2 f'$; $P'(x) = 6(2x-1)^2$

54. –

55. a) $f'(x) = \frac{2}{(x+1)^2}$ b) $f'(x) = \frac{3x^2-1}{x^2}$ c) $f'(x) = \frac{-11}{(3x+4)^2}$ d) $f'(x) = \frac{x^2-1}{2x^2}$

e) $g'(x) = \frac{1}{3x^2}$ f) $g'(x) = \frac{3}{(1-x)^2}$ g) $g'(x) = \frac{-1}{(x+1)^2}$ h) $g'(x) = \frac{-6x}{(1+x^2)^2}$

i) $v'(t) = \frac{-6}{(2t-1)^2}$ j) $v'(t) = \frac{t(3t-2)}{(3t-1)^2}$ k) $v'(t) = \frac{9t^2}{(t^3+2)^2}$ l) $v'(t) = \frac{1+3t^2}{(1-3t^2)^2}$

m) $w'(t) = \frac{t^2-2t+2}{(t-1)^2}$ n) $w'(t) = \frac{9t^2}{(1-2t^3)^2}$ o) $w'(t) = \frac{t-2}{2t\sqrt{t}}$ p) $w'(t) = \frac{2(1+t^2)}{(1-t^2)^2}$

56. –

57. –

58. –

59. a) $f''(x) = 60x^3$ b) $f''(x) = x^2 - x + 1$ c) $f''(x) = 50$

60. a) $\frac{\mathrm{d}^2 y}{\mathrm{d}x^2} = 24x$ b) $\frac{\mathrm{d}^2 y}{\mathrm{d}x^2} = 6x - 8$

61. a) $90x^4$ b) $\frac{1}{2}x^2 + \frac{1}{3}x - \frac{1}{4}$ c) $84x^5 - 120x^4 - 36x^2 + 120x$

d) $6a^2(ax - b)$ e) $-2t$ f) $6x$

62. –

63. a) i) $3x^2$; $6x$; 6; 0; 0 ii) $4x^3$; $12x^2$; $24x$; 24; 0

iii) $5x^4$; $20x^3$; $60x^2$; $120x$; 120 iv) $6x^5$; $30x^4$; $120x^3$; $360x^2$; $720x$

b) $f^{(5)}(x) = n(n-1)(n-2)(n-3)(n-4)x^{n-5}$

c) Yes

64. a) $y = \frac{x^3}{3}$ b) $y = \frac{x^4}{12}$ c) $y = \frac{x^5}{60}$

65. a) $y' = n \cdot x^{n-1}$; $y'' = n(n-1) \cdot x^{n-2}$; $y''' = n(n-1)(n-2) \cdot x^{n-3}$

b) $y' = n(n-1) \cdot x^{n-2}$; $y'' = n(n-1)(n-2) \cdot x^{n-3}$; $y''' = n(n-1)(n-2)(n-3) \cdot x^{n-4}$

c) $y' = \frac{x^{n-1}}{(n-1)!}$; $y'' = \frac{x^{n-2}}{(n-2)!}$; $y''' = \frac{x^{n-3}}{(n-3)!}$

66. –

67. a) No; $v'' = f''g + 2f'g' + fg''$ b) No; $v''' = f'''g + 3f''g' + 3f'g'' + fg'''$

68. a) $p'''(x) = 0$ b) $p^{(4)}(x) = 0$ c) $p^{(6)}(x) = 0$ d) $(n+1)$ times

69. a) $y' = 6(3x - 4)$ b) $y' = 6x(x^2 + 1)^2$ c) $y' = 4x(3x-2)(1-x^2+x^3)^3$

d) $y' = 8x(3 - x^2)^{-2}$ e) $y' = -12x^2(x^3 + 21)^{-5}$ f) $y' = \frac{1000x^9}{(10-x^{10})^{11}}$

70. a) $w'(x) = \frac{3}{\sqrt{6x+1}}$ b) $w'(x) = \frac{3x}{\sqrt{x^2-1}}$ c) $w'(x) = \frac{2}{\sqrt{x}} \cdot (\sqrt{x} + 1)^3$

d) $w'(t) = \frac{2t}{(2-t^2) \cdot \sqrt{2-t^2}}$ e) $w'(t) = \frac{1}{(t+1)\sqrt{t^2-1}}$ f) $w'(t) = \frac{1}{4 \cdot \sqrt{t} \cdot \sqrt{1+\sqrt{t}}}$

71. a) $f'(x) = \frac{4(x-1)}{(1+x)^3}$ b) $g'(t) = \frac{2 \cdot (t^2-1)}{(1+t^2)^2}$ c) $h'(u) = \frac{2(3u-2)}{(2u+1)^3}$

72. –

73. a) $p'(x) = 4(x + 1)^3$ b) $q'(x) = 4x(x^2 + 2)$

c) $r'(x) = 4x(x^2 + 1)$ d) $s'(x) = 4(x + 1)(x^2 + 2x + 2)$

74. a) $f'(x) = 90x^9 + 252x^5 + 98x$ b) $g'(x) = \frac{-24x}{(1-2x^2)^4}$

c) $f'(t) - 6t^2 - 140t^4 + 686t^6$ d) $g'(t) = \frac{-3(t+1)}{(l-1)^3}$

e) $f'(x) = \frac{(x+2)\sqrt{x+1}}{2(x+1)^2}$ f) $g'(t) = \frac{4t^2-1}{3t^2}$

75. a) $a_1 = 0, \ a_2 = 4$ b) $a = \frac{1}{64}$ c) $a_1 = \frac{16}{3}, a_2 = 3$

76. a) $P'(x) = 2 \cdot f(x) \cdot f'(x)$ b) $P'(x) = 3 \cdot f^2(x) \cdot f'(x)$ c) $P'(x) = n \cdot f^{n-1}(x) \cdot f'(x)$

77. $w' = \frac{1}{2\sqrt{x}}$

78. $f'(x) = u'(v(w(x))) \cdot v'(w(x)) \cdot w'(x)$ or $f' = (u' \circ v \circ w) \cdot (v' \circ w) \cdot w'$, respectively

79. a) $f'(x) = \frac{-v'(x)}{(v(x))^2}$ b) $f'(x) = \frac{u'(x)v(x)-u(x)v'(x)}{(v(x))^2}$

80. Both formulas are false.

a) $(f(-x))' = -f'(-x)$ b) $\left(f\left(\frac{1}{x}\right)\right)' = \frac{-f'\left(\frac{1}{x}\right)}{x^2}$

81. a) $f'(x) = \cos(x) + \sin(x)$ b) $f'(x) = 3\cos(x) - 5\sin(x)$ c) $f'(x) = \frac{1}{3}\cos(x)$

d) $f'(x) = \sin(x)$ e) $f'(x) = 1 - \sin(x)$ f) $f'(x) = \frac{\cos(x)}{\pi}$

82. a) $f'(t) = -5\cos(t) + 2t$ b) $f'(t) = -\sqrt{2} \cdot \cos(t)$ c) $f'(t) = \frac{1}{\sqrt{t}} + \frac{1}{3}\sin(t)$

d) $g'(t) = \sqrt{3} \cdot \cos(t)$ e) $g'(t) = \frac{-2}{t^3} - \pi \cdot \cos(t)$ f) $g'(t) = \frac{1}{\sqrt{2}} \cdot \sin(t)$

83. a) $y' = \sin(x) + x \cdot \cos(x)$ b) $y' = -2x^3 \cdot \sin(x) + 6x^2 \cdot \cos(x)$

c) $y' = -\frac{2}{x^3} \cdot \sin(x) + \frac{1}{x^2} \cdot \cos(x) + 3$ d) $y' = (2x - 3)\cos(x) - (x^2 - 3x + 1)\sin(x)$

e) $y' = \frac{1}{2\sqrt{x}}\cos(x) - \sqrt{x}\sin(x)$ f) $y' = \cos^2(x) - \sin^2(x) = 2\cos^2(x) - 1$

84. a) $y' = \frac{x\cos(x)-\sin(x)}{x^2}$ b) $y' = \frac{2\cos(x)+2x\sin(x)}{\cos^2(x)}$ c) $y' = \frac{2x\sin(x)+\cos(x)}{x \cdot \sqrt{x}}$

d) $y' = \frac{-\pi \cdot \cos(t)}{\sin^2(t)}$ e) $y' = \frac{(3t^2+2)\cos(t)+(t^3+2t)\sin(t)}{\cos^2(t)}$ f) $y' = \frac{-1}{\sin^2(t)} + 2t$

85. –

86. a) $\frac{\mathrm{d}}{\mathrm{d}t}f(t) = -\tan^2(t)$ b) $\frac{\mathrm{d}}{\mathrm{d}t}f(t) = \frac{\sin(t)\cdot\cos(t)+t}{\cos^2(t)}$ c) $\frac{\mathrm{d}}{\mathrm{d}t}f(t) = \frac{-1}{\sin^2(t)}$

87. a) $f'(x) = 2 \cdot \cos(2x)$ b) $f'(x) = (2x+1)\cos(x^2+x)$ c) $f'(x) = -\pi \cdot \sin(\pi \cdot x)$

d) $f'(x) = -2x \cdot \sin(x^2)$ e) $f'(x) = \frac{2x}{\cos^2(x^2)}$ f) $f'(x) = \frac{\cos(\sqrt{x})}{2\sqrt{x}}$

g) $f'(x) = -2\cos(x)\sin(x)$ h) $f'(x) = \frac{2\tan(x)}{\cos^2(x)}$ i) $f'(x) = \frac{\cos(x)}{2\sqrt{\sin(x)}}$

88. a) $f'(x) = \frac{1}{1+\cos(x)}$ b) $f'(t) = \frac{\cos(t)-\sin(t)-1}{(1-\cos(t))^2}$ c) $f'(u) = \frac{-3}{2} \cdot \cos^2(u) \cdot \sin(u)$

d) $f'(x) = \frac{x\cdot\sin(x)-\cos(x)}{x^2\cdot\cos^2(x)}$ e) $f'(t) = \sin(t^2)+2t^2\cdot\cos(t^2)$ f) $f'(u) = \sin\left(\frac{1-u}{1+u}\right) \cdot \frac{2}{(u+1)^2}$

89. a) $f'(x) = -2\mathrm{e}^{-2x}$ b) $g'(x) = 2x\mathrm{e}^{x^2}$ c) $h'(x) = -\frac{1}{x^2}\mathrm{e}^{\frac{1}{x}} + 2x$

d) $f'(t) = -3t \cdot \mathrm{e}^{-\frac{t^2}{2}}$ e) $g'(t) = t \cdot (2 - t) \cdot \mathrm{e}^{1-t}$ f) $h'(t) = \frac{1}{2} \cdot \sqrt{\mathrm{e}^t}$

90. a) $\frac{\mathrm{d}y}{\mathrm{d}x} = 6^x \cdot \ln(6)$ b) $\frac{\mathrm{d}y}{\mathrm{d}x} = -5^{-x} \cdot \ln(5)$ c) $\frac{\mathrm{d}y}{\mathrm{d}x} = 3^{2x} \cdot \ln(3)$

d) $\frac{\mathrm{d}y}{\mathrm{d}x} = 3 \cdot 10^x \left(1 + x \cdot \ln(10)\right)$ e) $\frac{\mathrm{d}y}{\mathrm{d}x} = \frac{2^x \cdot x\cdot\ln(2) - 2^x}{x^2}$ f) $\frac{\mathrm{d}y}{\mathrm{d}x} = (2\mathrm{e})^x \cdot (\ln(2) + 1)$

91. a) $y' = \frac{-3}{2-3x}$ b) $y' = \frac{1}{(x+1)\cdot\ln(10)}$ c) $y' = 6x + \frac{1}{2x\cdot\ln(2)}$

d) $y' = \frac{\ln(t)-1}{(\ln(t))^2}$ e) $y' = \frac{\ln(t)+2}{2\sqrt{t}}$ f) $y' = \frac{1}{2(t+2\sqrt{t})}$

92. a) $f'(x) = \frac{-\mathrm{e}^x}{(1+\mathrm{e}^x)^2}$ b) $g'(x) = -x \cdot \mathrm{e}^{-x}$ c) $h'(x) = -\frac{1}{x}$

d) $f'(u) = \frac{-1}{u(u-1)}$ e) $g'(u) = -\tan(u) - 3$ f) $h'(u) = 10^{\frac{2u}{u+1}} \cdot \frac{2\cdot\ln(10)}{(u+1)^2}$

g) $f'(x) = \frac{\mathrm{e}^x \cdot (x^2-4) - \mathrm{e}^x \cdot 2x}{(x^2-4)^2}$ h) $g'(x) = \frac{-\tan(x)}{\ln(3)}$ i) $h'(x) = \cos(x)\cdot\ln(x) + \frac{\sin(x)}{x}$

j) $f'(u) = (\mathrm{e} - u) \cdot u^{\mathrm{e}-1} \cdot \mathrm{e}^{-u}$ k) $g'(u) = \frac{u-1}{(u+1)(u^2+1)}$

l) $h'(u) = -\tan(u) \cdot \cos(\ln(\cos(u)))$

93. –

94. $(\ln(x))' = \frac{1}{x}$

95. $g'(x) = \frac{f'(x)}{f(x)}$

96. a) $y'(t) = A\,\omega \cdot \cos(\omega \cdot t)$ b) $y'(t) = \frac{A}{2\omega} \cdot \sin(\omega \cdot t) + \frac{A}{2}\,t \cdot \cos(\omega \cdot t)$

c) $y'(t) = -a\,\lambda \cdot \mathrm{e}^{-\lambda\cdot t}$

97. $y' = 0$

98. a) $f'(x) = x^x \cdot (\ln(x) + 1)$ b) $g'(x) = x^{\sin(x)} \cdot \left(\cos(x) \cdot \ln(x) + \sin(x) \cdot \frac{1}{x}\right)$

c) $h'(x) = x^{\frac{1}{x}} \cdot \frac{1-\ln(x)}{x^2}$ d) $i'(x) = x^{\sqrt{x}} \cdot \frac{\ln(x)+2}{2\sqrt{x}}$

e) $j'(x) = x^{(\mathrm{e}^x)}\mathrm{e}^x \left(\ln(x) + \frac{1}{x}\right)$ f) $k'(x) = \frac{-\ln(2)}{x\cdot\ln^2(x)}$

99. i) $f'(x) = (x + 1) \cdot \mathrm{e}^x$; $f''(x) = (x + 2) \cdot \mathrm{e}^x$ ii) $f^{(4)}(x) = (x + 4) \cdot \mathrm{e}^x$

iii) $f^{(n)}(x) = (x + n) \cdot \mathrm{e}^x$, $n \in \mathbb{N}$ iv) –

100. $f^{(n)}(x) = \mathrm{e}^x \left(x^2 + 2nx + n(n - 1)\right)$

101. i) $y^{(3)} = -\cos(x),\ y^{(4)} = \sin(x)$ and $y^{(5)} = \cos(x)$

ii) $y^{(11)} = -\cos(x),\ y^{(101)} = \cos(x)$ and $y^{(1001)} = \cos(x)$

iii) $(y)^{(1001)} = -\sin(x)$

102. $y' = x \cdot \cos(x) + \sin(x) \;\Rightarrow\; y'' = -x \cdot \sin(x) + 2\cos(x) \;\Rightarrow\; y''' = -x \cdot \cos(x) - 3\sin(x)$

$\Rightarrow\; y^{(4)} = x \cdot \sin(x) - 4\cos(x) \;\Rightarrow\; y^{(5)} = x \cdot \cos(x) + 5\sin(x) \;\Rightarrow\; y^{(6)} = -x \cdot \sin(x) + 6\cos(x)$

$y^{(12)} = x \cdot \sin(x) - 12\cos(x);\ y^{(13)} = x \cdot \cos(x) + 13\sin(x);\ y^{(25)} = x \cdot \cos(x) + 25\sin(x)$

103. a) $v(t) = v_0 - g \cdot t;\ a(t) = -g$

b) $v(t) = -C\,\omega_0\,\sin(\omega_0\,t + \varphi);\ a(t) = -C\,\omega_0^2\,\cos(\omega_0\,t + \varphi)$

104. a) $t \in [0; 6]$ b) $V'(t) = \frac{10^6}{8}(12t - 3t^2)$ c) –   d) –

105. a) $0\,\text{s},\ 16\,\text{s}$ b) $3\,\text{m/s},\ 1.5\,\text{m/s},\ -2.5\,\text{m/s}$ c) $-0.5\,\text{m/s}^2$

106. a) $v(1) - 10\,\text{km/h},\ v(2) = 20\,\text{km/h},\ v(3) \approx 24.5\,\text{km/h},\ v(7) \approx 28.8\,\text{km/h}$

b) $8.2\,\text{km/h},\ 3.9\,\text{km/h},\ 0.9\,\text{km/h}$

c) $a(t) = \frac{120t}{(t^2+2)^2};\ a(0.5) \approx 3.3\,\text{m/s}^2,\ a(1) \approx 3.7\,\text{m/s}^2,\ a(1.5) \approx 2.8\,\text{m/s}^2$ and $a(5) \approx 0.2\,\text{m/s}^2$

d) $\lim\limits_{t \to \infty} \frac{30t^2}{t^2 + 2} = 30\,\text{km/h}$

107. $C \to s,\ A \to v,\ B \to a$

108. $B \to s,\ C \to v,\ A \to a$

109. a) $2 \le s(t) \le 4,\ t \in \mathbb{R}$ b) $\dot{s}(t) = -\pi \cdot \sin(\pi \cdot t);\ \ddot{s}(t) = -\pi^2 \cdot \cos(\pi \cdot t)$

c) –   d) $\dot{s}(0) = 0,\ \dot{s}\left(\frac{1}{2}\right) = -\pi,\ \dot{s}(1) = 0,\ \dot{s}\left(\frac{3}{2}\right) = \pi$

110. a) 1000 bacteria   b) approx. 6 days and 6 hours   c) approx. 5200 bacteria

111. $-0.1\,\text{A/s};$ the current is decreasing.

112. a) $v(t) = s'(t) = \frac{\pi}{20}\left(100 + s^2(t)\right)$

b) $v_1 = 5\pi\,\text{m/s} \approx 15.7\,\text{m/s},\ v_2 = 10\pi\,\text{m/s} \approx 31.4\,\text{m/s}.$

113. a) $v''(t) = a'(t) = j(t)$ b) $j(t) = -\sin(t)$ c) $v(t) = 0.2t^3$

114. a) $s'''(t) = j(t)$ b) $j(t) = 0$ c) $s(t) = \frac{1}{8}t^4$

115. a) $j(t) = 0$ b) –

116. –

117. –

118. –

119. a) Yes   b) Yes   c) Yes   d) Yes   e) No   f) No

120. a) $k = 12$ b) $k = 5$ c) $k = \frac{3}{2}$ d) $k = -1$ e) $k \in \mathbb{R}$ f) $k = \pm 3$

121. a) $m = 21$ b) $m = -2$ c) $m = 12$ d) no sol.   e) $m = 1001$ f) $m = 4$

122. a) $s' - v = 0$; $v' - a = 0$; $s'' - a = 0$ b) $\frac{d}{dt}s = gt \Rightarrow \frac{d^2 s}{dt} = \frac{d}{dt}gt = g$
$$\Rightarrow s'' - g = 0$$

123. –

124. a) $t: y = 6x - 5$ b) $t: y = -4x$

125. $t(x) = 12x - 42$; $n(x) = \frac{-1}{12}x - \frac{23}{4}$

126. $P_1(2\,|\,-3)$ and $P_2(-2\,|\,21)$

127. $k = -20$

128. $t: y = -3x + 1$

129. $P(3\,|\,-5)$

130. $t_1: y = -6$ and $t_2: y = 21$

131. $P_1(3\,|\,-2)$, $P_2(9\,|\,2)$; $t_1(x) = 2x - 8$, $t_2(x) = 2x - 16$

132. a) $t_K: y = -5x + 3$ b) $E_1(0\,|\,1)$, $E_2\big(\frac{8}{3}\,\big|\,\frac{-229}{27}\big)$ c) $P_1\big(\frac{-1}{3}\,\big|\,\frac{14}{27}\big)$, $P_2 = (3\,|\,-8)$

133. $t: y = 2x - 1$

134. –

135. point of tangency $B(2\,|\,4)$

136. $a = \pm 3$

137. $x = \begin{cases} \left(\frac{1}{n}\right)^{\frac{1}{n-1}}, & \text{if } n \text{ is even} \\ \pm\left(\frac{1}{n}\right)^{\frac{1}{n-1}}, & \text{if } n \text{ is odd} \end{cases}$ and $x = \begin{cases} 1, & \text{if } n \text{ is even} \\ \pm 1, & \text{if } n \text{ is odd} \end{cases}$

138. a) $B(0\,|\,1)$ b) $B(0\,|\,3)$

139. $K(4\,|\,8)$

140. a) $y = 3x - 1$ b) $y = -10x + 11$ c) $y = 6x$ d) $y = -99$ e) $y = a_1 x + a_0$

141. a) $T(0\,|\,4)$; $N(0\,|\,-1)$ b) $S\big(-3\,\big|\,\frac{-5}{2}\big)$

142. $S(-1\,|\,-4)$

143. $x_S = \frac{u^2 - v^2}{2u - 2v} = \frac{u+v}{2}$

144. $l_{1,2}: y = \pm 6x$

145. a) $t_1: y = -4x$ b) $t_1: y = 4$ c) $t_1: y = 2x + 3$
     $\quad\;\; t_2: y = 4x$ $t_2: y = -8x - 12$ $t_2: y = 6x - 5$

146. a) $n: y = \frac{-1}{2}x + 3$ 　　　　b) $n: y = \frac{1}{4}x + 9$ 　　　　c) $n_{1,2}: y = \pm\frac{\sqrt{2}}{4}x + 5$

147. The angles are rounded to two decimal places.

　a) $\varphi_{1,2} = 75.96°$ 　　　　　　　　　b) $\varphi_{1,2} = 85.91°$

　c) $\varphi_1 = 80.54°$; $\varphi_2 = 45°$; $\varphi_3 = 50.19°$ 　　d) $\varphi_1 = 45°$; $\varphi_{2,3} = 0°$

148. a) $\left(1 \mid \frac{1}{3}\right)$ and $\left(-1 \mid \frac{-1}{3}\right)$ 　　　　b) $y' = x^2 \geq 0$ for all $x \in \mathbb{R}$

149. a) $90°$ 　　　　　　　　　　　　　　b) $2.7°$

150. a) $a = 1$ 　　　　b) $a = 1$ 　　　　c) $a = \frac{\sqrt{2}+1}{2}$

151. Both statements are true.

152. –

153. a) $t: y = -2x$ 　　　　b) $t: y = \frac{-x}{8} + \frac{3}{8} = \frac{1}{8}(3 - x)$ 　　c) $t: y = -\frac{1}{4}$

　　　　$n: y = \frac{x}{2} + \frac{5}{2} = \frac{1}{2}(x + 5)$ 　　$n: y = 8x - \frac{7}{3}$ 　　　　$n: x = -2$

154. a) $t: y = \frac{3}{4}x + \frac{5}{4}$ 　　b) $t: y = \frac{3}{16}x + \frac{3}{2}$ 　　c) $t: y = \frac{4}{3}x - \frac{7}{3} = \frac{1}{3}(4x - 7)$

　　　　$n: y = \frac{-4}{3}x + \frac{10}{3}$ 　　$n: y = \frac{-16}{3}x + \frac{283}{12}$ 　　$n: y = \frac{-3}{4}x + 6 = 6 - \frac{3}{4}x$

155. a) $t: y = 2x + 1 - \frac{\pi}{2}$ 　　b) $t: y = \frac{x}{2} + \frac{\sqrt{3}}{2} - \frac{\pi}{6}$ 　　c) $t: y = \frac{1}{2}$

　　　　$n: y = \frac{-x}{2} + 1 + \frac{\pi}{8}$ 　　$n: y = -2x + \frac{\sqrt{3}}{2} + \frac{2\pi}{3}$ 　　$n: x = \frac{\pi}{4}$

156. a) $t: y = 2x + 3$ 　　b) $t: y = \frac{1}{4}x - 1$ 　　c) $t: y = x - 1$

　　　　$n: y = \frac{-x}{2} + 3 = 3 - \frac{x}{2}$ 　　$n: y = -4x + 16$ 　　$n: y = -x + 1$

157. $t: y = 2 \cdot \sqrt{2}$

158. –

159. $M = \left(b \mid \frac{1}{b}\right) = B$

160. $t: y = x + \frac{1+\ln(3)}{3} \approx x + 0.70$

161. Intersection: $(1 \mid 0)$

162. Intersection: $x_0 = -\mathrm{e}^2$

163. $q = \sqrt{2}$

164. a) $t: y = -x$ 　　　　　　　　　　b) $t_1: y = 2x - 1$ and $t_2: y = \frac{1}{2}x - 4$

　c) $t: y = \frac{\mathrm{e}}{2}x$ 　　　　　　　　　　d) $t: y = \frac{1}{3}x + 2$

165. a) $(0 \mid 0)$, $45°$ 　　　　　　b) $(1 \mid 0)$, $26.57°$; $(-1 \mid 0)$, $26.57°$; $\left(0 \mid \frac{1}{2}\right)$, $90°$

　c) $(-1 \mid 0)$, $20.20°$; $(0 \mid 1)$, $26.57°$ 　　d) $\left(\frac{\pi}{2} \mid 0\right)$, $11.74°$; $\left(-\frac{\pi}{2} \mid 0\right)$, $78.26°$; $(0 \mid 1)$, $45°$

166. a) $71.6°$ twice 　　　　　　　　　b) $70.5°$

167. a) $90°$ b) $25.9°$ and $59.2°$ c) $22.6°$

168. $\frac{1}{2}\sqrt[4]{2} \approx 0.6$

169. a) $m = 1 + \sqrt{2}$ b) $-$

170. $-$

171. a) $\sigma_x \approx 69.8°$ b) $\sigma_1 \approx 63.4°$ and $\sigma_2 \approx 20.2°$

172. $-$

173. a) $-$ b) $n\colon y = \frac{-1}{f'(x_0)} \cdot x + f(x_0) + \frac{x_0}{f'(x_0)}$

174. a) False   b) False   c) True   d) False

175. a) $-$

     b) 1) increasing: $]0; 1.7[\,\cup\,]3.9; 8.8[$ 2) convexity: $]2.6; 4.8[\,\cup\,]6; 7.9[$
          decreasing: $]1.7; 3.9[\,\cup\,]8.8; 10[$ concavity: $]0; 2.6[\,\cup\,]4.8; 6[\,\cup\,]7.9; 10[$

        3) $x_1 = 1.7$, $x_2 = 8.8$ 4) $x_1 = 3.9$

        5) $x_1 = 2.6$, $x_2 = 4.8$, $x_3 = 6$, $x_4 = 7.9$ 6) does not exist

176. $(-5 \,|\, 119)$ and $(1 \,|\, 11)$

177. $T_1(-2 \,|\, -1)$, $T_2(2 \,|\, 3)$

178. a) Extremum at $x = -3$ b) $x_{\min} = -3$ and $x_{\max} = 1.5$

179. a) $H(-2 \,|\, 0)$, $T(0 \,|\, -4)$ and $W(-1 \,|\, -2)$ b) $\nearrow$ on $]-\infty; -2[\,\cup\,]0; \infty[$ and $\searrow$ on $]-2; 0[$
     c) $-$ d) $x_{\min} = 0$ and $x_{\max} = 2.5$

180. • $D_p = \mathbb{R}$ and $W_p = \mathbb{R}$ • roots: $x_1 = -2$ and $x_2 = 1$

     • no symmetry   • extrema: $x_3 = -2$, $x_4 = 0$

     • inflection points: $x_5 = -1$

181. a) $t_1\colon y = \frac{16}{9}x - \frac{5}{27}$ and $t_2\colon y = 1$ b) convex: $]-\infty; \frac{1}{3}[\,\cup\,]1; \infty[$; concave: $]\frac{1}{3}; 1[$

182. a) $f(a) = 0$; $f'(a) < 0$; $f''(a) > 0$ b) $f(a) > 0$; $f'(a) > 0$; $f''(a) = 0$
     c) $f(a) < 0$; $f'(a) = f''(a) = 0$ d) $f(a) > 0$; $f'(a) = 0$; $f''(a) < 0$

183. c.o.s. = change of sign

     a) monotonically decr. on $]-\infty; -2] \cup [-0.5; 2]$ b) monotonically decr. on $[-2; 2.5]$
       monotonically incr. on $[-2; -0.5] \cup [2; \infty[$ monotonically incr. on $]-\infty; -2] \cup [2.5; \infty[$

       minimum at $x = -2$ ($-/+$ c.o.s.)   maximum at $x = -2$ ($+/-$ c.o.s.)
       maximum at $x = -0.5$ ($+/-$ c.o.s.)   minimum at $x = 2.5$ ($-/+$ c.o.s.)
       minimum at $x = 2$ ($-/+$ c.o.s.)

184. a) convex on $]-2; 0.5[$ b) convex on $]4; \infty[$
       concave on $]-\infty; -2[\,\cup\,]0.5; \infty[$ concave on $]-\infty; 4[$

       inflection points at $x = -2$ and $x = 0.5$ inflection point at $x = 4$

185. $a = \frac{3}{4}$

186. for $a \geq 0$

187. The point $(x_0 \mid f(x_0))$ could also be a saddle point of f.

188. a) True   b) False   c) True

189. a) False   b) False   c) False   d) True

190. a) Yes   b) –

191. Ida

192. Nael

193. a) No   b) No

194. –

195. a) –   b) No

196. $S_0 = W = (0 \mid 1.5)$, $S_1 = (2.1 \mid 2.0)$, $S_2 = (-2.1 \mid 1.0)$

197. –

198. Without graphs:

a) $D_f = \mathbb{R}$, $W_f = \mathbb{R}$
roots: $x_1 = 0$, $x_{2,3} = \pm\sqrt{3}$
$T\left(1 \mid -\frac{2}{3}\right)$, $H\left(-1 \mid \frac{2}{3}\right)$
$W = (0 \mid 0)$
symmetric w.r.t. origin

b) $D_f = \mathbb{R}$, $W_f = [-1; \infty[$
roots: $x_1 = \frac{-4}{3}$, $x_2 = 0$
$T(-1 \mid -1)$
$W\left(-\frac{2}{3} \mid -\frac{16}{27}\right)$, saddle point: $(0 \mid 0)$

c) $D_f = \mathbb{R}$, $W_f = \left[\frac{103}{64}; \infty\right[$
roots: none
$H(0 \mid 2)$, $T_{1,2}\left(\pm\frac{\sqrt{5}}{4} \mid \frac{103}{64}\right) \approx (\pm 0.56 \mid 1.61)$
$W_{1,2}\left(\pm\frac{\sqrt{15}}{12} \mid \frac{1027}{576}\right) \approx (\pm 0.32 \mid 1.78)$
symmetric w.r.t. y-axis

d) $D_f = \mathbb{R}$, $W_f = \mathbb{R}$
roots: $x_1 = 0$
$H_1\left(\sqrt{2} \mid \frac{56}{15}\sqrt{2}\right) \approx (1.41 \mid 5.28)$,
$H_2\left(-\sqrt{6} \mid -\frac{8}{5}\sqrt{6}\right) \approx (-2.45 \mid -3.92)$,
$T_1\left(-\sqrt{2} \mid -\frac{56}{15}\sqrt{2}\right) \approx (-1.41 \mid -5.28)$,
$T_2\left(\sqrt{6} \mid \frac{8}{5}\sqrt{6}\right) \approx (2.45 \mid 3.92)$
$W_1(0 \mid 0)$, $W_2\left(2 \mid \frac{68}{15}\right) \approx (2 \mid 4.53)$,
$W_3\left(-2 \mid -\frac{68}{15}\right) \approx (-2 \mid -4.53)$
symmetric w.r.t. origin

e) $D_f = \mathbb{R}$, $W_f = \mathbb{R}$
roots: $x_1 = -1$, $x_2 = 2$
$H(-1 \mid 0)$, $T\left(1 \mid -\frac{2}{3}\right)$
$W\left(0 \mid -\frac{1}{3}\right)$

f) $D_f = \mathbb{R}$, $W_f = \mathbb{R}$
roots: $x_1 = 1$, $x_2 = -2$
$H(-2 \mid 0)$, $T(0 \mid -4)$
$W(-1 \mid -2)$

g) $D_f = \mathbb{R}$, $W_f = \mathbb{R}$

    roots: $x_1 = 0$, $x_2 = 1$

    $T(1 \,|\, 0)$, $H\!\left(\frac{1}{3} \,\middle|\, \frac{4}{27}\right)$

    $W\!\left(\frac{2}{3} \,\middle|\, \frac{2}{27}\right)$

h) $D_f = \mathbb{R}$, $W_f = \mathbb{R}_0^+$

    roots: $x_1 = 0$

    $T(0 \,|\, 0)$

    $W_1\!\left(1 \,\middle|\, \frac{7}{8}\right)$, $W_2(2 \,|\, 2)$

199. $a = -2$

200. a) $D_k = \mathbb{R}$; $W_k = \,]0; 4]$; no roots; $H(0 \,|\, 4)$; $W_{1,2}(\pm 1 \,|\, 3)$; symmetric w.r.t. y-axis; asymptote: $y = 0$

    b) $t_{1,2}$: $y = \mp 1.5x + 4.5$

201. a) $D_g = \mathbb{R} \backslash \{-1\}$; $W_g = \mathbb{R} \backslash \{1\}$; roots: $x_0 = 2$; Pole: $x_p = -1$; asymptote: $y = 1$

    b) –   c) t: $y = 3x - 2$; $\varphi \approx 18.4°$

202. a) roots: $x_{1,2} = -1$, $x_3 = -3$; poles: $x_4 = 0$, $x_5 = -4$; $E_1(1 \,|\, 0)$ and $E_2(6.52 \,|\, -6.06)$; $W(-1.32 \,|\, -1.28)$; vertical asymptotes: $x = 0$ and $x = -4$; oblique asymptote $y = \frac{x}{2} - \frac{3}{2}$

    b) $S\!\left(\frac{-3}{7} \,\middle|\, \frac{-12}{7}\right)$

203. a) roots: $x_1 = \frac{1}{2}$, $x_2 = 3$; $H\!\left(\frac{6}{7} \,\middle|\, \frac{25}{12}\right)$; asymptotes: $y = -2$, $x = 0$

    b) t_1: $y = 20x - 10$, t_2: $y = -\frac{5}{9}x + \frac{5}{3}$ c) $m_1 \approx 2.68807$, $m_2 \approx -0.153092$

204. a) $D_f = \mathbb{R}$; root: $x_1 = 0$; $W_1(0 \,|\, 0)$, $W_2(6 \,|\, 4.5)$, $W_3(-6 \,|\, -4.5)$; asymptote $y = x$; symmetric w.r.t. origin

    b) $d = \frac{6}{7}$

205. a) $0 < x < 4$ b) $x > 16$

206. $y' \neq 0$

207. a) $D_f = \mathbb{R}$, $W_f = [-1; 1]$

    root: $x_1 = 0$

    $H(1 \,|\, 1)$, $T(-1 \,|\, -1)$

    $W_1(0 \,|\, 0)$, $W_{2,3}\!\left(\pm\sqrt{3} \,\middle|\, \pm\frac{\sqrt{3}}{2}\right)$

    symmetric w.r.t. origin

    asymptote: $y = 0$

b) $D_f = \mathbb{R}$, $W_f = \,]0; 4]$

    roots: none

    $H(0 \,|\, 4)$

    $W_{1,2}\!\left(\pm\frac{\sqrt{3}}{3} \,\middle|\, 3\right)$

    symmetric w.r.t. y-axis

    asymptote: $y = 0$

    c) $D_g = \mathbb{R} \backslash \{-1\}$, $W_g = \mathbb{R} \backslash \{1\}$

    root: $t_1 = 0$

    pole: $t = -1$

    asymptote: $y = 1$

d) $D_g = \mathbb{R} \backslash \{0\}$, $W_g = \mathbb{R} \backslash \{1\}$

    root: $t_1 = -1$

    pole: $t = 0$

    asymptote: $y = 1$

e) $D_h = \mathbb{R}\backslash\{-2,2\}$, $W_h = \mathbb{R}$

root: $x_1 = 0$

$H\left(-2\sqrt{3}\mid -3\sqrt{3}\right)$, $T\left(2\sqrt{3}\mid 3\sqrt{3}\right)$

saddle point: $(0\mid 0)$

pole: $x = \pm 2$

symmetric w.r.t. origin

asymptote: $y = x$

f) $D_h = \mathbb{R}\backslash\{-1\}$, $W_h = \mathbb{R}_0^+$

root: $x_1 = 1$

$T(1\mid 0)$

$W\left(2\mid \frac{1}{9}\right)$

pole: $x = -1$

asymptote: $y = 1$

208. a) $D_f = \mathbb{R}_0^+$, $W_f = \mathbb{R}_0^+$

root: $x_1 = 0$

b) $D_f = \mathbb{R}_0^+$, $W_f = \mathbb{R}_0^+$

root: $x_1 = 0$

$W\left(\frac{1}{4}\mid \frac{9}{16}\right)$

c) $D_g = \mathbb{R}^+$, $W_g = \mathbb{R}$

root: $t_1 = \frac{\sqrt[3]{4}}{4}$

pole: $t = 0$

asymptote: $y = 4t$

d) $D_g = \mathbb{R}^+$, $W_g = \left[-\frac{1}{4};\infty\right[$

root: $t_1 = 1$

$T\left(4\mid -\frac{1}{4}\right)$, $W\left(\frac{64}{9}\mid -\frac{15}{64}\right)$

pole: $t = 0$

asymptote: $y = 0$

209. a) $D_f = \mathbb{R}$, $W_f = [-2.24; 2.24]$

roots: $x_k = 2.03 + k\cdot\pi$

$H_k(0.46 + k\cdot 2\pi\mid 2.24)$,

$T_k(3.61 + k\cdot 2\pi\mid -2.24)$

$W_k(2.03 + k\cdot\pi\mid 0)$

b) $D_f = \mathbb{R}$, $W_f = \left[-\frac{1}{2};\frac{1}{2}\right]$

roots: $x_k = k\cdot\frac{\pi}{2}$

$H_k\left(\frac{\pi}{4} + k\cdot\pi\mid \frac{1}{2}\right)$, $T_k\left(-\frac{\pi}{4} + k\cdot\pi\mid -\frac{1}{2}\right)$

$W_k\left(k\cdot\frac{\pi}{2}\mid 0\right)$

symmetric w.r.t. origin

c) $D_g = \mathbb{R}$, $W_g = \left[-\frac{9}{8};2\right]$

roots: $x_k = \pm\frac{\pi}{3} + k\cdot 2\pi$ and $\pi + k\cdot 2\pi$

$H_k(k\cdot 2\pi\mid 2)$ and $(\pi + k\cdot 2\pi\mid 0)$,

$T_k(\pm 1.82 + k\cdot 2\pi\mid -1.125)$

$W_k(\pm 0.87 + k\cdot 2\pi\mid 0.49)$ and

$(\pm 2.45 + k\cdot 2\pi\mid -0.58)$

symmetric w.r.t. y-axis

d) $D_g = \mathbb{R}\setminus\{k\cdot 2\pi\}$, $W_g = \left[-\frac{1}{2};\infty\right[$

roots: $x_k = \pm\frac{\pi}{2} + k\cdot\pi$

$T_k(\pi + k\cdot 2\pi\mid -0.5)$

poles: $x_k = k\cdot 2\pi$

symmetric w.r.t. y-axis

210. a) $D_f = \mathbb{R}$, $W_f = \left]-\infty;\frac{1}{e}\right]$

root: $x_1 = 0$

$H\left(1\mid \frac{1}{e}\right) \approx (1\mid 0.37)$

$W\left(2\mid \frac{2}{e^2}\right) \approx (2\mid 0.27)$

asymptote: $y = 0$

b) $D_f = \mathbb{R}$, $W_f = \mathbb{R}_0^+$

root: $x_1 = 1$

$H\left(-1\mid \frac{4}{e}\right) \approx (-1\mid 1.47)$, $T(1\mid 0)$

$W_1\left(-\sqrt{2}-1\mid 1.04\right)$, $W_2\left(\sqrt{2}-1\mid 0.52\right)$

asymptote: $y = 0$

c) $D_g = \mathbb{R}$, $W_g = [-0.43; 0.43]$

root: $t_1 = 0$

$T\left(\frac{-\sqrt{2}}{2}\,\middle|\,-0.43\right)$, $H\left(\frac{\sqrt{2}}{2}\,\middle|\,0.43\right)$

$W_1(0\,|\,0)$, $W_{2,3}\left(\frac{\pm\sqrt{6}}{2}\,\middle|\,\pm0.27\right)$

symmetric w.r.t. origin

asymptote: $y = 0$

d) $D_g = \mathbb{R}$, $W_g = \mathbb{R}$

root: $t_1 = -0.85$

asymptote: $y = \frac{t}{2}$

211. a) $D_f = \mathbb{R}^+$, $W_f = \left[\frac{-1}{e}; \infty\right[$

root: $x_1 = 1$

$T\left(\frac{1}{e}\,\middle|\,\frac{-1}{e}\right) \approx (0.37\,|\,-0.37)$

b) $D_f = \mathbb{R}^+$, $W_f = \left]-\infty; \frac{1}{e}\right]$

root: $x_1 = 1$

$H\left(e\,\middle|\,\frac{1}{e}\right) \approx (2.72\,|\,0.37)$

$W\left(\sqrt{e^3}\,\middle|\,\frac{3}{2\sqrt{e^3}}\right) \approx (4.48\,|\,0.33)$

pole: $x = 0$

asymptote: $y = 0$

c) $D_f = \mathbb{R}^+$, $W_f = \mathbb{R} \setminus [0; e[$

roots: none

$T(e\,|\,e) \approx (2.72\,|\,2.72)$

$W\left(e^2\,\middle|\,\frac{e^2}{2}\right) \approx (7.39\,|\,3.69)$

pole: $x = 1$

212. $f(x) = \frac{-1}{8}x^3 + \frac{3}{4}x^2 - \frac{3}{2}x + 2$

213. $f(x) = \frac{1}{3}x^4 - \frac{8}{3}x^3 + 6x^2$

214. $f(x) = x^3 - 3x^2 + 4$

215. $f(x) = \frac{1}{6}x^3 + \frac{1}{3}x + 2$

216. $y = f(x) = 3x^5 - 10x^3 + 15x$

217. $f(x) = \frac{1}{9}x^3 - 3x$

218. $f(x) = \frac{-1}{27}x^5 + \frac{5}{9}x^3$

219. $f(x) = \frac{-1}{2}x^4 + 3x^2$

220. $f(x) = \frac{-1}{27}x^5 + \frac{5}{9}x^3$

221. $f(x) = 0.04x^4 - 1.28x^2 + 10.24$

222. $82.9°$

223. $X_{1,2} = (\pm 1\,|\,0)$ and $X_{3,4} = \left(\pm\sqrt{5}\,\middle|\,0\right)$; $E_5 = (0\,|\,5) = H$ and $E_{6,7} = \left(\pm\sqrt{3}\,\middle|\,-4\right) = T_{1,2}$

224. a) –

b) four specifications

225. $W_2 = \left(-1 \mid \frac{-5}{2}\right)$; $H = (0 \mid 0)$ and $T_{1,2} = \left(\pm\sqrt{3} \mid \frac{-9}{2}\right)$

226. $f(x) = (x+2)(x-1)^2 = x^3 - 3x + 2$

227. $f(x) = \frac{-1}{2}x^3 + \frac{3}{2}x$

228. $f(x) = \begin{cases} \frac{1}{6}x^2 - \frac{2}{3} & \text{for } -2 \le x \le 2 \\ \frac{2}{3}x - \frac{4}{3} & \text{for } 2 < x \le 5 \end{cases}$

229. $y = v(x) = \frac{-1}{8}x^3 + \frac{3}{8}x^2 + \frac{1}{8}x - \frac{3}{8} = \frac{-1}{8}(x-3)(x^2-1)$

230. Show: $h(0) = H$, $h(T) = 0$ and $h'(0) = h'(T) = 0$.

231. a) $f(x) = 2x^3 - 2x^2 - 2x + 2$ b) $f(x) = -6x^5 + 14x^4 - 8x^3 - 2x + 2$

232. $a = 3$, $b = \frac{1}{4}$

233. a) $a = \frac{1}{4}e^2$ b) –

234. $f(x) = \frac{11(x-3)^2}{15(x+1)}$

235. a) $a = -16$, $b = 48$ b) pole: $x_{1,2} = 0$; $T(6 \mid -3)$; $W\left(9 \mid \frac{-5}{3}\right)$;   asymptote: $y = 9$

236. a) $a = 4$ b) $a = -3$

237. $f(x) = \frac{x(x-2)^2}{4(x+1)^2} = \frac{x^3 - 4x^2 + 4x}{4x^2 + 8x + 4}$

238. $b = 3$ and $a = 4b = 12$

239. a) $u = 1$, $v = -1$;   b) $u = e + 1$, $v = 1 - e^{-1}$;
     $f(x) = e^{x-1} - 1$ $f(x) = e^{x-e-1} + 1 - e^{-1}$

240. a) $F(x) = \frac{-1}{2}x^3 + 3x^2$;

x	0	1	2	3	4	5	6
$F(x)$	0	2.5	8	13.5	16	12.5	0

    b) $u(x) = -x^2 + 7x + x \cdot \sqrt{x^2 - 12x + 37}$

241. a) $B(v) = 0.125v^2 + \frac{2000}{v}$ b)

v in km/h	5	10	15	20	25	30	35	40
$B(v)$ in Fr./km	403	212	161	150	158	179	210	250

242. a) $s(x) = \frac{|x|}{2}$ b) $s(x) = 1$ c) $s(x) = 2x$
    d) $s(x) = |x|$ e) $s(x) = \left|\frac{x}{n}\right|$ f) $s(x) = |\tan(x)|$

243. Length $l = 20\,\text{cm}$ and width $b = 10\,\text{cm}$. It is a minimum.

244. a) $x = 2\,\text{cm}$ b) $x = 3\,\text{cm}$

245. a) height: $21\,\text{cm}$; width: $14\,\text{cm}$ b) height: $\sqrt{\frac{aA}{b}} + 2a\,\text{cm}$; width: $\sqrt{\frac{Ab}{a}} + 2b\,\text{cm}$

246. For variant 1:

    i) $u = 9\,\text{m}$; $v = 4.5\,\text{m}$ ii) $u = 12\,\text{m}$; $v = 9\,\text{m}$ iii)$u = 12\,\text{m}$; $v = 15\,\text{m}$

    For variant 2:

    i) $u = 12\,\text{m}$; $v = 3\,\text{m}$ ii) $u = 12\,\text{m}$; $v = 9\,\text{m}$ iii)$u = 13.5\,\text{m}$; $v = 13.5\,\text{m}$

247. $b = c = \sqrt{5}\,\text{m}$

248. $5\,\text{cm}$, $10\,\text{cm}$ and $15\,\text{cm}$; $V_{\max} = 750\,\text{cm}^3$

249. $2\,\text{dm}$ and $1\,\text{dm}$

250. a) $r = 10\sqrt[3]{\frac{1}{2\pi}} \approx 5.4\,\text{cm}$ b) $r = 10\sqrt[3]{\frac{1}{\pi}} \approx 6.8\,\text{cm}$

        $h = 10 \cdot \sqrt[3]{\frac{4}{\pi}} \approx 10.8\,\text{cm}$ $h = \frac{V}{\pi r^2} = 10\sqrt[3]{\frac{1}{\pi}} \approx 6.8\,\text{cm}$

251. a) $L_{\text{circle}} = \frac{\pi\ell}{4+\pi}$; $L_{\text{square}} = \frac{4\ell}{4+\pi}$ b) $L_{\text{circle}} = \ell$; $L_{\text{square}} = 0$

252. $u = 10$; $v = 150$

253. a) 3.00   b) 1.50   c) 0.00   d) $-1.00 \mathrel{\widehat{=}}$ price hike

254. The edge of the base measures $30\,\text{cm}$ and the height of the lampshade $60\,\text{cm}$.

255. The edge of the square floor plan measures $20\,\text{m}$ and the building height $30\,\text{m}$.

256. $x_{\max} = 1$; $x_{\min} = -1$

257. $A = 169$

258. maximum area $F = 16\sqrt{2} \approx 22.63$; length $4\sqrt{2} \approx 5.66$ and width 4

259. $y_h = 8$ or $y_h = 4$ (extremum at boundary)

260. $\frac{4 \cdot \sqrt{30}}{\sqrt{4+\pi}} \approx 8.20\,\text{m}$

261. a) $1 : 3$ b) $1 : 2$

262. $294°$

263. 36 area units

264. width $b = 10\sqrt{3}\,\text{cm} \approx 17.3\,\text{cm}$; height $h = 10\sqrt{6}\,\text{cm} \approx 24.5\,\text{cm}$

265. a) $a = 1$ b) $F_\Delta \leq \frac{2}{3}$

266. $r = \sqrt{\frac{2}{3}}R \approx 10.0\,\text{cm}$; $h = \frac{1}{\sqrt{3}}R \approx 7.1\,\text{cm}$

267. 26.4 units

268. The minimum construction cost is $480'000$ euros.

269. Fr. 75'915.–

270. $K_{1,2}\left(\pm\frac{\sqrt{2}}{2}\,\middle|\,\frac{1}{10}\right)$

271. $(-0.948\,|\,0.235)$

272. a) $P_{1,2}(\pm2\,|\,1)$ b) $P(2\,|\,2)$

273. a) $44.7\,\text{cm}$ b) $48.7\,\text{cm}$

274. Y_1 lies $2\sqrt{3}$ units above the x-axis and Y_2 3 units.

275. $F_{\max} = 769.0\,\text{m}^2$

276. $4\,\text{cm}$

277. $\varphi = 120°$

278. $\frac{13}{4}\sqrt{13} \approx 11.7\,\text{m}$

279. $\varphi \approx 54.7°$

280. length: $100.0\,\text{cm}$; width: $25.0\,\text{cm}$

281. $111°$, $137°$ and $111°$

282. 7 length units

283. $P(3\,|\,1)$

284. a) –   b) Satisfies the law.   c) $x_1 = \frac{100}{3} = 33.\overline{3}$ d) $x_1 = 100$ e) –

285. a) $G(x) = \frac{-1}{50}x^3 + 4x^2 - 96x - 4000$ b) $x = 120$ hectolitres
    c) $0 < x < 45.3$ or $x > 167.85$

286. a) $x = 250$ panels   b) $x = 181$ panels

287. a) 2 monetary units   b) $x = 11$ c) • $x = 30$, • $x = 14$, • $x \approx 12$

288. –

289. a) $S_1(-1\,|\,-1)$ and $S_2(3\,|\,-5)$ b) –

290. a) $v\colon x = \frac{1}{2}$ b) –
    c) With respect to the axis $x = \frac{1}{2}m$ and in the direction of the line g

291. a) –   b) –   c) –   d) $W_{1,2}\left(\pm\frac{2\sqrt{3}}{3}\,\middle|\,\frac{1}{3}\pm\frac{1}{3}\sqrt{3}\right)$
    e) G_f has oblique symmetry in the direction of the double tangent d with respect to the y-axis.
    f) $\overline{LW_1} : \overline{W_1W_2} = \frac{1}{2}\left(\sqrt{5}-1\right) = \tau \approx 0.618$

292. a) $y' = \frac{dy}{dx} = \frac{3x^2}{2y}$

    b) $y' = \frac{dy}{dx} = \frac{x-2y}{2x-y}$

    c) $y' = \frac{dy}{dx} = \frac{\cos(x)-2}{\sin(y)-3}$

    d) $y' = \frac{dy}{dx} = -\sqrt{\frac{y}{x}}$

293. $y' = \frac{dy}{dx} = \frac{x^2-y}{x-y^2}$

294. $m_P = \frac{-1}{4}$ and $m_Q = \frac{1}{2}$

295. $m = y' = \frac{-a}{b}$

296. a) –

    b) $m_1 = 8$ and $m_2 = \frac{7}{4}$

    c) $m_t = \frac{-x}{y}$

297. a) The equation holds when substituting any of the given coordinate pairs; k is a parabola.

    b) $y' = \frac{x-y-1}{x-y+1}$; $m_A = \frac{5}{4} = 1.25$, $m_C = \frac{3}{2} = 1.5$, $m_D = 2$, $m_F = 0$, $m_G = \frac{1}{2} = 0.5$, $m_I = \frac{3}{4} = 0.75$

    c) $t_A : y = \frac{5}{4}x + 5$; $t_G : y = \frac{1}{2}x - 1$

    d) $n_H : y = \frac{-3}{2}x + \frac{35}{2}$

    e) $n_D : y = \frac{-1}{2}x + \frac{9}{2}$

    f) $F(x, y) = F(y, x)$ holds.

    g) $S(\frac{1}{4} | \frac{1}{4})$; $m_S = -1$

    h) No

298. a) $(2 | 0)$

    b) $y' = \frac{-1}{y}$

    c) $\sigma \approx 63.4°$

    d) $Q(\frac{3}{2} | -1)$; $R(\frac{1}{2} | \sqrt{3})$

299. a) $x_1 = \frac{4}{5}$, $x_2 = \frac{333}{440}$

    b) initial guess $x_0 = 1$: $x_1 = 2$, $x_2 = \frac{33}{20}$; initial guess $x_0 = -1$: $x_1 = \frac{-6}{7}$, $x_2 = \frac{-9313}{11'340}$

    c) initial guess $x_0 = 2.5$: $x_1 = 2.512$, $x_2 = 2.51188\ldots$

300. $f(x) = x^3 - 4x^2 - 2$, $x_0 = 4$, $x_1 = \frac{33}{8} = 4.125$, $x_2 = \frac{3805}{924} = 4.1179\ldots$

301. a) root $\approx -2.3553\ldots$

    b) The method fails.

    c) root $\approx -2.3553\ldots$

302. $f(x) = e^x - 2$, $x_0 = 1$, $x_1 = 0.7357\ldots$, $x_2 = 0.6940\ldots$, $x_3 = 0.6931\ldots$ ($\ln(2) = 0.6931471\ldots$)

303. The method fails.

304. a) initial guess $x_0 = 2 \Rightarrow x_1 = -1.6268\ldots$, $x_2 = 0.8188\ldots$, $x_3 = -0.0946\ldots$, $x_4 = 0.00014\ldots$

    $\Rightarrow$ The approximations converge to the root $x = 0$.

    initial guess $x_0 = 2.5 \Rightarrow x_1 = -3.5502\ldots$, $x_2 = 13.8456\ldots$, $x_3 = -515'287.6282\ldots$

    $\Rightarrow$ The approximations drift apart, which is why the method fails.

    b) The initial guess x_0 cannot be chosen too far from the root $x = 0$.

305. a) True

    b) False

    c) True

306. a) Linearisation: $e^x \approx 1 + x$ for $x \approx 0$, $e^{0.01} \approx 1 + 0.01 = 1.01$

    b) Linearisation: $(1 + x)^3 \approx 1 + 3x$ for $x \approx 0$, $1.1^3 \approx 1 + 0.3 = 1.3$

    c) Linearisation: $\sqrt{x} \approx 1 + \frac{x}{4}$ for $x \approx 4$, $\sqrt{4.05} \approx 1 + \frac{4.05}{4} = 2.0125$

    d) Linearisation: $\sin(x) \approx \pi - x$ for $x \approx \pi$, $\sin(\frac{5\pi}{6}) \approx \pi - \frac{5\pi}{6} = \frac{\pi}{6}$

307. a) Linearisation of $\sqrt{x}$ at point $x_0 = 9$ yields $\sqrt{9.5} \approx 3.083$.

  b) Linearisation of x^3 at point $x_0 = 100$ yields $99.9^3 \approx 997'000$.

  c) Linearisation of $\sqrt{x}$ at point $x_0 = 4$ yields $\sqrt{3.98} \approx 1.995$.

  d) Linearisation of $\sqrt[3]{x}$ at point $x_0 = 1000$ yields $\sqrt[3]{1001} \approx 10.003$.

308. iii)

309. The linearisation of $f(x) = \sin(x)$ yields the most precise approximation.

310. a) $1 - 6x$ b) $2 - 2x$ c) $1 - \frac{x}{2}$ d) $4^{\frac{1}{3}} + \frac{x}{4^{\frac{2}{3}}}$

311. a) Linearisation of $\ln(x)$ at point $x_0 = 1$. The limit is -1.

  b) Linearisation of $\sin(x)$ at point $x_0 = 0$. The limit is 1.

  c) Linearisation of e^x at point $x_0 = 0$. The limit is 2.

  d) Linearisation of $(1 + x)^n$ at point $x_0 = 0$. The limit is 90.

312. approximation formula $\ell_1 \approx \ell_0 \left(1 + \frac{\gamma}{3} \cdot \Delta T \right)$ for $\Delta T \approx 0$

313. a) $\sqrt{50} \approx \frac{99}{14}$ b) $\sqrt{2} \approx \frac{99}{70}$ c) $\sqrt{50} \approx \frac{10}{3} \cdot \frac{17}{8} = \frac{85}{12}$

  d) The approximation from part a) is generally more precise.

314. $\ln(2) \approx \frac{2}{e} \approx 0.7$

315. a) $f(0) = f_1(0) = f_2(0) = f_3(0) = 2$. All graphs pass through $(0\,|\,2)$.

  b) $f'(0) = f_1'(0) = f_2'(0) = f_3'(0) = -1$. All graphs touch one another at $(0\,|\,2)$.

  c) $f''(0) = f_2''(0) = f_3''(0) = -10$. f, f_2 and f_3 have the same curvature at $(0\,|\,2)$.

  d) $f'''(0) = f_3'''(0) = 6$. f_3 is the best approximation of f.

  e) $f^{(4)}(0) = 72$, i.e. $3x^4 = \frac{f^{(4)}(0)}{4!} \cdot x^4$

  f) $f(x) = \frac{f^{(4)}(0)}{4!} \cdot x^4 + \frac{f'''(0)}{3!} \cdot x^3 + \frac{f''(0)}{2} \cdot x^2 + f'(0) \cdot x + f(0)$

316. –

317. $g'(x) = \left(x - \frac{x^3}{3!} + \frac{x^5}{5!} - \frac{x^7}{7!} \right)' = 1 - \frac{3x^2}{3!} + \frac{5x^4}{5!} - \frac{7x^6}{7!} = 1 - \frac{x^2}{2!} + \frac{x^4}{4!} - \frac{x^6}{6!}$

318. $g''(x) = -x + \frac{x^3}{3!} - \frac{x^5}{5!} \approx -\sin(x)$

319. $h(x) = x - \frac{x^3}{3!} + \frac{x^5}{5!} - \frac{x^7}{7!} + \frac{x^9}{9!}$ and $k(x) = x - \frac{x^3}{3!} + \frac{x^5}{5!} - \frac{x^7}{7!} + \frac{x^9}{9!} - \frac{x^{11}}{11!}$

320. $p(x) = 1 + x + \frac{x^2}{2!} + \frac{x^3}{3!} + \frac{x^4}{4!} + \frac{x^5}{5!} + \frac{x^6}{6!}$ and $e^x \approx 1 + x + \frac{x^2}{2!} + \frac{x^3}{3!} + \ldots + \frac{x^6}{6!}$

321. a) $T_2(x) = -2x^2 + 4$ b) $T_2(x) = 4x^2 - 16x + 16$

322. a) $f'(x) = \cos(x)$, $f''(x) = -\sin(x)$ and $f'''(x) = -\cos(x)$;
$f(0) = \sin(0) = 0$, $f'(0) = \cos(0) = 1$, $f''(0) = -\sin(0) = 0$ and $f'''(0) = -\cos(0) = -1$
$\Rightarrow T_3(x) = 0 + \frac{1}{1!}(x-0) + 0 + \frac{-1}{3!}(x-0)^3 = x - \frac{1}{6}x^3$

b) $f'(x) = -\sin(x)$, $f''(x) = -\cos(x)$, $f'''(x) = \sin(x)$ and $f^{(4)}(x) = \cos(x)$;
$f(0) = \cos(0) = 1$, $f'(0) = -\sin(0) = 0$, $f''(0) = -\cos(0) = -1$, $f'''(0) = \sin(0) = 0$ and
$f^{(4)}(0) = \cos(0) = 1 \Rightarrow T_4(x) = 1 + 0 + \frac{-1}{2!}(x-0)^2 + 0 + \frac{1}{4!}(x-0)^4 = 1 - \frac{1}{2}x^2 + \frac{1}{24}x^4$

The formulas of the function are identical, $y = 1 - \frac{x^2}{2!} + \frac{x^4}{4!}$.

323. $T_1(27) = 5.2$; $T_2(27) = 5.196$; $T_3(27) = 5.19616$; $T_4(27) = 5.196152$

324. $T_1\left(\frac{5}{4}\right) = \frac{3}{4} = 0.75$; $T_2\left(\frac{5}{4}\right) = \frac{13}{16} = 0.8125$; $T_3\left(\frac{5}{4}\right) = \frac{51}{64} = 0.796875$; $T_4\left(\frac{5}{4}\right) = \frac{205}{256} = 0.80078125$

325. a) $1, 3, 2, -1, a+b, 2a+h$ b) –

326. a) $2.46\,\text{cm}^2/\text{s}$ b) $a = 32.7\,\text{cm}$

327. a) $\frac{f(x+h)-f(x-h)}{2h} = 2x$; $\frac{g(x+h)-g(x-h)}{2h} = 2x+1$; $\frac{k(x+h)-k(x-h)}{2h} = 2ax+b$

b) $\frac{f\left(x+\frac{h}{2}\right)-f\left(x-\frac{h}{2}\right)}{h}$ c) –

328. a) $f'(4) = 6$; $f'(-3) = -8$; $f'(1-\sqrt{2}) = -2\sqrt{2}$

b) $q'(2) = 1$; $q'(-1) = \frac{1}{25}$; $q'\left(\frac{3}{2} - \sqrt{2}\right) = \frac{1}{8}$

c) $w'(2) = 1$; $w'\left(\frac{11}{7}\right) = \sqrt{7}$; $w'(4) = \frac{1}{\sqrt{5}} = \frac{\sqrt{5}}{5}$

329. $\mathrm{d}F = 2s_1(s_2 - s_1)$

330. a) True   b) True   c) True   d) False

331. –

332. a) False   b) True   c) False

333. a) $2ax + b$ b) $nx^{n-1} + n$ c) $e^x + e + \frac{1}{x^2}$

d) $t^5 + 2t^3x - 3tx^2$ e) 0 f) $2x - 2$

334. a) $F'(r) = 2\pi r \,\widehat{=}\,$ circumference of the circle   b) $\frac{\mathrm{d}F}{\mathrm{d}t} = 2\pi r \cdot r'$; $\frac{\mathrm{d}F}{\mathrm{d}t} = 12.6\pi + 8.82t$

335. $\frac{\mathrm{d}V}{\mathrm{d}t} = -0.13\,\text{m}^3/\text{s}$

336. $\frac{\mathrm{d}V}{\mathrm{d}T} = \frac{\mathrm{d}}{\mathrm{d}T}(s^2 \cdot h) = 2s\frac{\mathrm{d}s}{\mathrm{d}T} \cdot h + s^2 \cdot \frac{\mathrm{d}h}{\mathrm{d}T}$

337. a) $-1\,\text{cm/min}$; $-2\,\text{cm/min}$ b) $V = 94.0\,\text{m}^3$

338. –

339. –

340. a) True   b) False   c) True

341. 1

342. $f^{(99)}(x) = 3^{99} \cdot \sin(3x) + 100! \cdot x$

343. –

344. The shadow decreases with a velocity of $0.6\,\text{m/s}$.

345. Anna is right.

346. –

347. a) $X\left(1 - \frac{1}{m}\,\big|\,0\right)$; $Y\left(0\,\big|\,1 - m\right)$ b) $X(m+1\,|\,0)$; $Y\left(0\,\big|\,1 + \frac{1}{m}\right)$ c) $d = |F(\Delta_n) - F(\Delta_t)| = 2$

348. a) –   b) $m_t = 1$ c) $Q_1(-1\,|\,2) = P$, $Q_2(1\,|\,0)$

349. a) $y = \frac{-1}{2}x^2 + 3x$ b) $S^*(2\,|\,4)$; after 1 second

350. Both are wrong.

351. $\sqrt{2}$ units

352. –

353. a) $y = \frac{f(b) - f(a)}{b - a} \cdot (x - a) + f(a)$ b) $y = f'(a) \cdot (x - a) + f(a)$

354. –

355. a) $x = 1$ b) $x = \frac{3}{2}$ c) $x = \frac{s}{2}$

356. –

357. a) $p: y = -x^2 + 1$ b) $p: y = -ax^2 + c$ c) $g: y = \frac{1}{2}x + 1$ d) $g: y = \frac{b}{2}x + c$

358. $y = \frac{2}{3}x$

359. $x \in [-1, 3]$

360. at $m = c$

361. a) $[0, 5]$ b) $t = 5 + 5\sqrt{2}$ c) $f(t) = \frac{-1}{50}t + \frac{1}{10}$

362. a) $k: y = f(x) = \frac{1}{8}x^3 - \frac{3}{2}x$ b) $k: y = f(x) = \frac{-3}{128}x^5 + \frac{5}{16}x^3 - \frac{15}{8}x$

363. i) $F(4\,|\,8)$; $\overline{PF} = 15$ ii) $x = 4$

364. a) $X\left(2u^3 + u\,\big|\,0\right)$ b) –
    c) $S\left(-u - \frac{1}{2u}\,\big|\,u^2 + 1 + \frac{1}{4u^2}\right)$ d) $\overline{SP} = \frac{3}{2} \cdot \sqrt{3} \approx 2.60$

365. $C(-1\,|\,0)$

366. Emma is right.

367. Yes

368. $x = e$ with $e^{\frac{1}{e}} = 1.44466\ldots$

369. –

370. a) local maximum at $x = 2$, local minima at $x = 1$ and $x = 3$
b) local maximum at $x = 1$, local minimum at $x = 3$
c) local maximum at $x = 0$, local minimum at $x = 2$

371. –

372. $\varphi = 60°$

373. a) $x = 1$ b) $x = \frac{1}{2}$ c) $a = 3$

374. a) $\frac{\Delta y}{\Delta x} = \frac{f(0)-f(-2)}{0-(-2)} = \frac{4-(-6)}{0-(-2)} = 5$ b) $f'(2) = -9$

375. a) $v = 7.2\,\text{m/s};\quad 7.6\,\text{m/s};\quad 7.92\,\text{m/s};\quad 7.992\,\text{m/s}$ b) $8\,\text{m/s}$

376. a) $T(t_2) - T(t_1)$ b) $\frac{T(t_2)-T(t_1)}{t_2-t_1}$ c) $\lim_{t_1 \to t_2}\left(\frac{T(t_2)-T(t_1)}{t_2-t_1}\right)$

377. a) $-19'000$ b) $V'(t) = 400t - 20'000$ c) $-18'000$

378. a) 3 b) $\frac{1}{9}$

379. a) $f'(x) = 2x - 2$ b) $f'(x) = \frac{1}{(1-x)^2}$

380. a) $f'(x) = \frac{1}{(x-2)^2}$ b) $x_1 = 1,\ x_2 = 3$

381. Yes

382. –

383. a) False b) True c) False d) False

384. a) $f'(x) = 20x^4$ b) $f'(x) = 2x^5 - 2$ c) $f'(x) = \frac{6}{x^4}$
d) $f'(x) = \frac{2}{x^3} - \frac{1}{x^2}$ e) $f'(x) = \frac{2}{\sqrt{x}}$ f) $f'(x) = \frac{1}{3\cdot\sqrt[3]{x^2}}$

385. a) $y = \frac{1}{4}\cdot x^2 + 13$ b) $y = (-1)\cdot x^{-1}$ c) $y = \frac{1}{7}\cdot x^{\frac{1}{2}}$
d) $y = 6\cdot x^2 - 3\cdot x$ e) $y = x^2 - 4\cdot x + 4$ f) $y = \frac{5}{21}\cdot x^{-4} + \frac{1}{7}\cdot x$

386. a) $f'(x) = 4x^3 - 9x^2 + 2x$ b) $f'(x) = \frac{2}{(x+2)^2}$ c) $f'(x) = 3\sqrt{2x}$

387. –

388. a) $f'(x) = \cos(x) - \sin(x)$ b) $f'(x) = 1 - 2\cdot\cos(x)$ c) $f'(x) = 6x + 10\cdot\sin(x)$

389. a) $y' = \cos(x) - x\cdot\sin(x)$ b) $y' = 2\cdot\cos(2x+1)$ c) $y' = (x^2+2)\cos(x)$

390. a) $\frac{ds}{dt} = \frac{1}{\cos(t)-1}$ b) $\frac{ds}{dt} = \tan^2(t) - 1$ c) $\frac{ds}{dt} = \frac{1-\sin(t)}{\cos^2(t)}$

391. a) $y^{(4)} = y = 6 \cdot \cos(x)$ b) $y^{(4)} = y = -5 \cdot \sin(x)$
   $y^{(5)} = y' = -6 \cdot \sin(x)$ $y^{(5)} = y' = -5 \cdot \cos(x)$

392. a) $f'(x) = 3 \cdot e^{3x}$ b) $f'(x) = 1 - 2 \cdot e^x$ c) $f'(x) = \ln(2) \cdot 2^x$
   d) $f'(t) = \frac{3}{t}$ e) $f'(t) = \frac{2 + \ln(t)}{\sqrt{t}}$ f) $f'(t) = \frac{1}{t \cdot (t-1)}$

393. a) $\frac{dy}{dx} = 2 \cdot 10^x (1 + x \cdot \ln(10))$ b) $\frac{dy}{dx} = \frac{3^x}{x^2}(x\ln(3) - 1)$ c) $\frac{dy}{dx} = x \cdot 2^x(2 + x \cdot \ln(2))$

394. $d'(x_0) = -5$, $p'(x_0) = 14$, $q'(x_0) = -10$

395. $y' = 60x^2(2 + x^3)^3(1 + (2 + x^3)^4)^4$

396. a) False   b) False   c) True   d) False

397. a) True   b) True   c) False

398. a) $\frac{d}{dx}F_1(x) = \frac{d}{dx}F_2(x) = 1$ b) –

399. a) $t: y = 4x$ b) $t: y = -2x + 18$ c) $t: y = 8$
   $n: y = -\frac{1}{4}x$ $n: y = \frac{1}{2}x + 3$ $n: x = 4$

400. a) $t: y = -x + 1$ b) $t: y = \frac{7}{9}x + \frac{1}{9}$ c) $t: y = \frac{1}{2}x - \frac{7}{2}$
   $n: y = x + 1$ $n: y = -\frac{9}{7}x + \frac{89}{21}$ $n: y = -2x - 11$

401. a) $x = 1$ b) $x_1 = -1$, $x_2 = 3$ c) nowhere

402. $P(2 \mid -8)$

403. $t: y = \frac{1}{4}x + 1$; $n: y = 18 - 4x$

404. $a = -7$

405. a) $t_1: y = -9x$ and $t_2: y = 7x$ b) $t_1: y = 3x + 12$ and $t_2: y = -13x - 20$

406. a) $12.0°$ b) $63.4°$

407. a) $26.6°$ b) $18.9°$

408. $71.6°$

409. $T_1(0 \mid 0)$ and $T_2(2 \mid 0)$

410. $T(0 \mid 1)$, $H(-2 \mid -3)$

411. a) maximum point: none; minimum point: $T(-3 \mid -4.5)$; saddle point: $W(0 \mid 0)$
   b) $x_{\min} = -3$ and $x_{\max} = 1$

412. a) $D_f = \mathbb{R}$, $W_f = \mathbb{R}$

roots: $x_1 = 0$, $x_{2,3} = \pm\sqrt{3}$

$H(-1\,|\,2)$, $T(1\,|\,-2)$

$W(0\,|\,0)$

symmetric w.r.t. origin

b) $D_f = \mathbb{R}$, $W_f = [-1; \infty[$

roots: $x_1 = 0$, $x_{2,3} = \pm\sqrt{2}$

$H(0\,|\,0)$, $T_{1,2}(\pm 1\,|\,-1)$

$W_{1,2}\left(\pm\frac{\sqrt{3}}{3}\,\middle|\,-\frac{5}{9}\right)$

symmetric w.r.t. y-axis

c) $D_f = \mathbb{R}$, $W_f = \mathbb{R}$

roots: $x_1 = 0$, $x_2 = 3$

$T(0\,|\,0)$, $H(2\,|\,4)$

$W(1\,|\,2)$

d) $D_f = \mathbb{R}$, $W_f = \,]-\infty; \frac{27}{16}]$

roots: $x_1 = 0$, $x_2 = 2$

$H\left(\frac{3}{2}\,\middle|\,\frac{27}{16}\right)$

$W(1\,|\,1)$, saddle point: $(0\,|\,0)$

413. iv)

414. a) False   b) True   c) False

415. $a = b = 1$, $c = 0$

416. $f(x) = \frac{9}{16}x^3 - \frac{27}{4}x$

417. $f(x) = \frac{-1}{4}x^3 + 3x$

418. $y = k(x) = \frac{-1}{3}x^3 + 2x^2$; $\sigma \approx 85.2°$

419. n odd

420. $f(x) = x^4 - 7x^2 + 15$

421. $x = -1$

422. $2\,\text{cm}$

423. $25\,\text{m}^2$ per pair of bunnies

424. $(30\,|\,0)$

425. $42.0\,\text{cm}$ to $84.0\,\text{cm}$

426. $F_{\max} = 24$

427. height of cylinder: $R \cdot \sqrt{2}$

4 Integral Calculus

1. a) i)   e.g. $f(x) = x^2$
      iii) e.g. $f(x) = x^4$
      v)   e.g. $f(x) = \sin(x)$
      vii) e.g. $f(x) = \frac{-1}{x}$

   ii)   e.g. $f(x) = x^2 - x$
   iv) e.g. $f(x) = \frac{1}{n+1} \cdot x^{n+1}$
   vi) e.g. $f(x) = e^x$
   viii) e.g. $f(x) = \frac{1}{-m+1} \cdot x^{-m+1}$

   b) No

   c) i) and ii)

   d) $f(x) = x^2 + C, C \in \mathbb{R}$

   e) i) $f(x) = -\cos(x) + C$
      ii) $f(x) = x^3 + C$
      iii) $f(x) = e^x + x^2 + C$

2. a) $2x^2$ b) $\frac{x^4}{4}$ c) $-0.39x^6$ d) $\frac{x^{n+1}}{n+1}$
   e) $-\frac{t^9}{12}$ f) $1001t$ g) 0 h) $\frac{g}{2}t^2$

3. a) $2x^2 + 321x$ b) $2x^3 - 4x$ c) $3x + \frac{1}{x}$ d) $\frac{1}{4}x^4 - \frac{1}{2}x^{-2}$

4. a) $F(t) = \sqrt{t}$ b) $F(t) = -6\sqrt{t}$ c) $F(t) = 4t\sqrt[4]{t^3}$ d) $F(t) = 3\sqrt[3]{t}$

5. a) $-\cos(x) + \sin(x)$ b) $-3\cos(x) - 2\sin(x)$ c) $\tan(x)$
   d) $2e^x$ e) $\frac{1}{\ln(2)} \cdot 2^x$ f) $\frac{1}{\ln(a)} \cdot a^x, \ a > 0, a \neq 1$

6. a) $x^3 + 2x^2 - x$ b) $x^3 - x$ c) $\frac{1}{2-x} - \frac{1}{2}$

7. a) $f(x) = x^3 - 14x + 12$ b) $f(x) = x + \frac{1}{x} + \frac{2}{3}$ c) $f(x) = -\frac{1}{2}x^2 + 5x + 2$
   d) $f(x) = x - \cos(x) + 1$ e) $f(x) = 4 - 3\sin(x)$ f) $f(x) = \pi e^x + 1 - \pi e$

8. a) $F(x) = 2\sqrt{x} - 4$ b) $F(x) = \frac{1}{6}x^3 - x^2 + \frac{19}{6}$ c) $F(x) = x^2 + 5x - \frac{27}{2}$

9. $f(x) = 2x\sqrt{x} - 50$

10. a) Curve b b) Curve a

11. –

12. –

13. –

14. a) $2x^2 + C$ b) $x + C$ c) C
   d) $\frac{3}{2}x^4 + C$ e) $x^3 - x + C$ f) $x^4 + \frac{1}{3}x^3 - 7x + C$
   g) $x^5 + \frac{1}{8}x^4 - \frac{7}{9}x^3 + C$ h) $\frac{3}{4}x^4 + \frac{4}{3}x^3 - \frac{3}{2}x^2 - 4x + C$ i) $\frac{9}{5}x^5 - 2x^3 + x + C$

15. a) $\frac{-1}{x} + C$ b) $\frac{-1}{x^2} + C$ c) $\frac{-2}{x^3} + \frac{1}{x} + C$
   d) $6\ln|x| + \frac{5}{x} - \frac{2}{x^2} + C$ e) $\frac{1}{3}x^3 - x + 5\ln|x| + C$ f) $10\ln|x + 2| + C$

16. a) $\frac{2}{3}x\sqrt{x} + C$ b) $\sin(x) + C$ c) $\frac{1}{\ln(5)} \cdot 5^x + C$
   d) $\frac{2}{5}x^2\sqrt{x} + \frac{10}{3}x\sqrt{x} + C$ e) $e^x - \cos(x) + C$ f) $2\sqrt{x} + C$
   g) $\frac{1}{2}x^2 + \frac{4}{5}x\sqrt[4]{x} + C$ h) $-2\cos(x) - 3\sin(x) + C$ i) $\pi \cdot x\ln(x) - \pi \cdot x + C$
   j) $\tan(x) + C$ k) $\frac{1}{2}t^2 + \sin(t) + C$ l) $\frac{1}{2-2t}x^{2-2t} + C$

17. Taking the derivative of the right-hand side always yields the integrand on the left-hand side.

18. Taking the derivative of the right-hand side yields the integrand on the left-hand side.

19. a) Yes   b) Yes

20. a) True   b) False (C is missing)

21. a) $\int x\mathrm{e}^x\,\mathrm{d}x = x\mathrm{e}^x - \mathrm{e}^x + C$ b) $\int \ln(x)\,\mathrm{d}x = x\ln(x) - x + C$

22. a) $F(x) = \frac{1}{12}(2x+3)^6 + C$ b) $F(x) = \frac{1}{3}(3x-1)^7 + C$ c) $F(x) = \frac{-1}{4}(12-x)^4 + C$
   d) $F(x) = \frac{-1}{2}\left(3 - \frac{1}{2}x\right)^4 + C$ e) $F(x) = (3x-9)^4 + C$ f) $F(x) = \frac{-\sqrt{2}}{6}\left(3 - x\sqrt{2}\right)^3 + C$

23. a) $F(x) = \frac{-1}{4}(4x+3)^{-1} + C$ b) $F(x) = (1-x)^{-2} + C$ c) $F(x) = \frac{2}{3}\left(4 - \frac{1}{2}x\right)^{-3} + C$
   d) $F(x) = \frac{-1}{(2x-1)^2} + C$ e) $F(x) = \frac{3}{2\left(2 - \frac{1}{3}x\right)^4} + C$ f) $F(x) = \sqrt{2x+1} + C$

24. a) $F(t) = \frac{-1}{2}\cos(2t) + C$ b) $F(t) = \frac{1}{\pi}\sin(\pi t) + C$ c) $F(t) = \frac{1}{2}\sin(10t) + C$

25. a) $F(x) = -\mathrm{e}^{-4x} + C$ b) $F(x) = \mathrm{e}^x + \mathrm{e}^{-x} + C$ c) $F(x) = \frac{-3}{\ln(2)} \cdot 2^{-3x} + C$
   d) $F(x) = \frac{1}{2} \cdot \ln|2x-1| + C$ e) $F(x) = -\ln|3-2x| + C$ f) $F(x) = 12x\ln(3x) - 12x + C$

26. –

27. a) $v(t) = a \cdot t + v_0$, $s(t) = \frac{a}{2} \cdot t^2 + v_0 \cdot t + s_0$
   b) $v(1) = 5\,\mathrm{m/s}$, $s(1) = 2.5\,\mathrm{m}$, $t \approx 2.83\,\mathrm{s}$
   c) $v(1) = -9.81\,\mathrm{m/s}$, $s(1) \approx 15.10\,\mathrm{m}$, $t \approx 2.02\,\mathrm{s}$

28. –

29. a) $s(t) = \frac{1}{2}at^2$ b) $s(t) = \frac{1}{2}at^2 + v_0 t$

30. a) $s_{[0;14]} = 280\,\mathrm{m}$ b) $s_{[14;20]} = 60\,\mathrm{m}$
   c) $s_{[0;14]} =$ rectangular area $= 280\,\mathrm{m}$ d) $s_{[20;30]} \approx 60\,\mathrm{m}$
   $s_{[14;20]} =$ triangular area $= 60\,\mathrm{m}$

31. a) work $\hat{=}$ area beneath the curve of the force   b) $W \approx 212\,\mathrm{J}$

32. a) $x_0 = a = 0$, $x_1 = \frac{1}{2}$, $x_2 = 1$,   b) $O_4 = 3.75$
   $x_3 = \frac{3}{2}$, $x_4 = b = 2$; $\Delta x = \frac{1}{2}$
   c) $O_4 = 3.75$ d) $O_{50} = \frac{1717}{625} \approx 2.75$
   e) $O_n = \frac{4}{3} \cdot \left(1 + \frac{1}{n}\right) \cdot \left(2 + \frac{1}{n}\right)$ f) $F(0,2) = \frac{8}{3} \approx 2.67$

33. $U_4 = 12.25$, $O_4 = 16.25$

34. a) $U_6 = \frac{163}{54} \approx 3.02$ b) $O_6 = \frac{199}{54} \approx 3.69$ c) $3.02 \le A \le 3.69$

35. a) i) $U_5 = 16$ ii) $U_{10} = 20.25$ iii) $U_n = 25\left(1 - \frac{1}{n}\right)^2$
   b) i) $O_5 = 36$ ii) $O_{10} = 30.25$ iii) $O_n = 25\left(1 + \frac{1}{n}\right)^2$
   c) i) –   ii) $\lim_{n\to\infty} U_n = 25$, $\lim_{n\to\infty} O_n = 25 \Rightarrow A = 25$

36. a) $O_n = \frac{b^3}{12} \cdot \left(1 + \frac{1}{n}\right) \cdot \left(2 + \frac{1}{n}\right) + b$ b) $A = \frac{b^3}{6} + b$

37.

$n =$	3	5	10	20	100
a) $U_n =$	0.6167	0.6456	0.6688	0.6808	0.6907
$O_n =$	0.7833	0.7456	0.7188	0.7058	0.6957
b) $U_n =$	0.5326	0.5710	0.6010	0.6164	0.6290
$O_n =$	0.7433	0.6974	0.6643	0.6481	0.6353
c) $U_n =$	0.7152	0.8347	0.9194	0.9602	0.9921
$O_n =$	1.2388	1.1488	1.0765	1.0388	1.0078
d) $U_n =$	0.8404	0.8684	0.8891	0.8994	0.9075
$O_n =$	0.9756	0.9495	0.9297	0.9196	0.9116

38. a) $U_n = \pi r^2 \left(1 - \frac{1}{n}\right), \ O_n = \pi r^2 \left(1 + \frac{1}{n}\right)$ b) $\lim\limits_{n \to \infty} U_n = \lim\limits_{n \to \infty} O_n = \pi r^2, \ A(\text{Kreis}) = \pi r^2$

39. lower sum $= 55.0\,\mathrm{m}^2$, upper sum $= 83.8\,\mathrm{m}^2 \ \Rightarrow \ 55.0\,\mathrm{m}^2 < A < 83.8\,\mathrm{m}^2$

40. a) 8   b) $\frac{74}{3} = 24.\overline{6}$

41. a) $x_1 = -1$, $x_2 = 3$ b) $x \geq 1$ c) $U_{10} = 8.96$
    d) $U_n = 16 - \frac{8}{3}\left(1 + \frac{1}{m}\right)\left(2 + \frac{1}{m}\right)$, where $n = 2m$.

    e) $A = \frac{32}{3} = 10.\overline{6}$ f) $A = 2 \cdot \int\limits_{1}^{3} \left(-x^2 + 2x + 3\right) \mathrm{d}x$

42. i) 17   ii) -4 iii) -5

43. i) -5 ii) 3   iii) 0   iv) -8

44. i) $-p$ ii) $p + q$ iii) 0   iv) $-2q$

45. a) 7.5   b) -1 c) 6.5   d) 8.5

46. a) 3   b) -10.5 c) -7.5

47. a) $O_n = \frac{x^2}{2}\left(1 + \frac{1}{n}\right) - x$ b) $F_0(x) = \frac{x^2}{2} - x$
    c) $O_n = \frac{x^3}{6}\left(1 + \frac{1}{n}\right)\left(2 + \frac{1}{n}\right)$, $F_0(x) = \frac{x^3}{3}$

48. a) $F_0(x) = x$ b) $F_0(x) = 2x$ c) $F_0(x) = c \cdot x$
    d) $F_0(x) = \frac{x^2}{2}$ e) $F_0(x) = x^2$ f) $F_0(x) = \frac{m \cdot x^2}{2}$
    g) $F_0(x) = \frac{x^2}{2} + x$ h) $F_0(x) = x^2 + x$ i) $F_0(x) = \frac{m \cdot x^2}{2} + c \cdot x$

49. a) $f(t) = t^2$; $F_0(x) = \frac{1}{3}x^3$ b) $f(t) = t^3$; $F_0(x) = \frac{1}{4}x^4$ c) $f(t) = t^4$; $F_0(x) = \frac{1}{5}x^5$
    d) $f(t) = t^5$; $F_0(x) = \frac{1}{6}x^6$ e) $f(t) = \frac{1}{2}t^2 + 1$; $F_0(x) = \frac{1}{6}x^3 + x$

50. $\frac{\mathrm{d}}{\mathrm{d}x} F_0(x) = f(x)$

51. a) $\int\limits_0^2 x^2 \,\mathrm{d}x = \frac{8}{3}$ i) $\int\limits_0^2 (x^2 + 1)\,\mathrm{d}x = \frac{14}{3}$ ii) $\int\limits_0^2 (x + 1)^2 \,\mathrm{d}x = \frac{26}{3}$ iii) $\int\limits_{-1}^1 (x + 1)^2 \,\mathrm{d}x = \frac{8}{3}$

b) $\int\limits_0^4 \sqrt{x}\,\mathrm{d}x = \frac{16}{3}$ i) $\int\limits_0^4 \sqrt{x + 1}\,\mathrm{d}x = \frac{10}{3}\cdot\sqrt{5} - \frac{2}{3}$ ii) $\int\limits_{-1}^3 \sqrt{x + 1}\,\mathrm{d}x = \frac{16}{3}$ iii) $\int\limits_1^5 \sqrt{x - 1}\,\mathrm{d}x = \frac{16}{3}$

52. a) $\int\limits_{-2}^3 (-(x - 3)(x + 2))\,\mathrm{d}x = \frac{125}{6} = A$ b) $\int\limits_{-3}^3 (3 - x)^2 \,\mathrm{d}x = 72 = A$

c) $\int\limits_0^3 (3 - x)^2 \,\mathrm{d}x = 9 = A$ d) $\int\limits_{-2}^2 x(x - 2)(x + 2)\,\mathrm{d}x = 0 \neq A(= 8)$

e) $\int\limits_{-1}^2 (6 + 3x - 3x^2)\,\mathrm{d}x = \frac{27}{2} = A$ f) $\int\limits_{-1}^2 (6x + 3x^2 - 3x^3)\,\mathrm{d}x = \frac{27}{4} \neq A(= \frac{37}{4})$

53. a) 16   b) 12   c) 12   d) 0   e) -9 f) 24

54. a) $\frac{1}{4}$ b) -11.25 c) $3 - 2\ln(2) \approx 1.61$

d) $\frac{52}{3} - 3\ln(3) \approx 14.04$ e) $2\ln(\frac{3}{5}) + 2 \approx 0.98$ f) $3\ln(\frac{3}{2}) + \frac{13}{2} \approx 7.72$

55. a) 14   b) 2   c) 12   d) $\frac{2}{5}$ e) 43   f) $\sqrt{3}$

56. a) 1   b) 0   c) -1 d) 8   e) 1   f) $\frac{1}{2}\ln(2)$

57. a) $e^2 - 1 \approx 6.39$ b) $\frac{1}{e} - 2 \approx -1.63$ c) $\pi(e - 1) \approx 5.40$

d) $e^2 - \frac{2^{e+1}}{e+1} - 1 \approx 2.85$ e) $\ln(2) + e^2 - e \approx 5.36$ f) $\frac{9}{\ln(10)} \approx 3.91$

58. a) 1705   b) 0   c) 2

d) $\frac{1}{3}$ e) $\frac{\sqrt{2}}{\pi}$ f) 1

g) $\frac{1}{6}(e^{20} - 1)$ h) $e^{12} - 1$ i) 0

j) 2   k) $3\ln(10)$ l) 3

59. a) 13   b) 2

60. a) $3t^2 - \frac{3}{2}$ b) $3x^2 - 3x$

61. a) $k = 3$ b) $k = 5$

62. a) 2   b) -6 c) 10

63. a) $x = 0,\ x = 2$ b) $x = 3$; $x = k \cdot \frac{\pi}{3}$ für $k \in \mathbb{Z}$

64. a) Yes; $f(x) = 3x^2 - 4x + 1$ b) No

c) Only for $c = -2$; $f(x) = 3x^2 - 4x + 1$

65. $F(x) = -\frac{1}{2}\cos(2x) + C$, where:

a) $C > \frac{1}{2}$ b) $C = \pm\frac{1}{2}$

66. a) Car A is leading.   b) Lead of car A over B after 1 min.

c) Car B is leading.   d) after approx. 2 min.

67. a) True   b) False   c) False

68. a) 5   b) 2   c) -15

69. a) no roots, $A = 30$ b) roots: $x_{1,2} \approx \pm 5.57 \notin [-1; 5]$, $A = 144$

    c) root: $x \approx -2.67 \notin [-2; 0]$, $A = 90$ d) roots: $x_{1,2} \approx \pm 1.73 \notin [2; 6]$, $A = 3$

    e) root: $x = \frac{1}{4} \notin [4; 5]$, $A \approx 11.33$ f) root: $x = 2 \notin [3; 6]$, $A \approx 6.60$

70. a) $A = 36$ b) $A = 20$

    c) $A \approx 3.32$ d) $A = 1$

71. a) $A = 72$ b) $A = \frac{1}{2}$

    c) $A = \frac{4}{3}$ d) $A = \frac{128}{15} = 8.5\overline{3}$

    e) $A = \frac{784}{15} = 52.2\overline{6}$ f) $A = 4\sqrt{3} - \ln(7 + 4\sqrt{3}) \approx 4.29$

72. $a = 6$

73. a) $k - 8$ b) $k = \frac{3 + 3\sqrt{5}}{2} \approx 4.85$ c) $k = \sqrt[3]{4} \approx 1.59$ d) $k \approx 1.51$

74. a) $\frac{9}{2}$ b) 2   c) $\frac{37}{12}$ d) $\frac{8\sqrt{2}}{3}$ e) 24

75. $a = 9$

76. a) $A(t) = \frac{t}{2}$, $t = 16$ b) $A(t) = \frac{4}{3}t^3$, $t = 3$

77. $y = -2x^3 + 6\sqrt{2}x$

78. $y = -x^2 + 3x$

79. a) $y = -x^3 + 4x^2 = x^2(4 - x)$ b) $A = \frac{64}{3}$

80. $y = \frac{-1}{2}x^4 + \frac{9}{2}x^2$

81. a) $A = \frac{64}{3}$ b) $A = \frac{256}{3}$ c) $A = 9$ d) $A = \frac{5}{3}$

82. a) $A = 18$ b) $A = 18$

83. a) $A = \frac{39}{5} = 7.8$ b) $A = \frac{221}{12} = 18.41\overline{6}$

84. a) $A = \frac{8}{3}$ b) $A = \frac{1}{3}$

85. a) $x_{1,2} = \pm a$ b) $f_a(-x) = f_a(x)$, $h_a(-x) = h_a(x)$

    c) $x_{1,2} = \pm a$ d) $A(a)$ has a maximum at $a = 2$.

86. a) $A = 1 + \ln(3) \approx 2.10$ b) $A = 4\ln(2) \approx 2.77$ c) $A = 22 - 16\ln(2) \approx 10.91$

87. a) $A = \sqrt{3} - \frac{\pi}{3}$ b) $A = 2 - \frac{\pi}{2}$ c) $A = 3\sqrt{3} - \frac{\pi}{2} - 2$

88. a) $A = 2\sqrt{2} - 2$ b) $A = \frac{2}{\pi} + \frac{1}{6}$ c) $A = \frac{2}{\pi}$

89. a) $a = \frac{-\pi^2}{4}$, $A = 2 + \frac{\pi^3}{6}$ b) $a = e$, $A = \frac{4}{3}e - 2$

    c) $A = \frac{4}{3}$ d) $a = \frac{e^3 - e}{2}$, $b = \frac{3e - e^3}{2}$, $A = 2e$

90. $t_a: y = 2ax - a^2, \ t_b: y = 2bx - b^2, \ A = \frac{(b-a)^3}{12}$

91. $A = e^2 - 1$

92. $A = \frac{1}{12}$

93. $A = \frac{\pi^2}{4} - 2$

94. $A = \frac{25}{8}$

95. $A = \frac{11}{12}$

96. $x = e$

97. a) $a = \frac{1}{12}$ b) $m = \frac{1}{3}$

98. a) $A = \frac{4}{3}$ b) $A = \frac{b^3}{6}$

99. a) $p = \frac{1}{2(e-1)} \approx 0.29$ b) $q = \frac{1}{2(e-1)} \approx 0.29$ c) $r = \ln(\frac{3}{2}) \approx 0.41$ d) $p = q$

100. $x = \sqrt[3]{4}$

101. $m = 12 - 6\sqrt[3]{4} \approx 2.48$

102. $A_1 : A_2 = 1 : 1$

103. A_k is $45\,\%$ of the area of the rectangle.

104. $A : A_S = 1 : 18 = 0.0555\ldots$, the area of the region cut away is slightly more than $5\,\%$.

105. $a = 1$

106. $v = 3$, independent of the choice of a

107. a) $s = \frac{a}{h} \cdot x$ b) $Q(x) = \frac{a^2}{h^2} \cdot x^2$ c) $\Delta V = \frac{a^2}{h^2} \cdot x^2 \cdot \Delta x$
     d) $V_T = \sum \frac{a^2}{h^2} \cdot x^2 \cdot \Delta x$ e) $V_P = \frac{1}{3}a^2 h$

108. a) $V = \frac{56\pi}{3}$ b) $V = \frac{32\pi}{15}$

109. a) $V = \frac{512}{15}\pi \approx 107.23$ b) $V = \frac{324}{5}\pi \approx 203.58$ c) $V = \frac{128}{105}\pi \approx 3.83$
     d) $V = \frac{256}{315}\pi \approx 2.55$ e) $V = \frac{64}{3}\pi \approx 67.02$ f) $V = \frac{4}{3}\pi \approx 4.19$

110. a) $V = \frac{192}{35}\pi$ b) $V = 2\pi$

111. $V = \frac{191}{480}\pi \approx 1.25\,\text{cm}^3$

112. $V = \frac{16}{63}\pi$

113. $V = \frac{2448}{5}\pi \approx 1538.12$

114. $V = \frac{\pi^2}{8}$

115. a) $V = 2\pi(\tan(1) - 1) \approx 3.50$ b) $V = \pi \approx 3.14$

116. a) $V = \frac{9}{2}\pi$ b) $V = \frac{3}{8\ln(2)}\pi$ c) $V = (e^2 - 3)\pi$

117. a) $V = 20\pi$ b) $V = \frac{1}{3}\pi r^2 h$

118. a) $V = \frac{3}{2}\pi$ b) $V = \frac{4}{15}\pi$

119. a) $V = \frac{3}{10}\pi$ b) $V = 24\pi$

120. $V = \frac{\pi}{96}$

121. a) $V = \pi \int\limits_0^H R^2\,\mathrm{d}x, \ \ V = \pi \int\limits_0^H (R^2 - r^2)\,\mathrm{d}x$

b) $V = \pi \int\limits_0^H (\frac{R}{H}x)^2\,\mathrm{d}x, \ \ V = \pi \int\limits_{\frac{rh}{R-r}}^{\frac{Rh}{R-r}} (\frac{R-r}{h}x)^2\,\mathrm{d}x$

c) $V = \pi \int\limits_{-R}^{R} (R^2 - x^2)\,\mathrm{d}x, \ \ V = \pi \int\limits_{R-h}^{R} (R^2 - x^2)\,\mathrm{d}x$

122. $h = \sqrt{\frac{60}{\pi}} \approx 4.37$

123. a) $A_1 : A_2 = 45 : 19$ b) $V_1 - V_2 = \frac{316}{15}\pi \approx 66.18$ c) $h = 4 - 2\sqrt[3]{2} \approx 1.48$

124. a) –   b) $T_M = 12\,\mathrm{h}$

c) $y_M \cdot (b - a) = \int\limits_a^b f(x)\,\mathrm{d}x$ d) $T_M = 12$ h, same as in b)

125. a) $\overline{f} = \frac{16}{3}, \ x_M = \frac{4}{\sqrt{3}}$ b) $\overline{f} = 7, \ t_M = 3$

c) $\overline{f} = \frac{8}{3}, \ u_M = \frac{32}{9}$ d) $\overline{f} = \frac{e}{5}(e^5 - 1), \ x_M = 2\ln\left(\frac{e^5 - 1}{5}\right)$

126. a) $\overline{f} = 5$ b) $\overline{f} = 2$

127. a) $\overline{f} = 1$ b) $c_1 = 4, c_2 = 2$ c) –

128. e.g. $f(x) = c \cdot e^x$

129. Yes

130. a) $\overline{I} = 600$ boxes   b) $\overline{K} = 18$ francs

131. a) $\overline{I} = 300$ barrels   b) $\overline{K} = 6$ francs

132. a) $A(t) = \ln(t), \ V(t) = \pi(1 - \frac{1}{t})$

b) $\lim\limits_{t \to \infty} A(t)$ does not exist, $\lim\limits_{t \to \infty} V(t) = \pi$ exists.

c) A is infinite, V is finite.

133. a) $\frac{1}{3}$ b) $-\frac{1}{2}$ c) 4 d) $\frac{3}{32}$

134. a) 2 b) divergent   c) $-\frac{1}{2}$ d) $\frac{11}{18}$

135. a) 1   b) divergent   c) $\frac{e^3}{2}$ d) 2

136. a) $A = 2$ b) $A = \frac{1}{2}$ c) $A = \frac{2}{\sqrt{e}}$

137. a) $A = 2$ b) $A = 2e$

138. –

139. a) $A = 1$ b) $A = e$

140. $A = \frac{4}{\pi} + 2$

141. $A = \frac{2}{3}\sqrt{e} \approx 1.10$

142. a) $A = 1$ b) $x = \frac{3}{2}, \ x = 3$

143. a)

a	$\left(\int_1^a e^{-x}\, dx\right) \big/ \left(\int_1^\infty e^{-x}\, dx\right)$
2	$63.212\ldots\,\%$
5	$98.168\ldots\,\%$
10	$99.987\ldots\,\%$
20	$99.999\ldots\,\%$
50	$\approx 100\,\%$
100	$\approx 100\,\%$

b)

a	$\left(\int_1^a \frac{1}{x^2}\, dx\right) \big/ \left(\int_1^\infty \frac{1}{x^2}\, dx\right)$
2	$50\,\%$
5	$80\,\%$
10	$90\,\%$
20	$95\,\%$
50	$98\,\%$
100	$99\,\%$

144. $A = \frac{1}{2}$

145. i) $a = \frac{1}{4}e^2$ ii) –   iii) $A = \frac{1}{3}e^2 \approx 2.46$ iv) $V = \frac{\pi}{10}e^4 \approx 17.15$

146. $16\,\mathrm{m}$

147. $50\,\mathrm{m}$

148. $\int_0^\infty f(x)\, dx \ \hat{=} \ $ total population

$$\frac{\int_0^\infty x \cdot f(x)\, dx}{\int_0^\infty f(x)\, dx} \ \hat{=} \ \text{mean income per person in Switzerland}$$

149. a) $k: y = \frac{1}{2}x^3 - x^2 - x + 2; \ x_1 = 2, \ x_{2,3} = \pm\sqrt{2}$ b) $A_t : A_s = 1 : 16$

150. $V \approx 0.44\,\mathrm{cm}^3, \ d \approx 1.9\,\mathrm{cm}$

151. Decrease of $21.3\,\%$

152. i) $t_1: y = 2x - 2; \ S_1(1\,|\,0), \ S_2(-1\,|\,-4)$ ii) $t_2: y = 6x + 2; \ S_3(3\,|\,20)$
     iii) $t_3: y = 22x - 46; \ S_4(-5\,|\,-156)$ iv) $A_1 : A_2 = A_2 : A_3 = 1 : 16$

153. a) –   b) $A_n = \frac{1}{2} - e^{-n}, \ \lim\limits_{n \to \infty} A_n = \frac{1}{2}$

154. $A_1 : A_2 = 1 : 1$

155. $\alpha \approx 20.3°$

156. a) $V = 1280\,\ell$ b) $V \approx 632\,\ell$ c) $h \approx 43\,\text{cm}$

157. $s(t) = -3t^2 + 120t + s_0$ with $s_0 \leq 800\,\text{m}$

158. a) $v(t) = -2t^3 + 6t^2$ b) $s(3) = 13.5\,\text{m}$

159. a) $d_a = 69.\overline{4}\,\text{m}$ b) $d_b = 102.\overline{7}\,\text{m}$ c) $d_c = 60\,\text{m}$

160. a) $\overline{v} = 26.4\,\text{m/s} \approx 95.0\,\text{km/h}$

     b) $h(t) \approx (2785.71 - 50t - 285.71 \cdot e^{-0.175t})\,\text{m}$

     c) $A \,\hat{=}\,$ altitude difference of the two skydivers

161. $W = \frac{Q_1 Q_2}{4\pi\epsilon}\left(\frac{1}{r} - \frac{1}{R}\right), \quad a(r) - \frac{Q_1 Q_2}{4\pi\epsilon \cdot m} \cdot \frac{1}{r^2}$

162. $W(x) = 1000 \cdot x^2$ (in Nm)

163. a) $W = Gm_1 m_2 \left(\frac{1}{a} - \frac{1}{b}\right)$ b) $W \approx 8.496 \cdot 10^9$ J   c) $W \approx 6.262 \cdot 10^{10}$ J

     d) $v_0 = \sqrt{\frac{2GM}{R}}$ e) $v_0 \approx 11'200\,\text{m/s} \approx 40'000\,\text{km/h}$

     f) Never, the rocket cannot overcome Earth's gravity.

164. a) $t \approx 3.66\,\text{h}$ b) $t \approx 7.32\,\text{h}$

     c) $\frac{1}{30} \int\limits_0^{30} K(t)\,\mathrm{d}t \approx 13.11\,\text{ng/ml}$ d) Because $K(t) > 0$.

165. a) $h_{\max} = \frac{45}{4}b^2$ b) $\frac{8}{3}$ months   c) e.g. $g_{a,b}(x) = a \cdot f_b(x)$

166. $K(2000) = \text{Fr. } 38'000$

167. $E(x) = -8x^2 + 64x, \ E_{\max} = \text{Fr. } 128$

168. a) $p(x) = -0.2x + 1100$ b) $E(x) = -0.2x^2 + 1100x$

     c) $G'(x) = -0.4x + 800$ d) discount $= \text{Fr. } 200$

     e) $G(x) = -0.2x^2 + 800x - 138'000$

169. a) Yes   b) Yes

     c) Yes   d) No

     e) infinitely many   f) e.g. $y'' + 9y = 0$, $y^{iv} - 81y = 0$ etc.

     g) e.g. $y_1(x) = x^2$ and $y_2(x) = x^2 + 100$; no   h) e.g. $y_1(x) = e^x$ and $y_2(x) = 100e^x$; no

170. a) $y' = -x \cdot y$

     b) $y' = \frac{2y}{x}$, solution fulfils the equation $y = ax^2, \ a \in \mathbb{R}$.

     c) $y' = \frac{y}{x}$, solution fulfils the equation $y = mx, \ m \in \mathbb{R}$.

     d) $y' = \frac{-x}{y}$, solution fulfils the equation $x^2 + y^2 = r^2, \ r \in \mathbb{R}^+$.

171. a) $P(t) = A \cdot e^t, \ A \in \mathbb{R}$ b) $P(t) = A \cdot e^{2t}, \ A \in \mathbb{R}$
     c) $P(t) = A \cdot 2^t, \ A \in \mathbb{R}$ d) Doubles per unit of time

172. a) $-$ b) $-$ c) $-$ d) $-$ e) $k = \pm 3$ or $k = 0$ f) $-$

173. equilibrium solutions: $y = 0, \ y = 2, \ y = -2$

174. a) iv   b) ii   c) iii   d) i

175. $-$

176. For $m > 0$, the isoclines are circles around $(0\,|\,0)$ with radius $\sqrt{m}$.

177. a) $-$
     b) $-$
     c) i) faster   ii) faster   iii) higher   iv) approaches   v) directly
     d) $\frac{dT}{dt} = k(T - U)$
     e) It is a differential equation.

178. Differentiate, then substitute into the differential equation and verify the initial condition (evaluate at $t = 0$).

179. a) $\frac{dT}{dt} = k(T(t) - U)$ b) $\approx 6.61 \min$

180. a) $I(t+1) - I(t) = k \cdot (N - I(t))$ b) $\frac{dI}{dt} = k \cdot (N - I(t))$ c) $\approx 23 \,\text{days}$

181. ≈ 15 days

182. a) $\approx 3.23 \cdot 10^7 \,\text{kg}$ b) $\approx 1.5 \,\text{years}$

183. a) $-$
     b) $-$
     c)

P_0	0	100	200	300	400	500
$P(t) = \dfrac{S}{1 + \frac{S - P_0}{P_0} \cdot e^{-kSt}}$	0	$\frac{500}{1 + 4 \cdot e^{-t}}$	$\frac{500}{1 + \frac{3}{2} \cdot e^{-t}}$	$\frac{500}{1 + \frac{2}{3} \cdot e^{-t}}$	$\frac{500}{1 + \frac{1}{4} \cdot e^{-t}}$	500

184. ≈ 6.55 million

185. a) $u = x^2; \ \sin(x^2) + C$ b) $u = x^3 + 1; \ \frac{2}{9}\sqrt{(1 + x^3)^3} + C$
     c) $u = 1 + x^2; \ \frac{1}{2}\ln(1 + x^2) + C$ d) $u = h(x); \ \ln|h(x)| + C$
     e) $u = \cos(x); \ \ln|\cos(x)| + C$ f) $u = \cos(x); \ -\ln|\cos(x)| + C$

186. a) $F(x) = \frac{1}{21}(3x - 5)^7$ b) $F(u) = \frac{5}{3}\ln|3u - 4|$
     c) $G(u) = 2\sqrt{u + 2}$ d) $F(u) = \frac{4}{5a}\sqrt[4]{(au + b)^5}$

187. a) $F(x) = \frac{1}{8}(x^2 + 1)^4$ b) $G(x) = \frac{1}{3}\sqrt{(x^2 + 1)^3}$
     c) $F(x) = -3\sqrt[3]{(1 - x^2)^2}$ d) $H(t) = \frac{3}{4} \cdot \ln(1 + t^4)$

188. a) $F(x) = e^{x^2}$ b) $G(x) = -\frac{a}{2b}e^{-bx^2}$
     c) $H(t) = \ln(|\ln(t)|)$ d) $G(u) = au$

189. a) $F(x) = \frac{-1}{3} \cdot \cos^3(t)$ b) $G(z) = \frac{1}{2} \cdot \sin^2(z)$ oder $G(z) = \frac{-1}{2} \cdot (\cos(z))^2$
         c) $F(x) = -2\cos(\sqrt{x})$

190. a) $\frac{1}{18}(3t^3 - 4)^2 + C$ b) $\frac{1}{2}\ln(|u^2 - 2u + 2|) + C$ c) $2\sqrt{x^2 - x - 1} + C$

191. a) $e^{x^2+x} + C$ b) $2 \cdot e^{\sqrt{z}-1} + C$ c) $\frac{1}{2}(\ln(t))^2 + C$

192. a) $\frac{3}{8}$ b) $\frac{-3}{2}\ln(2)$ c) $\frac{49}{3}$

193. a) $\frac{\sqrt{3}+1}{4}$ b) $\frac{1}{2}(\frac{1}{e^6} - 1)$ c) $1 - \frac{1}{e^2}$

194. i) False   ii) False   iii) True   iv) False

195. $\int\limits_{1}^{3} \frac{\sin(2x)}{x}\,\mathrm{d}x = F(6) - F(2)$

196. –

197. The areas of all three regions are equal.

198. i) False   ii) True

199. $a_1 = 1, \ a_2 = 6$

200. False

201. a) Integrate both sides of the product rule and then rearrange.
         b) $-x \cdot \cos(x) + \sin(x) + C$
         c) $\frac{1}{2}x^2 \cdot \sin(x) - \int \frac{1}{2}x^2 \cdot \cos(x)\,\mathrm{d}x$. The new integral is more complicated than the initial one.

202. a) i) $F(x) = \frac{1}{2}x^2 \cdot \ln(x) - \frac{1}{4}x^2$ ii) $F(x) = \frac{1}{3}x^3 \cdot \ln(x) - \frac{1}{9}x^3$ iii) $F(x) = \frac{1}{4}x^4 \cdot \ln(x) - \frac{1}{16}x^4$
         b) $F(x) = \frac{1}{n+1}x^{n+1} \cdot \ln(x) - \frac{1}{(n+1)^2}x^{n+1}$ for $n \in \mathbb{Z} \setminus \{-1\}$
         c) The formula in b) also holds for $n = 0$, but not for $n = -1$.

203. i) $(x-1)e^x$ ii) $(x^2 - 2x + 2)e^x$
         iii) $(x^3 - 3x^2 + 6x - 6)e^x$ iv) $(x^4 - 4x^3 + 12x^2 - 24x + 24)e^x$

204. a) $\frac{1}{2} \cdot e^x(\sin(x) - \cos(x))$ b) $\frac{1}{2} \cdot e^x(\sin(x) + \cos(x))$

205. a) 1   b) 2   c) $\frac{-1}{2}$

206. $A = \frac{1}{16}$

207. $A = e$

208. a) –
         b) $\int\limits_{2}^{3} \ln(x)\,\mathrm{d}x = 3\ln(3) - 2\ln(2) - 1; \ \int\limits_{0}^{3} \sqrt{x}\,\mathrm{d}x = 2\sqrt{3}$
         c) Yes, the formula also holds for strictly decreasing functions.

209. a) –

b) –

c) • Partition the curve into 100 segments by the points $P_0, \ldots, P_{100}$.

• Measure the segments $\Delta s_k = \overline{P_k P_{k+1}}$, $k = 0, \ldots, 99$.

• total arc length $\hat{=} \sum\limits_{k=0}^{99} \Delta s_k$. Divide by a 100 to obtain a further measurement of b.

d) i) –   ii) –   iii) $L = \int\limits_a^b \sqrt{1 + (f'(x))^2}\,\mathrm{d}x$

e) $b = \frac{10\pi}{3} \approx 10.47\,\mathrm{cm}$. The values b_1 and $\frac{b_2}{2}$ should be close to the value b.

f) i) $\alpha = 60°$ ii) $b = \frac{10\pi}{3}$ iii) As in e).

210. a) –   b) $L = \frac{2(11\sqrt{22}-4)}{27} \approx 3.53$ c) –

211. a) $L = \frac{14}{3}$ b) $L = \mathrm{e}^3 - \frac{1}{\mathrm{e}^3}$

212. $w \approx 73.40\,\mathrm{cm}$

213. a) $f(x) = \frac{1}{3}x^3 + \frac{1}{2}x^2 + x + \frac{1}{3}$ b) $f(x) = \frac{1}{2}x^3 + \frac{5}{2}x^2 - 5x - 78$

c) $f(x) = 4x^2\sqrt{x} + 4x\sqrt{x} - 4x + 5$ d) $f(x) = \frac{1}{3}x^3 - \frac{16}{3}x + 5$

e) $f(x) = \frac{2}{x^2} - \frac{3}{2x} - \frac{x}{54} + \frac{1}{3}$

214. The results are equivalent, apart from the constant of integration.

215. No

216. –

217. All of them yield an antiderivative of f.

218. $F(x) = \begin{cases} \frac{1}{4}x^2, & 0 \le x \le 2 \\ x^2 - 3x + 3, & x > 2 \end{cases}$

219. a) $s = \frac{3}{2}t^2 - 4t$ b) $s = \frac{a}{2}t^2 + v_0 t$ c) $s = 7t$ d) $s = \frac{1}{3}t^3 - \frac{3}{2}t^2$

220. a) $v(t) = 9.81t + 3$ b) $v(t) = 4$ c) $v(t) = 0.75 \cdot t^2$ d) $v(t) = -\frac{3}{8}t^2 + 82$

$s(8) = 337.92$ $s(11) = 44$ $s(20) = 2000$ $s(4) = 320$

221. a) $O_n - U_n = \frac{b-a}{n} \cdot (f(b) - f(a))$

b) $n \ge 2 \cdot (b-a) \cdot (f(b) - f(a))$

c) 0

222. a) $\int\limits_a^b f(x)\,\mathrm{d}x$ b) e.g. f with $f(x) = mx + q\ (m, q > 0)$

223. a) True   b) False   c) False   d) False

224. a) False   b) True   c) False

225. f is odd, so $f(-x) = -f(x)$ for all x.

226. $\int_0^1 f'(x)\,\mathrm{d}x = 1$

227. Equations ii) and iii)

228. a) $k = \frac{2}{9}$ b) $k_1 = \frac{\pi}{6} + 2\pi n,\ k_2 = \frac{5\pi}{6} + 2\pi n, n \in \mathbb{Z}$

      c) $k = (2n+1)\pi,\ n \in \mathbb{Z}$ d) $k_{1,2} = \pm e^4$

      e) $k_1 = 0,\ k_2 = \ln(2)$

229. –

230. –

231. a) maximum point at $t_1 = 0$, minimum points at $t_2 = 1$ and $t_3 = -1$

      b) minimum boundary point at $x = -1$

      c) saddle points at $x_{1,2} = 0$, minimum point at $x_3 = \frac{3}{2}$

      d) minimum point at $u = 1$

232. a) True   b) False

233. a) $\frac{2}{3}$ b) $\frac{2}{3}$ c) 1   d) $\frac{1}{27}$

234. a) 2   b) 0   c) 2   d) 2

235. a) 4   b) 0   c) -4 d) 4

236. The area of the region between the graph of $y = e^x$ and the x-axis over the interval $[r; r+3]$ is e^r-times as large as the corresponding region over the interval $[0; 3]$.

237. a) -3 b) 3   c) 0

238. term i)

239. a) $g\colon y = \frac{1}{2}x$ b) $A = \frac{37}{4} = 9.25$

240. a) $m \approx 1.3\,\mathrm{t}$ b) $m \approx 222.3\,\mathrm{t}$

241. $A = \frac{24}{5}\sqrt{3}$

242. a) $A = \frac{2}{3}\pi - 2 \approx 0.094$ b) $A = \frac{1}{30}\pi^2 + \frac{8}{15}\pi - 2 \approx 0.005$

243. a) $\theta(x) = \arccos\left(\dfrac{\left(2.7+x\cos\left(\frac{\pi}{9}\right)\right)^2 + \left(9.3-x\sin\left(\frac{\pi}{9}\right)\right)^2 + \left(2.7+x\cos\left(\frac{\pi}{9}\right)\right)^2 + \left(1.8-x\sin\left(\frac{\pi}{9}\right)\right)^2 - 56.25}{2\cdot\sqrt{\left(2.7+x\cos\left(\frac{\pi}{9}\right)\right)^2 + \left(9.3-x\sin\left(\frac{\pi}{9}\right)\right)^2}\cdot\sqrt{\left(2.7+x\cos\left(\frac{\pi}{9}\right)\right)^2 + \left(1.8-x\sin\left(\frac{\pi}{9}\right)\right)^2}} \right)$

      b) maximum angle $\approx 48.5°$ at $x \approx 2.5\,\mathrm{m}$, minimum angle $\approx 21.6°$ at $x = 18\,\mathrm{m}$

      c) maximum angle $\approx 48.5°$

      d) mean angle $\theta_{\mathrm{mittel}} \approx 35.8°$, so $21.6° \leq \theta_{\mathrm{mittel}} \leq 48.5°$

244. $\int_0^r u(t)\,\mathrm{d}t = \pi r^2 \mathrel{\widehat{=}}$ area of circle; $\int_0^r A(t)\,\mathrm{d}t = \frac{4\pi}{3}r^3 \mathrel{\widehat{=}}$ volume of sphere

245. a) Rotation around the x-axis:

n	5	10	20	40	80
U_n	$\frac{2\pi}{5} \approx 1.26$	$\frac{9\pi}{20} \approx 1.41$	$\frac{19\pi}{40} \approx 1.49$	$\frac{39\pi}{80} \approx 1.53$	$\frac{79\pi}{160} \approx 1.55$
O_n	$\frac{3\pi}{5} \approx 1.88$	$\frac{11\pi}{20} \approx 1.73$	$\frac{21\pi}{40} \approx 1.65$	$\frac{41\pi}{80} \approx 1.61$	$\frac{81\pi}{160} \approx 1.59$

Rotation around the y-axis:

n	5	10	20	40	80
U_n	$\frac{354\pi}{3125} \approx 0.36$	$\frac{15'333\pi}{100'000} \approx 0.48$	$\frac{281'333\pi}{1'600'000} \approx 0.55$	$\frac{4'805'333\pi}{25'600'000} \approx 0.59$	$\frac{79'381'333\pi}{409'600'000} \approx 0.61$
O_n	$\frac{979\pi}{3125} \approx 0.98$	$\frac{25'333\pi}{100'000} \approx 0.80$	$\frac{361'333\pi}{1'600'000} \approx 0.71$	$\frac{5'445'333\pi}{25'600'000} \approx 0.67$	$\frac{84'501'333\pi}{409'600'000} \approx 0.65$

b) The volumes are different: $V_x = \frac{\pi}{2} \neq V_y = \frac{\pi}{5}$.

246. $13 - 6 \cdot \ln(3)$

247. shift of $a = \sqrt{2}$; area $A = \sqrt{2} \cdot 2\pi$

248. $A = \frac{4}{3}\sqrt{a} \cdot a$, independent of u

249. $b_1 = 1$ (trivial solution), $b_{2,3} = 1 \pm \sqrt{3}$

250. a) $V = \pi \int\limits_{-a}^{a} (b^2 - \frac{b^2}{a^2}x^2)\,\mathrm{d}x$, $V = \pi \int\limits_{-b}^{b} (a^2 - \frac{a^2}{b^2}y^2)\,\mathrm{d}y$

b) $V = \int\limits_{0}^{H} (\frac{x}{H})^2 \cdot G\,\mathrm{d}x$, $V = \int\limits_{a}^{a+h} (\frac{x}{H})^2 \cdot G\,\mathrm{d}x$, $a = \frac{\sqrt{D}}{\sqrt{G}-\sqrt{D}} \cdot h$, $a+h = \frac{\sqrt{G}}{\sqrt{G}-\sqrt{D}} \cdot h$

c) $V = \pi \int\limits_{-r}^{r} ((R + \sqrt{r^2 - x^2})^2 - (R - \sqrt{r^2 - x^2})^2)\,\mathrm{d}x$

251. $\frac{15}{4}$

252. $A = \frac{1}{2}$ for all $a \in \mathbb{R}$

253. a) $\int\limits_{0}^{1} \ln(x)\,\mathrm{d}x = -1$ b) $\lim\limits_{x \to 0^+} (x \ln(x)) = 0$

254. a) $12.4\,°C$ b) ≈ 94.5 minutes

255. a) $\frac{\mathrm{d}K}{\mathrm{d}t} = 0.075 \cdot (20 - K(t))$ b) $K(t) = 20 - 10 \cdot \mathrm{e}^{-0.075t}$ c) $\lim\limits_{t \to \infty} K(t) = 20\,\mathrm{mg}$

256. a) $F(x) = \frac{1}{3}x^3$ b) $F(x) = x^4 - x^2 + x$ c) $F(x) = \frac{1}{n+1} \cdot x^{n+1}$ d) $F(x) = 2\ln|x|$
e) $F(t) = 2 \cdot \sin(t)$ f) $F(t) = \pi \cdot \mathrm{e}^t$ g) $F(t) = \frac{1}{\ln(3)} \cdot 3^t$ h) $F(t) = t^{\frac{3}{2}}$

257. a) $f(x) = x^3 - 41x + 9$ b) $f(x) = 3x + 2 - \frac{1}{x}$
c) $f(t) = 2 - 2 \cdot \cos(t)$ d) $f(t) = \frac{1}{2}t^2 + 4t - 8$

258. a) $3x^4 + C$ b) $14x - x^3 + C$ c) $x^4 - x^3 + 2x^2 - 3x + C$
d) $\frac{-2}{x} + C$ e) $\frac{-1}{x^3} + \frac{1}{x^2} + C$ f) $\frac{1}{3}x^3 + 4x - 5\ln|x| + C$
g) $2t\sqrt{t} + C$ h) $\sin(t) - \cos(t) + C$ i) $\frac{1}{\ln(12)} \cdot 12^t + C$

259. a) True b) True

260. a) True　　　　　　　b) True　　　　　　　c) False

261. –

262. a) $F(x) = \frac{1}{12} \cdot (4x + 3)^3$　　　b) $F(x) = \frac{-2}{5}(3 - \frac{1}{2}x)^5$　　　c) $F(x) = \frac{-1}{5(5x+43)}$

d) $F(x) = \frac{3}{2} \cdot (3 - \frac{x}{3})^{-2}$　　　e) $F(t) = \frac{-1}{7} \cdot \cos(21t)$　　　f) $F(t) = 3 \cdot \sin(\frac{t}{3})$

g) $F(x) = -e^{-2x}$　　　　h) $F(x) = \frac{1}{3} \ln|3x - 2|$

263. $U_6 = 13$, $U_{12} = 15.625$ and $O_6 = 22$, $O_{12} = 20.125$

Statement: $U_6 < U_{12} \leq A \leq O_{12} < O_6$ or, numerically, $15.625 \leq A \leq 20.125$

264. i) 3　　　　　　ii) $3\sqrt{5}$　　　　　　iii) -3　　　　　　iv) -3

265. a) 12　　　　　　b) 10　　　　　　c) -10　　　　　　d) -7

266. $F_0(x) = \frac{x^2}{2} + 2x$ for all $x \in \mathbb{R}$

267. a) 170　　　　　b) 54　　　　　c) $\frac{13}{2} = 6.5$　　　　　d) 0

268. a) 3　　　　　b) 10　　　　　c) $2\sqrt{3} \approx 3.46$

269. a) $e^\pi - 1$　　　　b) $e^2 - 1 \approx 6.39$　　　　c) $\frac{6}{\ln(3)} \approx 5.46$　　　　d) 1

270. a) Yes　　　　b) Yes　　　　c) No　　　　d) Yes

271. a) False　　　　b) True　　　　c) False

272. a) True　　　　b) True　　　　c) True　　　　d) False

273. a) A, like F, is an antiderivative of f, with the condition $A(0) = 0$.

b) $\int (3x^2 - 4x + 10)\,dx = x^3 - 2x^2 + 10x + C$

274. $a = -6$, $a = 4$

275. –

276. formula i)

277. Both the integrand and the differential are non-positive $\Rightarrow$ $- \cdot - = +$.

278. a) for $a = k \cdot \pi$ with $k \in \mathbb{Z}$　　　　　b) for all $a \in \mathbb{R}$

279. –

280. maximum point $H(-1\,|\,2)$, minimum point $T(1\,|\,-2)$

281. a) $\frac{40}{3} = 13.\overline{3}$　　　　b) $\frac{27}{4} = 6.75$　　　　c) $\frac{131}{4} = 32.75$

282. It is the lead of Susanne over Katharina after the first minute.

283. a) $V = \frac{8}{3} \cdot \pi$　　　　b) $V = 32\pi$　　　　c) $V = \frac{8}{3} \cdot \pi$　　　　d) $V = 32\pi$

284. $c = \pm\frac{1}{3}$

285. Lili is right.

286. a) $\overline{f} = 3$; $x_M = \sqrt{3} \approx 1.73$ b) $\overline{f} = \frac{2}{\pi} \approx 0.64$; $t_M = \arcsin\left(\frac{2}{\pi}\right) \approx 0.69$

287. a) 3   b) 2   c) $\frac{1}{8}$

288. $A = \frac{3}{8}$

289. $V = \frac{3\pi}{4}$

290. –

291. –

292. a) exponential growth: $\frac{\mathrm{d}P}{\mathrm{d}t} = k \cdot P(t)$; the relative growth rate $\frac{P'(t)}{P(t)} = k$.

b) Suitable for populations with unbounded resources (food and habitat).

c) The solutions are given by $P(t) = P_0 \cdot \mathrm{e}^{kt}$, where P_0 is the initial population.

293. a) differential equation for logistic growth: $\frac{\mathrm{d}P}{\mathrm{d}t} = k \cdot P(t) \cdot (S - P(t))$

b) Suitable for populations with limited resources (food and habitat).

294. a) True   b) True

X Functions

1. a) $0, 25, \frac{1}{25}, 2$ b) $2, \frac{3}{4}, \frac{41}{20}, \frac{-\sqrt{2}}{4} + 2$ c) $1, \frac{-1}{624}, \frac{625}{624}, \frac{-1}{3}$

d) $1, 4, \frac{\sqrt{10}}{5}, \sqrt{1 + 3\sqrt{2}}$ e) $0, \ln(6), \ln\left(\frac{4}{5}\right), \ln\left(\sqrt{2} + 1\right)$ or $0, 2.77, -0.22, 0.88$

f) $\frac{1}{243}, 1, 3^{-\frac{26}{5}}, 3^{\sqrt{2}-5}$ or $0.0041, 1, 0.0033, 0.0195$

2. a) 0.1   b) 250

3. a) $f(2) = \frac{1}{2}$ b) $f\left(\frac{1}{3}\right) = 3$ c) $f(0.2) = 5$ d) $f(2.5 \cdot 10^{12})$ $= 4 \cdot 10^{-13}$

e) $f\left(\frac{3}{2}\right) = \frac{2}{3}$ f) $f(4) = 0.25$ g) $f\left(-\frac{4}{7}\right) = \frac{-7}{4}$ h) $f(-2 \cdot 10^{-7})$ $= -5 \cdot 10^6$

4. a) $f\left(\frac{p}{q}\right) = \frac{q}{p}$ b) $f\left(\frac{n}{2}\right) = \frac{2}{n}$ c) $f(f(8)) = 8$ d) $f\left(f\left(-\frac{1}{5}\right)\right) = -\frac{1}{5}$

5. a) 25   b) a^2 c) $\frac{a^2}{b^2}$ d) $4a^2$

e) $(n + 2)^2$ f) 32   g) $2 \cdot x^4$ h) 7

i) $m^2 + 3$ j) $x^2 - 20x$ k) 99   l) $a \cdot (x - c)^2 + d$

6. a) $m - 1, 2x - 1, -x - 1, x^2 - 1, x, h, \frac{1-x}{x}, \sqrt{x} - 1$

b) $m^2, 4x^2, x^2, x^4, x^2 + 2x + 1, 2xh + h^2, \frac{1}{x^2}, x$

c) $m^2 - 1, 4x^2 - 1, x^2 - 1, x^4 - 1, x^2 + 2x, 2xh + h^2, \frac{1-x^2}{x^2}, x - 1$

d) $1 + \frac{1}{m}, 1 + \frac{1}{2x}, 1 - \frac{1}{x}, 1 + \frac{1}{x^2}, 1 + \frac{1}{x+1}, \frac{1}{x+h} - \frac{1}{x}, 1 + x, 1 + \frac{\sqrt{x}}{x}$

7. a)

x	-2	-1	0	1	2.2	10
$f(x)$	-6	-3	0	3	6.6	30

b)

x	-6	-3	0	1.7	3	6
$f(x)$	-4	-2	0	$\frac{17}{15}$	2	4

c)

x	-1	-0.5	0	1	1.5	2
$f(x)$	-9	-4	0	5	6	6

d)

x	-4	-2	0	1	2	3
$f(x)$	0	0	0	2	4	6

e)

x	-1	0	2	3	6	8
$f(x)$	0	1	$\sqrt{3}$	2	$\sqrt{7}$	3

8. –

9. –

10. a) $(0\,|\,{-2})$; $\left(-\frac{2}{9}\,\middle|\,0\right)$ b) $(0\,|\,{-25})$; $(\pm 5\,|\,0)$ c) $(0\,|\,1)$
d) $(7\,|\,0)$ e) $(0\,|\,\ln(2))$; $(-1\,|\,0)$ f) $(0\,|\,2)$

11. a) $\left(\frac{5}{3}\,\middle|\,\frac{7}{3}\right)$ b) $\left(-\frac{9}{5}\,\middle|\,-\frac{22}{5}\right)$, $(1\,|\,4)$
c) $(0\,|\,8)$ d) $(-1\,|\,3)$, $\left(\frac{9}{10}\,\middle|\,\frac{131}{50}\right)$
e) $\left(\frac{\sqrt{11}+4}{2}\,\middle|\,\frac{\sqrt{11}+1}{2}\right) \approx (3.66\,|\,2.16)$ f) $x = \frac{\lg(4)}{2\lg(5)-\lg(2)}$ $\Rightarrow$ $(0.55\,|\,5.85)$

12. a) $a = \frac{1}{4}$ b) $a = -20$ c) $a \in \mathbb{R}$ arbitrary
d) $a \geq 0$ (a is the base of a power with real exponent)   e) $a = 5$
f) $a = 0$

13. -71

14. $a = 1$, $a = 2$

15. a) $D = \mathbb{R}$ b) $D = [3;\infty[$ c) $D = \mathbb{R}\backslash\{-3, 3\}$ d) $D = \mathbb{R}$

16. a) $W = {]-\infty; 4]}$ b) $W = [-5; 4]$ c) $W = {]-\infty; -12]}$

17. a) $W = {]-2;\infty[}$ b) $W = {]\frac{-17}{9}; 2]}$ c) $W = {]-2; 1]}$

18. a) $[4; 25]$ b) $[0; 16]$ c) $[0; 16]$ d) $[0; 9]$ e) $[0; 9[$ f) $\mathbb{R}_0^+$

19. a) $D = \mathbb{R}$, $W = [0;\infty[$ b) $D = \mathbb{R}$, $W = \mathbb{R}$
c) $D = \mathbb{R}\backslash\{-1, 1\}$, $W = \mathbb{R}\backslash[0; 1[$ d) $D = [-\frac{1}{3};\infty[$, $W = [0;\infty[$
e) $D = {]-1;\infty[}$, $W = \mathbb{R}$ f) $D = \mathbb{R}$, $W = \mathbb{R}^+$

20. Possible solutions:
a) $f(x) = \sqrt{x}$ b) $f(x) = \sqrt{x-3}$ c) $f(x) = x^2$ d) $f(x) = \frac{1}{(x+1)(x-2)}$

21. a) 4 b) 9 c) any prime d) 1

e) 7 f) 4 g) 4 h) (a prime)^2

i) $D = \mathbb{N}$ j) –

22. a) 3, 0, 2, 3 b) $W_d = \{0, 1, 2, 3\}$ c) $d(n)$, 3, 1 d) $s(n) = 6$

23. a) $C \in [-273.15; \infty[$ b) $F \in [-459.67; \infty[$

c) increase in Fahrenheit per degree Celsius d) temperature in Fahrenheit at $0\,°$Celsius

e) $C = \frac{5}{9}(F - 32)$

24. a)

t	0	0.5	1	1.5	2
$h(t)$	1	4.5	5.5	4	0

b) – c) after 2 seconds d) $D = [0; 2]$, $W = [0; 5.5125]$

25. a) $6.9\,°$C, $13.8\,°$C, $20.9\,°$C, $15.3\,°$C b) $13\,$h, $19\,$h

c) $[12\,\text{h}; 21\,\text{h}]$ d) $[0\,\text{h}; 12\,\text{h}]$ and $[21\,\text{h}; 24\,\text{h}]$

e) $D = [0\,\text{h}; 24\,\text{h}]$, $W = [3.2\,°\text{C}; 21.2\,°\text{C}]$

26. a) $G(d) = 95 - 0.7d$ b) $53\,$kg c) 30 tins d) for 108 tins

27. 1d, 2b, 3c, 4a

28. –

29. –

30. $d(t) = \sqrt{61} \cdot t$

31. $\frac{\pi}{12}h^3$

32. a) $y = 5x$ b) $y = 6x + 2$ c) $y = \frac{5}{6}x - \frac{5}{2}$ d) $y = \frac{8}{3}x$

e) For example, place the vertex at $(-3\,|-8)$. Setting $y = f(x) = a(x + 3)^2 - 8$ and $f(3) = 8$ then yields $y = \frac{4}{9}(x + 3)^2 - 8$.

33. a) No b) No

34. –

35. –

36. –

37. –

38. a) Strictly increasing on $\left[-\frac{\pi}{2} + k \cdot 2\pi; \frac{\pi}{2} + k \cdot 2\pi\right]$, $k \in \mathbb{Z}$; strictly decreasing on $\left[\frac{\pi}{2} + k \cdot 2\pi; \frac{3\pi}{2} + k \cdot 2\pi\right]$, $k \in \mathbb{Z}$.

b) Strictly decreasing on $]-\infty; 2]$ and strictly increasing on $[2; \infty[$.

c) Monotonically increasing on $\mathbb{R}$.

39. a) –   b) –   c) e.g. $f\colon x \mapsto y = \sin^2(x)$

40. The function is continuous in the following intervals: $[-5; -2[,\ ]-2; 2[,\ ]2; 4[,\ ]4; 6[,\ ]6; 8[$.

41. a) continuous   b) continuous   c) discontinuous   d) discontinuous

42. a) continuous   b) discontinuous   c) continuous

43. True

44. a) discontinuous   b) continuous   c) continuous   d) discontinuous

45. a) $x = 0$ b) none   c) $x = \pm 1$

46. Yes

47. $c = \frac{2}{3}$

48. a) $m(x) = 3x + 2$ b) $m(x) = 2x - x^2$ c) $m(x) = (x + 2) \cdot \frac{1}{x} = 1 + \frac{2}{x}$
     d) $m(x) = x^2 : \frac{1}{x} = x^2 \cdot x = x^3$ e) $m(x) = 2x^2$ f) $m(x) = (2x)^2 = 4x^2$

49. a) $(2x + 7)^2$; $2x^2 + 7$ b) $\sqrt{3 - 4x}$; $3 - 4\sqrt{x}$ c) $\frac{2}{x^2 + 3}$; $\frac{4}{x^2} + 3$

50. a) $1 - x$ b) $2x^2 - x^4$ c) $\sin^2(x)$ d) $\cos(1 - x^2)$
     e) $\sin(x)$ f) $\cos(1 - x)$ g) $2x - x^2$ h) $1 - \sin^4(x)$

51. The solutions are not unique.
     a) $f(x) = x^3$, $g(x) = 3x + 10$ b) $f(x) = \frac{1}{x}$, $g(x) = 2x + 4$
     c) $f(x) = \sqrt{x}$, $g(x) = x^2 - 3x$ d) $f(x) = 6 \cdot \sin(x)$, $g(x) = 1 - x$
     e) $f(x) = \frac{10}{x^3}$, $g(x) = 3x - x^2$ f) $f(x) = \cos(x)$, $g(x) = \sqrt{x} + 1$
     g) $f(x) = x^5$, $g(x) = \tan(x)$ h) $f(x) = 8 \cdot e^x$, $g(x) = \sqrt{x}$
     i) $f(x) = 5 \cdot \ln(x) - 14$, $g(x) = \tan(x)$

52. –

53. a) $g(x) = 4x + 4$ b) $g(x) = \frac{1}{2}(x - 1)^2$
     c) $g(x) = \sqrt{-\frac{x}{4}} + 9$ or $g(x) = -\sqrt{-\frac{x}{4}} + 9$ d) $g(x) = \frac{1}{6}\left(e^{\frac{x-1}{10}} - 13\right)$

54. a) e.g. $g(x) = x^2 - 1$, $h(x) = x^3$, $f(x) = h(g(x))$
     b) e.g. $g(x) = x^2$, $h(x) = x - 1$, $i(x) = x^3$, $f(x) = i(h(g(x)))$
     c) e.g. $g(x) = x^2$, $h(x) = x - 1$, $i(x) = x^{\frac{1}{3}}$, $j(x) = x^9$, $f(x) = j(i(h(g(x))))$

55. Composition with t_1 shifts the graph to the right by 1.5 units, i.e. in the direction of the positive x-axis, and composition with t_2 scales it by a factor of 2 along the direction of the x-axis.

56. a) $D = \mathbb{R}_0^+$ and $W = \mathbb{R}_0^+$ b) $D = \mathbb{R}\backslash\{0\}$ and $W = \mathbb{R}\backslash\{0\}$
     c) $f(g(x)) = \sqrt{\frac{1}{x}} \Rightarrow D = \mathbb{R}^+$ and $W = \mathbb{R}^+$

57. –

58. –

59. a) $x - 12, \quad x - 4$ b) $9x - 36, \quad 9x - 12$
   c) $\frac{x-12}{9}, \quad \frac{x-36}{9}$ d) 0

60. a) $P_1\left(2 \mid \frac{1}{3}\right), P_2\left(\frac{1}{3} \mid \frac{1}{3}\right), P_3\left(\frac{1}{3} \mid \frac{3}{4}\right), P_4\left(\frac{3}{4} \mid \frac{3}{4}\right), P_5\left(\frac{3}{4} \mid \frac{4}{7}\right)$
   b) $P_1(x_0 \mid f(x_0)), P_2(f(x_0) \mid f(x_0)), P_3(f(x_0) \mid f(f(x_0))), P_4(f(f(x_0)) \mid f(f(x_0))),$
     $P_5(f(f(x_0)) \mid f(f(f(x_0))))$
   c) –

61. $-12, -8, -4, 0$

62. $\frac{-1}{1999}$

63. a) Translation by -3 in direction of the y-axis   b) Translation by 3 in direction of the x-axis
     Translation by 3 in direction of the y-axis   Translation by -3 in direction of the x-axis

   c) Reflection at the x-axis   d) Scaled by the factor 3 in direction of the
     Reflection at the y-axis y-axis
     Scaled by the factor $\frac{1}{3}$ in direction of the x-axis

64. a) Shortend by a factor 3 in direction of the y-axis
   b) Translated by 5 to the left
   c) Translated upward by 3
   d) Stretched by a factor 3 in direction of the x-axis
   e) Reflected across the y-axis
   f) Reflected across the x-axis and shortened by a factor 2 in direction of the x-axis

65. • Translation by 2 in direction of the x-axis
   • Reflection across the x-axis
   • Stretched by the factor 3 in direction of the y-axis
   • Translation by 7 in direction of the y-axis

66. –

67. a) $w(x+6) = \sqrt{(x+6) - 8} = \sqrt{x-2}$ b) $w(-x) + 1 = \sqrt{-x-8} + 1$
   c) $2.5 \cdot w(x) - 3 = 2.5\sqrt{x-8} - 3$ d) $-w(8x) = -\sqrt{8x-8}$

68. $-\sqrt{\frac{1}{2}x}$

69. As $p(x) = q(4x+1)$, the graph is
   • first shifted left 1 unit
   • and then stretched by the factor $\frac{1}{4}$ in direction of the x-axis.

70. a) Shifted left by 3 units, strechted by the factor 2 in direction of the y-axis and reflected across the x-axis, shifted up by 5 units

b) Shifted right by 2 units, strechted by the factor 3 in direction of the x-axis, reflected across the x-axis

71. –

72. –

73. a) $y = x^3$ b) $y = -x^3$ c) $y = -x^3$ d) $y = \begin{cases} -\sqrt[3]{-x} = -\sqrt[3]{|x|}, & \text{for } x < 0 \\ \sqrt[3]{x}, & \text{for } x \geq 0 \end{cases}$

74. a) $f(x) = \frac{1}{2}(x + 1)^2$ b) $f(x) = -(x - 3)^3 + 2$

c) $f(x) = \frac{1}{4}x^4 - 4$ d) $f(x) = 3 \cdot \sqrt{-x + 3}$

e) $f(x) = 2 \cdot 2^x - 5$ f) $f(x) = -\ln(x + 4)$

75. –

76. a) even   b) even   c) neither

d) odd   e) neither   f) odd

g) even   h) even   i) odd

77. –

78. –

79. For $x < 0$, we have:

a) $y = \sqrt{-x}$ b) $y = -\sqrt{-x}$

80. a) even   b) even   c) even   d) odd   e) neither   f) odd

81. f_1 even, f_2 odd

82. $y = f(x) = 0$

83. –

84. Mia is right.

85. –

86. $g(x) = \frac{x^2}{x^2 - 1}$ and $u(x) = \frac{x^3}{x^2 - 1}$, where $x \neq \pm 1$.

87. a) odd, period 2π b) odd, aperiodic

c) even, period π d) neither even nor odd, period π

e) odd, period 2π

88. a) $T = 3\pi$ b) $T = \frac{2\pi}{3}$ c) $T = \pi$ d) No T exists.

89. $f(2021) = f(7 \cdot 288 + 5) = f(5) = f(-2) = (-2)^3 = -8$

90. $a = \frac{6-9}{2} = -\frac{3}{2}$, $b = \frac{2\pi}{12} = \frac{\pi}{6}$, $c = 3$, $d = \frac{9+6}{2} = 7.5$

91. $h(t) = 30 - 25 \cdot \cos\left(\frac{2\pi}{8} \cdot t\right) = 30 - 25\cos\left(\frac{\pi}{4} \cdot t\right)$ (time t in minutes)

92. a) $y(t) = 5 \cdot \sin\left(\pi\left(t + \frac{1}{2}\right)\right) = 5\sin\left(\pi t + \frac{\pi}{2}\right) = 5\cos(\pi t)$

    b) after $\frac{3}{2} = 1.5$ seconds   c) Yes, as $5\cos\left(\frac{7}{4}\pi\right) = 5 \cdot \frac{\sqrt{2}}{2} > \frac{5}{2}$

93. a) 14 solutions   b) 6 solutions   c) 24 solutions   d) 18 solutions

94. $g(x) = 2\cos\left(3x - 4 + \frac{\pi}{2}\right)$

95. $g(x) = f(ax) = f(ax + T) = f\left(a\left(x + \frac{T}{a}\right)\right)$

96. –

97. a) $V(h) = \frac{\pi}{4}h \approx 0.785h$; $h(V) = \frac{4}{\pi}V \approx 1.27V$

    b) $V(r) = \frac{\pi}{3}r^2 \approx 1.05r^2$; $r(V) = \sqrt{\frac{3V}{\pi}} \approx 0.977\sqrt{V}$

    c) $h(r) = \frac{100}{\pi r^2} \approx \frac{31.8}{r^2}$; $r(h) = \frac{10}{\sqrt{\pi h}} \approx \frac{5.64}{\sqrt{h}}$

98. a) invertible: b) not invertible   c) invertible:

    $f^{-1}(x) = -\frac{2}{3}x$

$$f^{-1}(x) = \begin{cases} \sqrt[3]{x}, & x \geq 0 \\ -\sqrt[3]{|x|}, & x < 0 \end{cases}$$

    d) invertible: e) not invertible   f) invertible:

    $f^{-1}(x) = x^2$, $x \geq 0$ $f^{-1}(x) = f(x) = \frac{1}{3x}$

99. a) $f^{-1}(x) = 4x$ b) $f^{-1}(x) = \sqrt[3]{x}$ c) $f^{-1}(x) = \frac{1}{3}x - \frac{2}{3}$

    d) $f^{-1}(x) = 1 - x$ e) f is not invertible.   f) $f^{-1}(x) = -\sqrt{x}$, $x \geq 0$

100. –

101. The functions are invertible on any subsets of the domains given below.

    a) $D_f = [0; \infty[\;\Rightarrow\; f^{-1}(x) = \sqrt{x}$ or $D_f =]-\infty; 0] \;\Rightarrow\; f^{-1}(x) = -\sqrt{x}$

    b) $D_f = [2; \infty[\;\Rightarrow\; f^{-1}(x) = 2 + \sqrt{x}$ or $D_f =]-\infty; 2] \;\Rightarrow\; f^{-1}(x) = 2 - \sqrt{x}$

    c) $D_f = \mathbb{R} \;\Rightarrow\; f^{-1}(x) = \frac{3}{2}x + \frac{1}{2}$

    d) $D_f = [4; \infty[\;\Rightarrow\; f^{-1}(x) = x^2 + 4$

    e) $D_f = \mathbb{R}^+ \;\Rightarrow\; f^{-1}(x) = \mathrm{e}^x$

    f) $D_f = \mathbb{R} \;\Rightarrow\; f^{-1}(x) = \log_2(x)$

    g) $D_f = [0; \infty[\;\Rightarrow\; f^{-1}(x) = x$ or $D_f =]-\infty; 0] \;\Rightarrow\; f^{-1}(x) = -x$

    h) $D_f = \left[-\frac{\pi}{2}; \frac{\pi}{2}\right] \;\Rightarrow\; f^{-1}(x) = \arcsin(x)$

102. $D_1 = \{1, 2, 3, 4\}$, $D_2 = \{99, 100, 101, 102\}$, in general for $n \in \mathbb{N}$: $D = \{n, n+1, n+2, n+3\}$.

103. –

104. a) $w(x) = 12x + 1$ b) $w^{-1}(x) = \frac{x-1}{12}$ c) $w^{-1}(x) = v^{-1}(u^{-1}(x))$

105. $w^{-1} \circ v^{-1} \circ u^{-1}$

106. Examples: $f: y = x$, $f: y = \frac{1}{x}$ on $]0; \infty[$, $f: y = -x + c$, $f: y = \sqrt{1 - x^2}$ on $[0; 1]$.

107. a) $(2 \mid 2)$ b) $\left(\frac{-1+\sqrt{5}}{2} \mid \frac{-1-\sqrt{5}}{2}\right)$ and $\left(\frac{-1-\sqrt{5}}{2} \mid \frac{-1+\sqrt{5}}{2}\right)$ c) $(1 \mid 1)$.

108. a) False   b) False   c) False

109. a) Yes; $f(x) = -\frac{1}{2}x^4 - 2x^2 + x - \frac{9}{7}$ b) Yes; $f(x) = \frac{5}{2}x^2 + x$ c) No

     d) Yes; $f(x) = 4$ e) Yes; $f(x) = 4x^2 + 4 \cdot \sqrt{2} \cdot x + 2$ f) No

     g) No   h) Yes; $f(x) = x^2 - 1$ i) No

110. a) $Q(x) = 8x^3 + 4x^2 + 2x + 1$ b) $Q(x) = -x^3 + x^2 - x + 1$

     c) $Q(x) = x^6 + x^4 + x^2 + 1$ d) $Q(x) = x^3 - 2x^2 + 2x$

111. a) 7   b) 6   c) 14   d) 8   e) 5

112. a) 24   b) 12   c) 12   d) 9

113. $P \cdot Q$ has degree $m + n$ and the degree of $P + Q$ is at most as large as the larger of the two numbers m and n.

114. a) $n = 1$ and $n = 4$ b) $n = 3$

115. a) $a = 0$, even   b) $a = 0$, odd

     c) $a = -8$, even   d) a can be any odd number, odd.

116. a) $1, 0, -1, 0, 3, 0, 0$ b) $x^4 - 1$

117. a) $1, 0, -3, 0$ b) $1, -1, -3, 0, 0, 0$

     c) $1, 0, -1, 0, -3, 0, 0$ d) $1, -2, -5, 6, 9, 0, 0$

118. a) $p(x) = 2x^2 - 7x + 1$ b) $p(x) = -\frac{1}{2}x^2 + \frac{5}{2}x$

     c) $p(x) = \frac{1}{2}x^3 - x^2$ d) $p(x) = x^3 + x^2 - 4x + 2$

119. Type 1: $f(x) = 4x^{12} - 32x$, Type 2: $f(x) = -x^8 + x$, Type 3: $f(x) = x^5 - 2x^4 + 56x^3 + 2x^2 + 2x + 2$, Type 4: $f(x) = -0.1x^9 + 13x^7 - 90$

120. a) Type 4   b) Type 1   c) Type 3   d) Type 2

121. –

122. a) $4x^2 + 5x - 1$ b) $4x - 5$

     c) $3x^2 - 2x + 4$ d) $x^2 - x + 1$

     e) $4x^2 - 5x + 6$ (rest 3)   f) $x^4 - x^3 + x^2 - x + 1$

     g) $5x^2 + 4x - 2$ (rest -16)   h) $x^4 + x$ (rest $x - 1$)

123. p_3

124. a) $x_2 = -3$, $x_3 = -5$ b) $x_2 = -7$, $x_3 = 2$

125. roots of the function f: $\{-2, 1, 3\}$, roots of the function g: $\{-2, 1, 3\}$

126. $20x^2 + 4x - 3$

127. Roots: $\{-4, -1, 2, 3\}$ $\Rightarrow$ $f(x) = (x+4)(x+1)(x-2)(x-3)$

128. a) $p(x) = x^2 + 5x$ b) $p(x) = x^3 - 5x^2 + 5x + 3$

     c) $p(x) = 3x^3 - 6x^2 - x + 2$ d) $p(x) = x^4 - 7x^3 + 12x^2 + 4x - 16$

129. a) Roots: $\{-4, -2, 1\}$ $\Rightarrow$ $f(x) = (x-1)(x+4)(x+2)$

     b) Roots: $\{-2, -1, 2\}$ $\Rightarrow$ $g(x) = (x+2)(x-2)^2(x+1)$

130. a) $x_1 = 2$ (simple root)   b) $x_1 = -1$ (triple root)

        $x_2 = -5$ (quadruple root) $x_2 = 0$ (triple root)

                                           $x_3 = 1$ (simple root)

     c) $x_1 = -4$ (double root)   d) $x_1 = 0$ (quintuple root)

                                              $x_2 = 1$ (simple root)

     e) $x_1 = -5$ (simple root)   f) $x_1 = -1$ (simple root)

        $x_2 = 1$ (double root) $x_2 = 0$ (simple root)

                                              $x_3 = 1$ (simple root)

     g) $x_1 = -2$ (simple root)   h) $x_1 = -2$ (double root)

        $x_2 = 1$ (double root) $x_2 = 1$ (double root)

                                              $x_3 = 5$ (simple root)

131. a) $y = -x^5$ b) $y = x(x+1)(x-1)(x-2)$ c) $y = x(x+2)(x-2)^2$

     d) $y = -x(x+1)^2(x-2)$ e) $y = x^5(x+1)(x-1)$ f) $y = -x(x-1)(x^2+2)$

132. $p(x) = x^3 + 2x - 3 = (x-1)(x^2+x+3)$

     a) False   b) False   c) True   d) False   e) False

133. a) $y = x^2 + 5x - 6$ b) $y = \frac{1}{2}x^2 - \frac{3}{2}x$

     c) $y = x^3 - 14x^2 + 61x - 84$ d) $y = 2x^4 - 8x^3 + 2x^2 + 12x$

     e) $y = -2x^5 + 8x^4 + 30x^3 - 212x^2 + 392x - 240$

134. a) Roots at -2, 5 and 9; degree $n = 3$ b) Roots at -2, 0 and 2; degree $n = 6$

135. a) $\rightarrow 4)$ b) $\rightarrow 6)$ c) $\rightarrow 3)$ d) $\rightarrow 1)$ e) $\rightarrow 5)$ f) $\rightarrow 2)$

136. a) $D = \mathbb{R}\backslash\{-9\}$, pole without change of sign (c.o.s.) at $x = -9$

     b) $D = \mathbb{R}\backslash\{-1, 0, 1\}$, poles with c.o.s. at $x = \pm1$, removable singularity at $x = 0$

     c) $D = \mathbb{R}\backslash\{-2, 3\}$, poles with c.o.s. at $x = -2$ and $x = 3$

     d) $D = \mathbb{R}\backslash\{3\}$, pole without c.o.s. at $x = 3$

     e) $D = \mathbb{R}\backslash\{-8, 2\}$, pole with c.o.s. at $x = -8$, removable singularity at $x = 2$

     f) $D = \mathbb{R}\backslash\{-2, 2\}$, poles with c.o.s. at $x = -2$ and $x = 2$

     g) $D = \mathbb{R}\backslash\{3\}$, pole with c.o.s. at $x = 3$

     h) $D = \mathbb{R}\backslash\{-3, 4\}$, pole with c.o.s. at $x = 4$, removable singularity at $x = -3$

137. a) $y = \frac{1}{-2} = -\frac{1}{2}$ b) $y = 0$ c) $y = x - 1$

d) $y = \frac{1}{2}x$ e) $y = 2x$ f) $y = -x$

138. a) $y = -10$ b) $y = \frac{5}{3}$ c) $y = x^2 - 3x + 1$ d) $y = 2x + 1$

139. a) $f(x) = 4x + 5 + \frac{1}{(x-3)^2}$ b) $f(x) = \frac{(x-1)(x+1)^2}{x^2(x-2)}$

140. a) $f(x) = \frac{1}{(x-2)^2} + 2$ b) $f(x) = \frac{1}{x+(-2)} + 1$ c) $f(x) = x + 1 + \frac{1}{4(x-1)^2}$

141. a) a odd,   b) a even and b odd,   c) a odd and b even,

e.g. $f(x) = \frac{1}{x-1}$ e.g. $f(x) = \frac{x}{(x-2)^2(x+2)}$ e.g. $f(x) = x + \frac{1}{(x+2)(x-1)^2}$

142. $a = -1$ and $b = 1$

143. –

144. a) $y = \frac{1}{x+4} - 2$ b) $y = \frac{1}{(x+2)^2} + 2$

c) $y = \frac{x}{(x+2)^2(x-2)}$ d) $y = \frac{x}{(x+1)(x-1)^2}$

e) $y = -x - \frac{1}{x+2}$ f) $y = \frac{1}{2}x - 1 + \frac{1}{(x-2)x}$

145. Let $k \in \mathbb{Z}$ for all answers.

a) G_f is symmetric with respect to the axis $x = \frac{\pi}{2} + k\pi$, and to the point $(k\pi \,|\, 0)$.

b) G_f is symmetric with respect to the axis $x = k\pi$, and to the point $\left(\frac{\pi}{2} + k \cdot \pi \,\middle|\, 0\right)$.

c) G_f ist and to the point $\left(k \cdot \frac{\pi}{2} \,\middle|\, 0\right)$.

d) G_f is symmetric with respect to the axis $x = \frac{\pi}{4} + k\pi$, and to the point $\left(\frac{3\pi}{4} + k\pi \,\middle|\, 0\right)$.

e) G_f ist and to the point $(0 \,|\, 0)$.

f) G_f is symmetric with respect to the axis $x = k \cdot \pi$, and to the point $\left(\frac{\pi}{2} + k \cdot \pi \,\middle|\, 0\right)$.

146. a) odd   b) even   c) even   d) odd

147. –

148. –

149. –

150. –

151. a) $f(t) = 2 \cdot \sin\left(\frac{4t}{3}\right)$ b) $f(t) = 2 \cdot \cos\left(t + \frac{\pi}{4}\right) = 2 \cdot \sin\left(t + \frac{3\pi}{4}\right)$ c) $f(t) = 2 + 4 \cdot \sin(2t)$

152. a) $f(t) = \frac{1}{2} + \frac{1}{2}\sin\left(t + \frac{3\pi}{4}\right)$ b) $f(t) = \frac{1}{2} + \frac{1}{2}\cos\left(t + \frac{\pi}{4}\right)$

153. a) $x = k\pi$, $x = \frac{\pi}{2} + k2\pi$, $x = -\frac{\pi}{2} + k2\pi$; $k \in \mathbb{Z}$ b) $x = \frac{1}{k\pi}$, $x = \frac{1}{\frac{\pi}{2} + k2\pi}$, $x = \frac{1}{-\frac{\pi}{2} + k2\pi}$; $k \in \mathbb{Z}$

c) –

154. i) $f_1 \rightarrow$ (4)   ii) $f_2 \rightarrow$ (3)   iii) f_3

iv) $f_4 \rightarrow$ (5)   v) $f_5 \rightarrow$ (1)   vi) $f_6 \rightarrow$ (2)

155. a) $x = 2$ b) $x = \frac{1}{2}$

c) $u = \ln\left(\frac{2e^2}{e^2-1}\right) \approx 0.83856$ d) $t = \frac{3}{e^3-1} \approx 0.15719$

156. a) $D = \mathbb{R}_0^+$, $W = [1;\infty[$ b) $D = [1;\infty[$, $W = \mathbb{R}_0^+$ c) $D = \mathbb{R}^+$, $W = \mathbb{R}$

d) $D = \mathbb{R}\backslash\{0\}$, $W = \mathbb{R}$ e) $D = \mathbb{R}^+$, $W = \mathbb{R}_0^+$ f) $D =]1;\infty[$, $W = \mathbb{R}$

157. a) $f(x) = \frac{1}{2} \cdot 2^x$ b) $f(x) = 32 \cdot 4^x$ c) $f(x) = 3 \cdot \left(\frac{1}{3}\right)^x$

158. a) $f(t) = e^{\ln(2)\cdot t}$ b) $f(t) = \frac{1}{32} \cdot e^{\ln(2)\cdot t}$ c) not possible

159. a) $f^{-1}: x \mapsto \frac{\log(x)}{\log\left(\frac{1}{2}\right)} + 2$ b) $f^{-1}: x \mapsto \frac{\log(2x)}{\log(5)}$

c) $f^{-1}: t \mapsto \frac{\log\left(\frac{t}{4}\right)}{3\log(2)}$ d) $f^{-1}: t \mapsto \sqrt{e^{\frac{t}{2}} + 1}$